QUANTUM ELECTRODYNAMICS

PROCEEDINGS OF THE
IV. INTERNATIONALE UNIVERSITÄTSWOCHEN
FÜR KERNPHYSIK 1965 DER KARL-FRANZENS-UNIVERSITÄT
GRAZ, AT SCHLADMING (STEIERMARK, AUSTRIA)
25th FEBRUARY—10th MARCH 1965
(ACTA PHYSICA AUSTRIACA / SUPPLEMENTUM II)

SPONSORED BY
BUNDESMINISTERIUM FÜR UNTERRICHT
BUNDESMINISTERIUM FÜR HANDEL UND WIEDERAUFBAU
STEIERMÄRKISCHE LANDESREGIERUNG AND
THE INTERNATIONAL ATOMIC ENERGY AGENCY

EDITED BY

PAUL URBAN
GRAZ

WITH 20 FIGURES

1965
SPRINGER SCIENCE+BUSINESS MEDIA, LLC

Organizing Committee:

Chairman: Prof. Dr. PAUL URBAN
Vorstand des Institutes für Theoretische Physik, Universität Graz

Committee Members: Dr. P. KOCEVAR, Graz
Dozent Dr. H. ZINGL, Graz

Secretary: A. SCHMALDIENST

ISBN 978-3-211-80734-7 ISBN 978-3-7091-7649-8 (eBook)

DOI 10.1007/978-3-7091-7649-8

Softcover reprint of the hardcover 1st edition 1965

Library of Congress Catalog Card Number 50-557
Title-No. 9159

Preface

Whereas there has been significant progress in various branches of elementary particle physics during the last decade or so, Quantum Electrodynamics seems to be in a somewhat stagnating phase. For people not involved in sophisticated questions of relativistic field theories the situation might appear as not very distinct from the time, when Schwinger, Feynman and Dyson published their famous articles in the early 1950's. Only some arguments against the possibility of a finite Quantum Electrodynamics (1953) has caused a significant reaction and initiated a rather vehement discussion. In connection with these questions about the consistency of Quantum Electrodynamics there arises the problem of the high energy behaviour of the theory, since higher and higher energies will be available in future experiments.

For these reasons I thought it quite useful to choose Quantum Electrodynamics at high energies as subject of this year's meetings and hope to meet the demands of both experimental and theoretical physicists working in this field.

Graz, October, 1965. **P. Urban**

Contents

Ladies and Gentlemen:

As organizer of this meeting I welcome you and hope that your stay will be pleasent and profitable for you. Many of you have attended one of the past Schladming-meetings and will have noticed that we were able to expand this meeting from a rather small scale event to a respectable conference. There is, however, always the serious problem, how such a meeting should be arranged especially when one considers that the number of participants has almost doubled every year. This year we have tried to split the meeting into two parts: the first part is ment to outline the general background of the more sophisticated lectures of the second part. The success of this meeting will show whether this was a good idea.

Now let me say a few words about this year's Topic:

Quantumelectrodynamics. It is the only field theory we have today, for which the theoretical predictions and the numerical values are in excellent agreement with the experimental results. The more abstract and formal approach which has been developed recently will be the main subject of this symposium. Whether this very abstract mathematical formulation will help us to gain further insight into nature cannot be said at the present time. It is, however my sincere hope that you will be stimulated by this meeting. You will have noticed that we succeeded in bringing lecturers who are outstanding experts in their particular fields to Schladming, many of them from foreign countries. This was made possible only through the financial and moral support from many institutions, whom we thank very heartily. I close my speech with the best wishes for the time you spend here in the wonderful landscape of Styria.

P. URBAN, Graz

An Introduction to Theories of Integration over Function Spaces*

By

L. Streit

Universität Hamburg

With 1 Figure

1. Introduction

To begin with let me say some words about what the talk I am going to give is aiming at. In theoretical physics one happens again and again and in a great variety of fields upon functional formulations. The idea to use functional calculus for the description of physical systems depending on an infinite number of variables dates back to a paper written already in 1914 by VOLTERRA [VOLTERRA 1914], where he proposed to make use of a functional formulation for the dynamics of many particle systems and continuous media.

Milestones of the introduction of functional integration to theoretical physics were the papers by WIENER (c.f. e.g. [PALEY 1930]) treating Brownian motion, those by FEYNMAN on a new approach to quantum mechanics [FEYNMAN 1948, 1955] and the work done by FRIEDRICHS [FRIEDRICHS 1951, 1953, 1957, 1959], SEGAL [SEGAL 1953, 1954, 1958, 1959], GÅRDING + WIGHTMAN [GÅRDING 1954 a, b] on the representation problem in quantum field theory. Many other applications are collected e.g. in the bibliography of the review articles by GELFAND + YAGLOM [GELFAND 1960] and by BRUSH [BRUSH 1961] but the three groups of papers mentioned above have in common that the authors devised new types of integrals to cope with a special class of physical problems.

They were followed by "second generations" of works concerned with the rigorization of the newly defined notions, setting them into the frame of general integration theory, giving computational rules, generalizations and so forth. Most of them appeared in mathematical journals.

It is this second generation of mathematical results that my talk is supposed to give a collection of. For lack of time almost nothing will

---

* Lecture given at the IV. Internationalen Universitätswochen für Kernphysik, Schladming, 25 February—10 March 1965.

be proven but examples will be given of the kind of problems, theorems and computational devices one encounters with theories of functional integration, so that you get an idea what all this means, what it is good for, and where to get the mathematical details.

2. Some basic notions

When Volterra in the years about 1890 began to develop the theory of functionals he characterized them as "function of lines" or functions depending on other functions [Volterra 1930]. We will distinguish the notions of function and functional by the spaces on which they are defined as mappings. We will call a scalar, vector, operator valued *function* any mapping f from a *finite*-dimensional point space like the space R^n of real n-tuples into a space of numbers, vectors, operators etc.

$$f: \quad (x_1, \ldots, x_n) \to f(x_1, \ldots, x_n)$$

whereas by *functional* we mean a mapping F whose domain of definition is a given set of *functions*, say

$$F: \quad x(t) \to F[x]$$

where F again may be a scalar, vector, operator or whatever you like.

The essential difference is that the set of indices characteristic for the dimension of the point x is no more finite, like $1, \ldots, n$ but infinite

$$a \leqslant t \leqslant b$$

so that in *functional* calculus we have variables of *infinite dimension.*

For the set $\{x(t)\}$ of all functions of a real variable t the dimension is that of the continuum: to define x, you must give $x(t)$ for all real t.

On the other hand the set of all continuous functions is of countable dimension, you only have to know $x(t)$ for all rational t to know it everywhere.

Another example of the latter kind is the (separable) Hilbert space L_2 of square integrable functions, any element $x \in L_2$ is given by its expansion in terms of a countable orthonormal set $\varphi_n(t)$, i.e. by the countable set of x_n in

$$x(t) = \sum x_n \varphi_n(t)$$

Certain notions may be carried over more or less directly from the finite dimensional case, e.g. partial differentiation[1]

$$\frac{\partial}{\partial x_\nu} \to \frac{\delta}{\delta x(t)}$$

[1] For details on the functional derivative, with generalizations to operator valued $x(t)$ cf. [Rohrlich 1964].

via the definition:

$$\frac{\delta F[x]}{\delta x(t)} = \lim_{\varepsilon \to 0} \frac{F[x + \varepsilon\,\delta(\cdot - t)] - F[x]}{\varepsilon} = F_t'$$

The analogue of the Taylor expansion

$$f(x + x_0) = \sum_n \frac{f^{(n)}(x_0)}{n!}\, x^n$$

is then the Volterra expansion

$$F[x + x_0] = \sum_n \frac{1}{n!} \int dt_1 \ldots dt_n\, F^{(n)}_{t_1 \ldots t_n}[x_0]\, x(t_1) \ldots x(t_n)$$

where $x_0 = x_0(t)$ is a function now.

On the other hand it is by no means obvious how to define integration over infinite dimensional spaces (the so called continuous or functional integration). In the case of finite integration one feels that Lebesque integration is the obvious thing because of a very strong uniqueness property of the measure associated with it: if we look for measures on R^n with the following weakened invariance property under translations (the so called quasi-invariance):

$$\mu(M) = 0 \Rightarrow \mu\,\{m + x_0 | m \in M \subseteq R^n\} = 0$$

for any fixed $x_0 \in R^n$, that is sets of measure zero transform into sets of measure zero under any translation, then we find that the set of quasi-invariant measures on R^n exactly exhausts the class of measures equivalent to the Lebesque measure, where two measures μ, ν are called equivalent if they have the same sets of measure zero.

$$\mu(M) = 0 \Leftrightarrow \nu(M) = 0$$

That is: any measure on R^n that just has this weak form of translational invariance may be expressed in terms of Lebesque measure (via the theorem of Radon and Nikodym), c. f. [Gelfand **1964**, p. **322** ff.]. We will see that the situation is quite different for infinite dimensional integrals.

3. The Wiener Integral

Probably the most well-known infinite dimensional integral is the Wiener integral over the space of continuous functions.

Before giving its definition let me just remind you with a few words of how one may obtain the Lebesque measure by a direct generalization of the Riemann content of intervals in R^n, since we will be doing the same in function space afterwards.

The function

$$\mu(J) = \prod_{j=1}^{n} (b_j - a_j)$$

defined on the set algebra K of intervals

$$J = \{(x_1, \ldots, x_n) \in R^n | a_\nu \leqslant x_\nu < b_\nu, \quad \nu = 1 \ldots n\}$$

is a σ-additive content, i.e.

$$\mu\left(\sum_{\lambda=1}^{\infty} J_\lambda\right) = \sum_{\lambda=1}^{\infty} \mu(J_\lambda) \qquad \text{if only} \qquad J_{\lambda 1} \cap J_{\lambda 2} = 0 \qquad \forall \lambda_1 \neq \lambda_2$$

Since it is σ-additive, it defines a unique measure on the smallest σ-algebra or "Borel algebra" over K through

$$\mu(M) = \lim_{M \subset \bigcup_n J_n} \sum_n \mu(J_n) \qquad \text{for} \qquad M \in B \qquad \text{with} \qquad J_n \in K$$

$\mu(M)$ is the Lebesque measure and

$$\int f(x)\, d\mu = \sum_a a \cdot \mu\, [f(x) = a]$$

For the details of this see any modern textbook on general integration theory.

Instead of R^n we now have the space

$$C = \{x(t) | x(0) = 0, \quad 0 \leqslant t \leqslant 1\}$$

of continuous functions and with metric $\varrho(x, y) = \max_t |x(t) - y(t)|$.

To obtain Wiener measure we see that we need something like intervals and a σ-additive content for them.

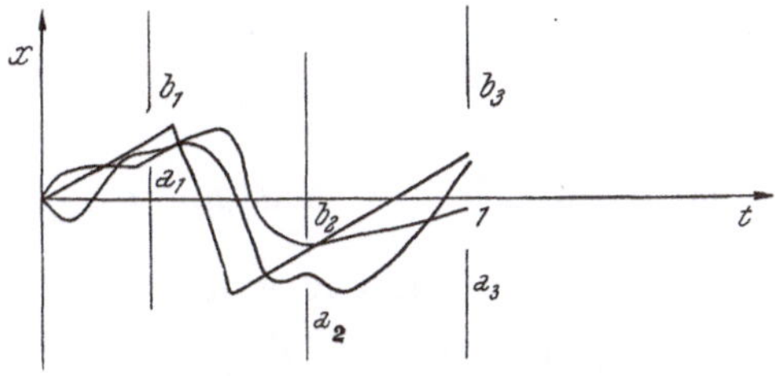

Fig. 1

Given three sets of numbers a, b, t

$$0 < t_1 < \ldots < t_n = 1 \qquad a_j \leqslant b_j \qquad j = 1, \ldots, n$$

we will call quasi-interval the set

$$S = \{x(t) \in C | a_j \leqslant x(t_j) \leqslant b_j \qquad j = 1, \ldots, n\}$$

i.e. all the continuous functions whose graphs start from zero and pass through the x-intervals $a_j \ldots b_j$ at the points t_j.

The content of S

$$\mu(S) = \frac{1}{\sqrt{\pi^n t_1(t_2 - t_1) \ldots (t_n - t_{n-1})}} \cdot$$

$$\cdot \int_{a_1}^{b_1} \ldots \int_{a_n}^{b_n} dx_1 \ldots dx_n \, e^{-\frac{x_1^2}{t_1} - \sum_j^{n-1} \frac{(x_{j+1} - x_j)^2}{t_{j+1} - t_j}}$$

being σ-additive, may be extended to the Borel closure of the set of quasi-intervals in C and is called the Wiener measure μ_w in C. It gives rise to the Wiener integral

$$\int_C F[x] \, d_w x \qquad \text{and} \qquad \int_{E \subseteq C} \cdot /. = \int_C F[x] \, \chi_E(x) \, d_w x$$

where $\chi_E(x)$ is the characteristic function of the set E:

$$\chi_E(x) = \begin{cases} 1 & x \in E \\ 0 & x \notin E \end{cases}$$

For an arbitrary open set $G \subseteq C$ (where open refers to the topology introduced by ϱ) and

$$\Gamma_n = \{(x_1, \ldots, x_n) \in R^n | \exists \, x \in G \quad \text{with} \quad x(t_j) = x_j, \quad j = 1, \ldots, n\}$$

we have

$$\mu_w(G) = \lim_{n \to \infty} \frac{1}{\sqrt{\pi^n t_1(t_2 - t_1) \ldots (t_n - t_{n-1})}} \cdot$$

$$\cdot \int_{\Gamma_n} dx_1 \ldots dx_n \, e^{-\frac{x_1^2}{t_1} - \sum_j^{n-1} \frac{(x_{j+1} - x_j)^2}{t_{j+1} - t_j}}$$

where lim is to be taken such that $\max_j |t_j - t_{j-1}| \underset{n}{\to} 0$.

For the exponent we may formally write

$$\sum \left(\frac{\Delta_i x}{\Delta_i t}\right)^2 \Delta_i t \sim \int \dot{x}^2(t) \, dt$$

and for the measure

$$d_w x \sim e^{-\int_0^1 \dot{x}^2 dt} \prod_0^1 dx(t)$$

so that we see: it is not the x that is weighted with a Gaussian distribution but rather the infinitesimal increment $\Delta x / \Delta t \sim \dot{x}(t)$ is taken as Gaussian distributed variable. We may e.g. identify $x(t)$ with the path of a particle undergoing Brownian motion, where (as generally in simple Markoff processes) the transition probability from x at time t to $x + \Delta x$ at $t + \Delta t$ depends only on Δx, Δt

$$p(x + \Delta x, t + \Delta t | x, t) = \frac{1}{2\sqrt{\pi \Delta t D}} e^{-\frac{(\Delta x)^2}{4 \Delta t D}}$$

[MONTROLL 1952]

(We have set the diffusion constant $D = 1/4$ for simplicity's sake, which is always possible with a suitable time scale). So that

$$\int_G d\mu_w = \lim_{n \to \infty} \int_\Gamma p(x_n, t_n | x_{n-1} t_{n-1}) \ldots p(x_1 t_1 | 00) \, dx_1 \ldots dx_n$$

This illustrates that the Wiener integral may be interpreted as an averaging over paths, indicates its usefulness in statistical mechanics, and suggests at once another definition of the Wiener integral, namely: Be $F[x]$ a functional on C, $x^{(n)}(t)$ the piecewise linear function that coincides with x at $t = t_j$, put $F[x^{(n)}] = F(x_1 \ldots x_n)$ and finally

$$J(F) = \lim_{n \to \infty} \int_{R^n} F(x_1 \ldots x_n) \prod_{r=1}^{n} p(x_r t_r | x_{r-1} t_{r-1}) \, dx_r$$

FOMIN [FOMIN 1958] has shown that if $F[x] \geqslant 0$ and $J(F)$ exists, then

$$J(F) \geqslant \int F[x] \, d_w x$$

and the rhs. exists. If F is bounded and continuous, the equality sign holds. But the two integrals are *not* equivalent since there exists an example of a functional given by MAIKOV [MAIKOV 1958] with

$$\int F \, d_w x < J(F)$$

Still other methods to introduce the Wiener integral would be through the theorem of Kolmogorov or following Wiener's original method [PALEY 1930] by giving an explicit mapping of C onto the interval $[0, 1]$ where one has the Lebesque integration.

One might also ask at what kind of measure one arrives when one drops the continuity condition in the definition of the quasi-intervals. In fact one obtains thus a measure on the space of *all* real valued functions but this is of no obvious use to physics since the sub-set of continuous functions that one would like to identify with particle paths proves to be non-measurable in this space [KAC 1959].

4. Change of variables in the Wiener integral

It is only with the knowledge of a sufficient amount of algorithmic rules and other devices of computation that the Wiener integral becomes a useful mathematical tool. The first question in this direction shall be:

what happens to the Wiener integral under a translation of the argument by a fixed function $a(t) \in C$?

$$y(t) \to x(t) + a(t)$$

The formal representation mentioned above

$$\int F\, d_w y = \frac{1}{N} \int_{-\infty}^{+\infty} \int F[y]\, e^{-\int_0^1 \dot{y}^2(t)\, dt} \prod_0^1 dy(t)$$

indicates after inserting $x + a$ on the rhs,

$$\int_C F\, d_w y = e^{-\int_0^1 \dot{a}^2(t)\, dt} \int_C F[x + a]\, e^{-2\int_0^1 \dot{a}\, dx(t)}\, d_w x$$

The rigorous proof of this relation for bounded and continuous F and for $\dot{a}(t)$ of bounded variation may be carried out through the representation J. For generalizations see [CAMERON 1951, SUNOUCHI 1951, MARUYAMA 1950, SEGAL 1958]. It is a remarkable property of the Wiener integral that the allowed $a(t)$ are of measure zero in C [PALEY 1930]. Cameron [CAMERON 1954] has shown that almost all $a \in C$ transform measurable sets into non measurable ones and that there is no measure μ on C with

a) invariance against all translations in C,

b) μ absolutely continuous to μ_w or vice versa.

Note that setting $F \equiv 1$ in the above formula ($\int d_w y = 1$) yields a first example for a solvable Wiener integral, namely

$$\int_C e^{\int_0^1 \dot{a}(t)\, dx(t)}\, d_w x = e^{4\int_0^1 \dot{a}^2(t)\, dt}$$

The next operation we want to look at are linear transformations of the kind

$$y(t) = x(t) + \int_0^1 K(t, s)\, x(s)\, ds$$

Making use again of our formal representation and replacing the Jacobi determinant of the finite dimensional problem by its counterpart, the Fredholm determinant D, we find

$$\int_C F[y]\, d_w y = |D| \int_C F\left[x(t) + \int_0^1 K(t, s)\, x(s)\, ds\right] \times$$

$$\times\, e^{-\int_0^1 \left\{\frac{\partial}{\partial t} \int_0^1 K(t,s)\, x(s)\, ds\right\}^2 dt}\; e^{-2\int_0^1 \frac{\partial}{\partial t}\left(\int_0^1 K(t,s)\, x(s)\, ds\right) dx(t)}\, d_w x$$

Various conditions on K and F for this relation to hold may be found in the literature [CAMERON **1945**, SUNOUCHI **1951**, SEGAL **1958**, SEIDMAN **1959**, WOODWARD **1961**].

For non linear variable transformations see the review articles by Gelfand + Yaglom [GELFAND **1960**] and by Kovalchik [KOVALCHIK **1963**] and [CAMERON **1949**, **1953**].

5. The indefinite Wiener integral

Let us now introduce the notion of indefinite Wiener integral [CAMERON **1951** a]

$$G[u] = \int\limits_{x(t) \leqslant u(t)} F[x]\, d_w x$$

where u must not necessarily be in C but should be Borel measurable. If one sets further

$$\delta F[x, y] = \frac{\partial}{\partial h} F[x + h\, y]\big|_{h=0}$$

one has for well-behaved F and $y \in C$ (see [KOVALCHIK **1963**] for details)

$$\delta G[u, y] = \int\limits_{x<u} \delta F[x, y]\, d_w\, x - 2 \int\limits_{x<u} F(x) \left[\int\limits_0^1 \dot{y}(t)\, dx(t) \right] d_w x$$

We set

$$u(t) = u_0(t) = \infty, \qquad \delta G[u, y] = 0$$

and we have

$$\int\limits_C \delta F[x, y]\, d_w x = 2 \int\limits_C F[x] \left\{ \int\limits_0^1 \dot{y}(t), dx(t) \right\} d_w\, x$$

a result we also would have obtained, at least formally, through the translation formula with $z = x + h\, y$

$$\int\limits_C F(z)\, d_w z = e^{-h^2 \int \dot{y}^2 dt} \int\limits_C F[x + h\, y]\, e^{-2h \int\limits_0^1 \dot{y}\, dx(t)}\, d_w\, x$$

Partial differentiation with respect to h at $h = 0$ results in the above equation. Replacing now $F[x]$ by $F[x]\, G[x]$ one immediately gets a formula for integration by parts in function space

$$\int\limits_C F[x]\, \delta G[x, y]\, d_w x = \int\limits_C G[x] \left\{ 2\, F[x] \int\limits_0^1 \dot{y}\, dx(t) - \delta F[x, y] \right\} d_w x$$

obviously both formulas are a very useful tool to obtain further classes of Wiener integrals that are solvable in closed form.

6. The Fourier-Wiener transform

Be $K = \{x + i\,y | x, y \in C\}$ the complexification of C and $F[x + i\,y]$ Wiener summable for $y \in K$ then we will call

$$\tilde{F}[y] = \int_C F[\sqrt{2}\,x + i\,y]\, d_w x$$

the Fourier-Wiener-transform of $F[x]$. At first sight this does not look at all like a Fourier transform, but through the exponential in the Wiener measure one has an inversion rule

$$\tilde{\tilde{F}}[x] = F[-x]$$

and a Parseval equation

$$\int_C F_1[x]\, F_2[-x]\, d_w x = \int_C \tilde{F}_1[y]\, \tilde{F}_2[y]\, d_w y$$

holds for the square integrable functionals over C [CAMERON 1945 a, 1947].

7. The Wiener integral and differential equations

We will now give some important relations between the theory of differential equations and Wiener integrals. They are useful both ways: on the one hand very interesting results on the spectra and eigenfunctions of second order differential operators may be obtained [RAY 1954] through Wiener integral representations, on the other hand in the case of differential equations that are solvable in closed form the evaluation of the corresponding Wiener integral is reduced to an ordinary one dimensional integration.

To illustrate this we first introduce the so called conditional Wiener measure by the following slight change of definitions

$$C_{x_1 t_1,\, x_0 t_0} = \{x(t) | x(t_0) = x_0 \qquad x(t_1) = x_1\}$$

That is we now

a) take the continuous functions over the interval $[\tau_0, \tau_1]$ instead of $[0, 1]$,

b) set $x(\tau_0) = x_0$ instead of $x(0) = 0$ or in the language of paths the particle now starts at point x_0, time t_0 and, most important,

c) we keep the end point fixed, i.e. we average over all paths leading from x_0, τ_0 to x_1, τ_1.

The resulting changes in the definition of measure are obvious

$$\mu_{x_1 \tau_1,\, x_0 \tau_0}(S) = \frac{1}{\sqrt{\pi^n (t_1 - t_0) \ldots (t_n - t_{n-1})}} \int_{a_i}^{b_i} \prod^{n-1} dx_i\, e^{-\sum_1^{n-1} \frac{(x_{j+1} - x_j)^2}{t_{j+1} - t_j}}$$

is the new quasi-interval content and $C_{x_1 t_1, x_0 t_0}$ is now to be taken instead of C.

For the conditional integral we write

$$\int\limits_{C_{x_1 \tau_1, x_0 \tau_0}} F[x]\, d_{w(x_1 \tau_1, x_0 \tau_0)} x = \psi_F(x_1 t_1, x_0 t_0)$$

To get back to the original Wiener integral one uses the obvious relation

$$\int\limits_{C} F[x]\, d_w x = \int \psi_F(x\,1; 0\,0)\, dx$$

For $F = 1$ $\psi(x\,t, x_0\,\tau_0)$ describes the diffusion probability of a particle from $x_0\,\tau_0$ to $x\,t$ and obeys the diffusion equation

$$\frac{\partial \psi}{\partial t} = D \frac{\partial^2 \psi}{\partial x^2}$$

with (as mentioned above) $D = 1/4$ and the boundary condition

$$\psi(x\,\tau_0, x_0\,\tau_0) = \delta(x - x_0)$$

As a more general and very important result Kac [KAC 1949] showed that for

$$F = e^{-\int\limits_{\tau_0}^{t} V(x(t))\, dt}$$

$$\frac{\partial \psi}{\partial t} = \frac{1}{4} \frac{\partial^2 \psi}{\partial x^2} - V(x)\, \psi$$

with the above boundary condition, so that solving this differential equation one also has the Wiener integral over F.

The review articles bei Brush [BRUSH 1961] and by Gelfand + Yaglom [GELFAND 1960] give various examples for the usefulness of the Wiener integral method applied to the differential equations of theoretical physics, especially to obtain theorems on the number and distribution of the eigenvalues of the associated differential equations. A typical example here is the paper of Feynman on the eigenstate of lowest energy in the polaron model [FEYNMAN 1955].

Another important representation for the equation

$$a \frac{\partial \psi}{\partial t} = \frac{\partial^2 \psi}{\partial x^2} + p(x, t)\, \psi(x, t) \qquad \lim_{t \to +0} \psi(x, t) = \varphi(x) \qquad \text{for almost all } x$$

has been derived by Cameron [CAMERON 1954 a], namely

$$\psi(x, t) = \int\limits_{C} e^{\frac{\tau}{a} \int\limits_{0}^{1} p\left(2\sqrt{\frac{t}{a}}\, y(s) + x,\, t(1-s)\right) ds} \varphi\left(2\sqrt{\frac{t}{a}}\, y(1) + x\right) d_w y.$$

Further papers on the relation between differential equations and continuous integration are [Cameron **1956**, Yeh **1960**, Daletzkii **1960**, **1961**, **1961** a].

8. The Wiener integral and integral equations

Given

$$y(t) \equiv F[x, t] = x(t) + \int K(t, s)\, x(s)\, ds$$

the solution

$$x(t) = |D|\, e^{-\int_0^1 (\dot{y}(s))^2 ds} \int_C u(t)\, e^{2\int_0^1 \dot{y}(s)\, dF[u,s] - \varphi[u]}\, d_w u$$

$$\exp(-\varphi[u]) = e^{-\int_0^1 \left\{ \frac{\partial}{\partial t} \int_0^1 K(t,s)\, u(s)\, ds \right\}^2 dt - 2 \int_0^1 \left\{ \int_0^1 \frac{\partial K(t,s)}{\partial t} u(s)\, ds \right\} du(t)}$$

may be obtained with the help of the formulas describing the transformation of continuous integrals under linear transformations on C. For details and similar formulas for the resolvent kernel see [Ostrom **1949**] and [Kovalchik **1963**]. A formula for the Fredholm determinant results immediately by setting $F \equiv 1$ in the formula for general linear transformations (cf. § 4)

$$|D| = \left\{ \int_C e^{-\varphi[u]}\, d_w u \right\}^{-1}$$

The connection with integral equations gives us yet another method to gain exact expressions for certain classes of Wiener integrals.

Besides these there are several approximation methods [Brush **1961**] which we will just enumerate without giving the details here, since they are of a more technical nature. Cameron [Cameron **1951** b] provided the first results in this direction (rectangle rule and Simpson's rule) which consist in calculating an nth-order integral J_n with the property that $|J_n - J| \sim n^{-2}$, where J is the exact result, that is convergence is not so good.

A comparatively very simple method has been suggested by Brush, it consists in parametrization of a subset of paths and only integrating over the parameters.

A more accurate approximation

$$\int_C F[x]\, d_w x \approx \sum_{-\infty}^{+\infty} \lambda_i F(x_i)$$

has been given by Vladimirov [Vladimirov **1960**], where the λ_i may be so selected that the equation holds exactly e.g. for all odd functionals.

For approximation with the help of electronic computers see a paper by GELFAND + CHENTSOV [GELFAND 1956].[2]

Finally there results a useful approximation for $\int\limits_C e^{\alpha F[x]}\, d_w x$ by expanding $F[x]$ into a Volterra series $F[x] = F[x_0] + \frac{1}{2}\delta^2 F[x_0] + \dots$ around a path $x_0(t)$ that makes $F[x]$ extremal and by setting

$$\int\limits_C e^{\alpha F[x]}\, d_w x \sim e^{\alpha F[x_0]} \int\limits_C e^{-\alpha^2 \delta^2 F[x_0]}\, d_w x$$

which reduces the exponent to a quadratic functional in $x(t)$. For the relation of this method to that of steepest descent and to the WKB-method see MONTROLL [MONTROLL 1952] and MORETTE [MORETTE 1950].

9. The Feynman integral

We have already noticed the relation between the conditional Wiener integral

$$\int e^{-\int\limits_0^t \left\{\frac{\dot{x}^2}{4D} + V(x(\tau))\right\} d\tau} \prod_0^t dx(\tau)$$

and the heat flow equation

$$\frac{\partial \psi}{\partial t} = D \frac{\partial^2 \psi}{\partial x^2} - V \psi$$

out of which we can get the Schrödinger equation

$$-i\hbar\dot{\psi} = \frac{\hbar^2}{2}\psi'' - v\psi$$

if we set

$$D = \frac{1}{2} i\hbar \qquad V = \frac{i}{\hbar} v$$

Now let us see what happens to the exponent of the Wiener integral under this substitution. It now reads

$$\frac{i}{\hbar}\int\limits_0^t \left\{\frac{1}{2}\dot{x}^2 - v(x)\right\} d\tau = \frac{i}{\hbar}\int\limits_0^t L[\dot{x}, x]\, d\tau = \frac{i}{\hbar} S[x]$$

where L is the Lagrangian and S the classical action along the path $x(\tau)$. This is the result Feynman [FEYNMAN 1948] got when he formulated his space — time approach to quantum mechanics. The general idea is the following: when you describe a classical transition $\langle x, t|0, 0\rangle$

---

[2] For some approximation methods cf. also [ROSEN 1963].

of a randomly mowing particle you may subdivide the time interval by $0 < t_1 < \ldots < t_n < t$, multiply the Markoff chain of probabilities $p(x_i, t_i | x_{i-1}, t_{i-1})$ and integrate over the $x_i = x(t_i)$

$$p(x\,t|0\,0) = \int dx_1 \ldots dx_n\, p(x, t|x_n, t_n) \ldots p(x_1, t_1|0\,0)$$

whereas in quantum mechanics what you have to multiply is not a chain of (real, normed etc.) probabilities but rather a so called quantum chain of *complex* transition elements [Montroll 1952] which leads to the representation derived above

$$\psi(x, t, x_0, t_0) = \int\limits_{C_{xt,\,x_0t_0}} e^{i/\hbar\, S[x]} \prod_{t_0}^{t} dx(\tau)$$

the so called Feynman integral.

Now this is one of the objects of which we know very well the physical meaning but which has no mathematical meaning at all as it stands. The reason is that instead of an exponentially damped integrand we now have an undamped, oscillating one. An obvious attempt to get a well defined continuous integral representation for the Schrödinger wave function was to use a complex $\hbar + i\,\delta$ the imaginary part of which would provide a damping factor and to perform the limit $\delta \to 0$ only in the end [Gelfand 1960]. But it has been shown by Cameron [Cameron 1960] that this is not sufficient to make the Feynman integral well defined. Instead he recently proposed a different definition along the following lines [Cameron 1960]: The first step is to equate

$$\int\limits_{C} F[\lambda^{-1/2}\, x]\, d_w x = \int\limits_{0}^{\infty} e^{-\lambda s}\, df(s)$$

where $f(s)$ is the "Inverse Laplace Stieltjes Transform Of Wiener's" or "Ilstow" integral. The rhs then may be continued to imaginary λ $\lambda \to -i\,q$ which Cameron denotes as the Feynman integral with parameter q

$$\int\limits_{C}^{f_q} F[x]\, dx \equiv \int\limits_{0}^{\infty} e^{iqs}\, df(s)$$

Here the class of integrands may be enlarged by introducing a damping factor in the rhs resulting in the "limiting Feynman integral"

$$\int\limits_{C}^{\to f_q} F[x]\, dx \equiv \lim_{\eta \to +0} \int\limits_{0}^{\infty} e^{s(iq-\eta)}\, df(s)$$

CAMERON then shows that under suitable boundedness and integrability conditions on the functions φ, V and with

$$C[0, t] = \left\{ x(\tau) \,|\, x\left(\frac{\tau}{t}\right) \varepsilon\, C \right\}$$

the limiting Feynman integral

$$\psi(x, t) = \int_{C[0,t]}^{\to f_1} e^{i\int_0^t V(t-s,\, y(s)+x)ds} \varphi(y(t)+x)\, dy$$

exists and satisfies for $t > 0$ the Schrödinger equation

$$-i\frac{\partial\psi}{\partial t} = \frac{1}{2}\frac{\partial^2\psi}{\partial x^2} + V(t, x)\,\psi$$

and the boundary condition. (For a similar representation of the solution of the heat equation see the above paragraph on Wiener integrals and differential equations.)

You will find further rigorous approaches to the Feynman integral in articles by [SCHWEBER 1962], [BABBITT 1963], [NELSON 1964], [MCSHANE 1963] and the literature given therein.

10. Hilbert space integrals

In this last paragraph let me say some words about integration of functionals over Hilbert spaces, because it has found twofold application in quantum field theory. For its significance in the functional formulation of dynamics see e.g. the articles of SYMANZIK [SYMANZIK 1954, 1964] and BRUSH [BRUSH 1961] and the long lists of further references you will find there. Besides I think several lectures pointing in this direction will be given in the course of this school.

The other application is in the theory of representations of the canonical commutation relations [GÅRDING 1954, 1954 a], [SEGAL 1953, 1954, 1958, 1959], [HAAG 1960], [COESTER 1960] where the representation problem is connected with an ansatz for the Hamiltonian [FRIEDRICHS 1951, 1953, 1957, 1959], [GELFAND 1964].

In quantum field theory states may be classified as eigenvectors of an infinite number of commuting operators e.g. of the field $A(\vec{x}, t)$ for all $\vec{x}$ at a given t. That is the state is no more given as a function of a finite number of eigenvalues x_ν

$$\psi = \psi(x_1, \ldots, x_n)$$

but rather as a functional of the infinite set x

$$\psi = \psi[x]$$

so that the scalar product in Hilbert space is no more

$$\int |\psi(\chi_1, \ldots, \chi_n)|^2 d\chi_1 \ldots d\chi_n$$

but a continuous integral

$$\int |\psi[\chi]|^2 \, d\mu(\chi)$$

of some sort. The importance of this kind of representation lies in the fact that the task of finding the inequivalent representations of the canonical commutation relations may be "reduced" to finding certain classes of inequivalent measures on Hilbert space.

There are several basic difficulties one encounters when looking for Hilbert space measures:

a) it can be shown that there are no quasi-invariant σ-additive measures in the sense that $\mu(m) = 0 \Rightarrow \mu(m + x) = 0$ for all subsets and all elements x of a given Hilbert space. (The fact that the measure is given on a Hilbert space is not necessary for the argument, the same is true for vast classes of other infinite dimensional topological spaces [GELFAND **1964**, p. **326—334**].) On the other hand there do exist quasi-invariant measures if x varies only in a certain dense subset of H [SUDAKOW **1959**], [MITJAGIN **1961**], [GELFAND **1964**, p. **327**].

b) countable additivity and unitary invariance are incompatible. That is given any countably additive measure on a separable Hilbert space H with the properties that every $x \in H$ lies in a subset of H with arbitrarily small diameter and finite measure and such that $\mu(Um) = \mu(m)$ for all measurable m and all unitary U, then it follows

$$\mu(H - \{0\}) = 0 \qquad \text{[McSHANE 1963]}$$

This being the case there have been various inequivalent definitions of Hilbert space integrals e.g. by FRIEDRICHS, GELFAND, SEGAL, J. SCHWARTZ [DUNFORD **1948**, p. **402** ff.] and many others. See also [GROSS **1960, 1962, 1963**], [KOVACS **1963**]. Collections of them are given in DUNFORD's book and the review article by McSHANE. We will treat here only two examples, one of the unitary invariant type, because it seems to be quite useful and not much work has been done on it yet, and one of the countable additive type because it is closely related to the Wiener integral.

When one wants to introduce a measure in a separable Hilbert space one may take as a realization the set of sequences (x_i) with the property $\Sigma^\infty x_i^2 < \infty$. Now we could think of introducing an infinite product Πr of Gaussian measures

$$r(m_i) = \frac{1}{\sqrt{\pi}} \int_{m_i} e^{-x_i^2} \, dx_i$$

This would be countably additive on the space of *all* sequences but

$$\Pi \, r(H) = 0$$

This failure shows the necessity of a somewhat more sophisticated approach. Given any set $m \subseteq H$ and an orthogonal n-dimensional projection operator P on H such that

$$Px \in m \Leftrightarrow x \in m$$

we will call m a cylinder set. Since PH is n-dimensional m is isomorphic to a set of n-dimensional vectors $(x_1, \ldots, x_n)$ and we may introduce the measure

$$\mu_n(m) = \pi^{-n/2} \int_m e^{-(x_1^2 + \ldots + x_n^2)} \, dx_1 \ldots dx_n$$

Since

$$\mu_{n+n'}(m \oplus R^{n'}) = \mu_n(m) \qquad \text{and} \qquad \mu_n(m) = \mu_n(Um)$$

for all measure-preserving linear rotations U, $\mu_n = \mu$ is independent of the chosen projector and of the representation of m in R^m so that μ furnishes us with a non negative finitely additive rotational invariant function on the cylinder sets on which a finitely additive integration theory may be based. Incidentally no use has been made of separability in this definition.

To enlarge the space of integrable functions one now proceeds as follows [DUNFORD 1958]: The quantity

$$||F|| = \varliminf_{P, \varepsilon > 0} \ \varlimsup_{\substack{T \\ |T - TP| < \varepsilon}} \int_H |F[T_x]| \, d\mu$$

where T is any bounded linear mapping of H into a finite dimensional subspace and P any finite dimensional projector, can be shown to be a norm on the μ-integrable F. The completion of the latter set with respect to this norm is "integrable in the extended sense": since the mapping

$$F \to \int F \, d\mu$$

is uniformly continuous with respect to the norm topology, it may be uniquely extended to the limits of μ-integrable Cauchy sequences. Rotation invariance now reads: For any $F[x]$ integrable in the extended sense and any linear norm preserving mapping $U: H \to H$, $F[Ux]$ is integrable (in the extended sense) and the integrals of $F[x]$ and $F[Ux]$ over H are equal. In the special case of separable H, we may map it norm-preservingly onto the square-convergent real sequences (x_i) and have

$$\mu(\{(x_i) \in L_2 | a_i \leqslant x_i \leqslant b_i \quad i = 1 \ldots n\}) =$$

$$= \pi^{-n/2} \int_{a_i}^{b_i} \ldots \int dy_1 \ldots dx_1 \, e^{-\sum_1^n y_\nu^2}$$

(A similar technique — definition of an integral over function alsbased on cylinder sets: $F[x] = F[Px]$ with finite dimensional P and completion with respect to a norm — has been used by Symanzik in his definition of the Friedrichs-Shapiro integral which he uses to solve the coupled differential equations for the Green's Functions of Euclidean quantum field theory [SYMANZIK **1964**, App. 1]. But the norm he gives depends on basic systems of projections and the completion i.e. integrability is in general not independent of the chosen system.)

To conclude let me just make a very short remark on a countably additive but non rotational invariant measure in the Hilbert space $L_2(0, 1)$. With the mapping

$$x(t) = \int_0^t f(\tau)\, d\tau$$

the Hilbert space vectors $f(t)$ are mapped into the space C of Wiener measure, which in turn induces a measure on H that is countably additive on a certain extension (corona) of H. (For the details see DUNFORD or MCSHANE).

It is well worth while to note this correspondence between Wiener and Gaussian Hilbert space integration since a large amount of well established results pertaining to the Wiener integral may serve to develop similar ones for the Hilbert space integral.

An example for this is the definition of the Fourier-Wiener transform (see e.g. [MCSHANE **1963**])

$$\tilde{F}[y] = \int F[\sqrt{2}\, x + i\, y]\, dx$$

that reads exactly alike in both theories of integration.

With these remarks that link up the theory of the Wiener integral and that of Gaussian integrals over separable Hilbert spaces I have come to the end of my "introductory survey" of continuous integration, knowing that it is incomplete but hoping nevertheless that I have succeded in conveying to you a vague idea of what some of the titles given in the list of references might possibly mean.

References

D. G. BABBITT, A summation procedure for certain Feynman integrals. Journ. Math. Phys. **4**, **36** (**1963**).

S. G. BRUSH, Functional integrals and statistical physics. Rev. Mod. Phys. **33**, 79 (1961).

R. H. CAMERON, W. T. MARTIN, Transformation of Wiener integrals under a general class of linear transformations. Trans. Amer. Math. Soc. **58**, **184** (**1945**).

R. H. CAMERON, W. T. MARTIN, Fourier-Wiener transform of analytic functionals. Duke Math. J. **12**, **489** (**1945 a**).

R. H. CAMERON, W. T. MARTIN, Fourier transforms of functionals belonging to L_2 over the space C. Duke Math. J. **14**, **99** (**1947**).

R. H. CAMERON, W. T. MARTIN, The transformation of Wiener integrals by nonlinear transformations. Trans. Amer. Math. Soc. **66**, **253** (**1949**).

R. H. CAMERON, R. E. GRAVES, Additive functionals on a space of continuous functions. Trans. Amer. Math. Soc. **70**, **160** **(1951)**.

R. H. CAMERON, The first variation of an indefinite Wiener integral. Proc. Amer. Math. Soc. **2**, **914** (1951 a).

R. H. CAMERON, A Simpson's rule for the numerical evaluation of Wiener's integrals in function space. Duke Math. J. **18**, **111** (1951 b).

R. H. CAMERON, R. E. FAGEN, Nonlinear transformations of Volterra type in Wiener space. Trans. Amer. Soc. **75**, 552 (1953).

R. H. CAMERON, The translation pathology of Wiener space. Duke Math. J. **21**, 623 (1954).

R. H. CAMERON, The generalized heat flow equation and a corresponding Poisson formula. Ann. of Math. (2) **59**, **434** (1954 a).

R. H. CAMERON, Nonlinear Volterra functional equations and linear parabolic differential systems. J. Analyse Math. **5**, **136** **(1956)**.

R. H. CAMERON, A family of integrals serving to connect the Wiener and Feynman integrals. J. Math. and Phys. **39**, 126 (1960).

R. H. CAMERON, The Ilstow and Feynman integrals. J. Analyse Math. **10**, 287 (1963).

F. COESTER, R. HAAG, Representation of states in a field theory with canonical variables. Phys. Rev. **117**, 1137 (1960).

J. L. DALETSKII, On the representation of the solutions of operator equations in the form of continuous integrals. Dokl. Akad. Nauk **134**, 1013 (1960).

J. L. DALETSKII, Continuous integrals connected with certain differential equations. Dokl. Akad. Nauk **137**, 268 (1961).

J. L. DALETSKII, Continuous integrals and characteristics connected with a group of operators. Dokl. Akad. Nauk **141**, 1290 (1961 a).

N. DUNFORD, J. SCHWARTZ, Linear operators I, p. 402 ff. Interscience New York (1958).

R. P. FEYNMAN, Space-time approach to non-relativistic quantum mechanics. Rev. Mod. Phys. **20**, 367 (1948).

R. P. FEYNMAN, Slow electrons in a polar crystal. Phys. Rev. **97**, 660 (1955).

S. V. FOMIN, On the inclusion of the Wiener integral in general Lebesque integral theory. Nauchn. Dokl. Vyssh. Shkoly Fiz.-Mat. Nauki **2**, 83 (1958).

K. O. FRIEDRICHS, H. N. SHAPIRO, Integration over Hilbert space and outer extensions. Proc. Math. Acad. Sci. U.S. **43**, 336 (1951).

K. O. FRIEDRICHS, Mathematical aspects of the quantum theory of fields. Interscience, New York (1953).

K. O. FRIEDRICHS, H. N. SHAPIRO et al., Integration of functionals. CIMS seminar notes, New York University (1957).

K. O. FRIEDRICHS, Remarques sur l'intégration des fonctionelles dans l'espace d'Hilbert. (Les problèmes mathématiques de la théorie quantique des champs. Paris 1959, p. 139.)

L. GARDING, A. WIGHTMAN, Representations of the commutation relations. Proc. Nat. Acad. Sci. U.S. **40**, 622 (1954).

L. GARDING, A. WIGHTMAN, Representation of the anticommutation relations. Proc. Nat. Acad. Sci. U.S. **40**, 617 (1954).

J. M. GELFAND, N. N. CHENTSOV, On the numerical computation of continuous integrals. JETP **31**, 1106 (1956).

J. M. GELFAND, A. M. YAGLOM, Integration in functional spaces and its application in quantum physics. Journ. Math. Phys. **1**, 48 (1960).

J. M. GELFAND, N. J. WILENKIN, Verallgemeinerte Funktionen, Bd. IV. (Hochschulbücher für Mathematik Bd. 50, Berlin, 1964).

L. Gross, Integration and nonlinear transformations in Hilbert space. Trans. Amer. Math. Soc. **94**, **404** (**1960**).

L. Gross, Measurable functions on Hilbert space. Trans. Amer. Math. Soc. **105**, **372** (**1962**).

L. Gross, Harmonic analysis on Hilbert space. Memoire Amer. Math. Soc. No. **46** (**1963**).

R. Haag, Canonical commutation relations in field theory and functional integration in Lectures in theoretical physics. Vol. III, Boulder **1960**, p. **353**.

M. Kac, Probability and related topics in physical science. (Interscience, London, 1959: Lectures in applied mathematics Vol. I).

J. M. Kovalchik, The Wiener integral. Russian Math. Surveys **18**, **97** (**1963**).

E. V. Maikov, On the inequivalence of two definitions of continuous integral. Nauchn. Dokl. Vyssh. Shkoly Fiz.-Mat. Nauk. **3**, 85 (1958).

G. Maruyama, Notes on Wiener integrals. Kodai Math. Sem. Rep., p. **41** (**1950**).

E. J. McShane, Integrals devised for special purposes. Bull. Amer. Math. Soc. **69**, 597 (1963).

B. S. Mitjagin, Eine Bemerkung über die quasiinvarianten Maße. Uspekhi Mat. Nauk. **26**: 5 (101), **191**, (**1961**).

E. W. Montroll, Markoff chains, Wiener integrals, and quantum theory. Comm. Pure Appl. Math. V, p. **415** (1952).

C. Morette, On the definition and approximation of Feynman's path integrals. Phys. Rev. **81**, **848** (1950).

E. Nelson, Feynman integrals and the Schrödinger equation. Journ. Math. Phys. **5**, 332 (1964).

T. G. Ostrom, The solution of linear equations by means of Wiener integrals. Bull. Amer. Math. Soc. **55**, **343** (**1949**).

R. E. Paley, N. Wiener, Fourier-transforms in the complex domain. Amer. Math. Soc. Colloquium Publications XIX (1930).

D. Ray, On spectra of second order differential operators. Trans. Amer. Math. Soc. **77**, 299 (1954).

F. Rohrlich, Functional differential calculus of operators. Journ. Math. Phys. **5**, 324 (1964).

Gerald Rosen, Approximate evaluation of Feynman functional integrals. J. Math. Phys. **4**, **1327** (**1963**).

S. S. Schweber, On Feynman quantization. Journ. Math. Phys. **3**, **831** (1962).

J. E. Segal, Ann. Math. **57**, **401** (**1953**).

J. E. Segal, Tensor algebras over Hilbert spaces. Trans. Amer. Math. Soc. **81**, 106 (1964).

J. E. Segal, Distributions in Hilbert space and canonical systems of operators. Trans. Amer. Math. Soc. **88**, **12** (1958).

J. E. Segal, Charactérisation mathématique des observables en théorie quantique des champs et ses consequences pour la structure des particules libres. (Les problèmes mathématiques de la théorie des champs. Paris **1959**, p. **57**.)

T. J. Seidman, Linear transformations of a functional integral. Comm. Pure Appl. Math. **12**, **611** (**1959**).

N. N. Sudakow, Lineare Mengen mit quasiinvariantem Maß. Dokl. Akad. Nauk SSSR **127**, 524 (1959).

G. Sunouchi, Harmonic analysis and Wiener integrals. Tohoku Math. J. (2) **3**, 187 (1951).

K. Symanzik, Über das Schwingersche Funktional in der Feldtheorie. Zeitschr. f. Naturforschung **9**, 809 (**1954**).

K. Symanzik, A modified model of Euclidean quantum field theory. NYU-preprint JMM-NYU **327** (**1964**).

V. S. Vladimirov, The approximate evaluation of Wiener integrals. Uspekhi Mat. Nauk **15**, No. **4** (**94**), p. **129** (**1960**).

V. Volterra, Les problèmes qui resortent du concept de fonction de ligne. Sitzgsber. Berliner Math. Gesellsch. XIII. Jg. **6**. Sitzg. **1914**.

V. Volterra, Theory of functionals and of integral and integro-differential equations (**1930**).

D. A. Woodward, A general class of linear transformations of Wiener integrals. Trans. Amer. Math. Soc. **100**, **459** (**1960**).

J. Yeh, Nonlinear Volterra functional equations and linear parabolic differential systems. Trans. Amer. Math. Soc. **95**, **408** (**1960**).

High Energy Quantum Electrodynamics I*

By

H. Mitter

Max Plank Institut für Physik und Astrophysik, München

Yet the thing is not altogether desperate
I. Newton

1. Introduction

Quantum electrodynamics is up to the present day the only example of a quantised field theory the predictions of which agree with experiment to a high degree of precision. In the following lecture we shall adopt the somewhat optimistic standpoint, that this agreement is not accidential and that it is therefore worthwile to spend time on explaining the formal structure of the theory also in regions, where the connection with the experiment is not a close one. We shall along with it stress the field aspect: we shall study expectation values of products of field operators (Green's functions), which are the quantum analog to classical field functions, and shall study these Green's functions as response functions to small, external perturbations: the analogy to the classical concept of a field strength and test charge is evident.

Thus we start from a Lagrangian density describing an electromagnetic field A_μ resp. $F_{\mu\nu}$ and an electron-positron field ψ [1] in interaction with each other and with external, classical source s[2] J_μ, A_μ^e (which we imagine switched on in the remote past and switched off in the remote future).

---

* Lecture given at the IV. Internationalen Universitätswochen für Kernphysik, Schladming, 25 February—10 March 1965. Dedicated to Prof. Urban on the occasion of his 60th birthday.

[1] The metric is $g^{\mu\nu} = \begin{pmatrix} -1 & & & \\ & 1 & & \\ & & 1 & \\ & & & 1 \end{pmatrix}$, the γ's are $\gamma^0 = \beta = \gamma^{0+}$, $\gamma_k = -\gamma_k{}^+$

$\{\gamma^\mu, \gamma^\nu\} = -2\, g^{\mu\nu}$, the charge of the bare electron is $-e_0$.

[2] We will neither assume, that A^e and J are related, nor that J_μ is conserved in order that all components are independent, so that we can perform functional differentiation with respect to J and A^e. The condition $\partial^\mu J_\mu = 0$ has to be imposed after all functional differentiations are carried out.

$$L = L_M + L_e + L_i + L_s \tag{1}$$

$$\begin{aligned} L_M &= -\frac{1}{2} F_{\mu\nu}(\partial^\mu A^\nu - \partial^\nu A^\mu) + \frac{1}{4} F_{\mu\nu} F^{\mu\nu} \\ L_e &= -\bar{\psi}\left(\frac{1}{i}(\gamma \cdot \partial) + m_0\right)\psi \\ L_i &= +e_0 j_\mu A^\mu \\ L_s &= J_\mu A^\mu + e_0 j^\mu A_\mu^e. \end{aligned} \tag{2}$$

For the current of the electron-positron-field we take for the moment the usual expression

$$j_\mu = \frac{1}{2}[\bar{\psi}, \gamma_\mu \psi] \tag{3}$$

(later we shall give a more sophisticated form). We consider F and A as independent variables. The field equations can be deduced by standard methods and read

$$\begin{aligned} \left(\frac{1}{i}\gamma \cdot \partial + m_0\right)\psi + e_0 \gamma_\mu \psi(x)\,(A^\mu(x) + A^{\mu e}(x)) &= 0 \\ \Box A_\mu + e_0 j_\mu + J_\mu &= 0 \\ F^{\mu\nu} &= \partial^\mu A^\nu - \partial^\nu A^\mu. \end{aligned} \tag{4}$$

We shall impose the standard canonical commutation relations, which read e.g. for the Dirac field

$$\begin{aligned} \{\psi(x), \bar{\psi}(y)\}_{x_0 = y_0} &= -\gamma_0\,\delta(\vec{x} - \vec{y}) \\ \left[A_\mu(x), \frac{\partial A_\nu(y)}{\partial y_0}\right]_{x_0 = y_0} &= i\, g_{\mu\nu}\,\delta(\vec{x} - \vec{y}). \end{aligned} \tag{5}$$

2. Transformation to "Bound Interaction Picture"

We want to get rid of the source terms in the field equations. We will therefore transform the state vectors and field operators in such a way that[3]

(I) the new operators fulfill the source-free-equations,

(II) the new states vary in time according to

$$\mathfrak{H}_s^{(\text{int})}(t) = \int H_s^{(\text{int})}(x)\, d^3x.$$

This is accomplished by a unitary transformation. Point (II) means

$$|\text{int}, t_2\rangle = U_s(t_2, t_1)\,|\text{int}, t_1\rangle, \tag{6}$$

---

[3] We shall characterize the Heisenberg operators *for the moment* by the superscript (H) and the new operators by the superscript (int).

where

$$i \frac{\partial}{\partial t} U_s(t, t') = \mathfrak{H}_s^{(\text{int})}(t)\, U_s(t, t'). \tag{7}$$

If we impose the boundary condition $U_s(t, t) = 1$ we have by standard time-dependent perturbation theory

$$U_s(t_1, t_2) = T \exp\left(i \int_{t_1}^{t_2} \mathfrak{H}_s^{(\text{int})}(t)\, dt\right) = T \exp\left(i \int_{t_2}^{t_1} L_s(x)\, d^4x\right), \tag{8}$$

where U_s has the group property and is unitary:

$$U_s(t_1, t_2) = U_s(t_1, t_3)\, U_s(t_3, t_2)$$

$$U_s^{-1}(t_1\, t_2) = U_s^{+}(t_1\, t_2) = U_s(t_2\, t_1).$$

We identify the Heisenberg states with the interaction states at $t = 0$

$$\frac{1}{\sqrt{N}} |\text{int}, t = 0\rangle = |H\rangle.$$

N is some normalisation constant. Then any operator B has to be transformed according to

$$B^{(\text{int})} = U_s(t, 0)\, B^{(H)}\, U_s^{-1}(t, 0)$$

and we can easily verify (using eq. (7) and its hermitian adjoint)

$$i \frac{\partial}{\partial t} B^{(\text{int})} = [B^{(\text{int})}, \mathfrak{H}^{(\text{int})} - \mathfrak{H}_s^{(\text{int})}]$$

which is postulate (I).

We consider now some important expectation values. At first we express the Heisenberg vacuum by the interaction vacua at remote times. Consider the Heisenberg state describing the vacuum before the sources are switched on:

$$|0_H^{\text{in}}\rangle = \frac{1}{\sqrt{N}} |0_{\text{int}}, t = 0\rangle = U_s(0, -\infty)\, |0_{\text{int}}, t = -\infty\rangle \frac{1}{\sqrt{N}}.$$

Similarly we find for the vacuum after the sources are switched off

$$|0_H^{\text{out}}\rangle = \frac{1}{\sqrt{N}} U_s(0, +\infty)\, |0_{\text{int}}, t = +\infty\rangle.$$

The total change in the vacuum, induced by the sources, is then

$$V = \langle 0_H^{\text{out}} | 0_H^{\text{in}}\rangle = \frac{1}{N} \langle 0_{\text{int}}, t = +\infty |\, U_s(+\infty, -\infty)\, | 0_{\text{int}}, t = -\infty\rangle.$$

In order to avoid too many superscripts we shall use a more compact notation: we write simply $|\text{out}\rangle$ resp. $|\text{in}\rangle$ for the vacuum state characterizing a situation at remote times, drop all superscripts (H), (int) and indicate the representation by subscripts s (= presence of sources) and/or i (= presence of interaction). Thus for instance

$$\langle\text{out}|\, B\, |\text{in}\rangle_{s,i} = \text{denotes the Heisenberg expectation value} = \\ = \langle 0_H{}^{\text{out}}|\, B^{(H)}\, |0_H{}^{\text{in}}\rangle,$$

while the expectation value in the bound interaction picture is

$$\langle\text{out}|\, B\, |\text{in}\rangle_i = \text{in explicit notation} = \langle 0_{\text{int}}, t = +\infty|\, B^{(\text{int})}\, |0_{\text{int}}, t = -\infty\rangle.$$

Thus our equation for V reads in short notation

$$V = \langle\text{out}|\text{in}\rangle_{s,i} = \frac{1}{N}\langle\text{out}|\, U_s(+\infty, -\infty)\, |\text{in}\rangle_i \tag{9}$$

in words: the total change due to the sources may be calculated by means of operators fulfilling equations without sources, but containing interaction. The normalization will be correct, if we take

$$N = \langle\text{out}|\text{in}\rangle_i,$$

i.e. the change in the vacuum without sources. In perturbation theory N is the sum of all vacuum diagrams

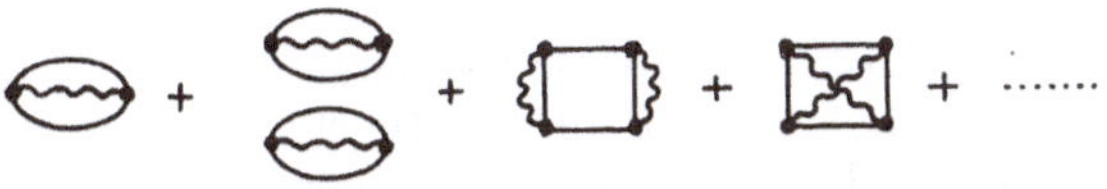

We now consider the vacuum expectation value of a time-ordered product of Heisenberg operators

$$\langle\text{out}|\, T\, B(t_1)\ldots B(t_n)\, |\text{in}\rangle_{s,i} \equiv \langle 0_H{}^{\text{out}}|\, T\, B^{(H)}(t_1)\ldots B^{(H)}(t_n)\, |0_H{}^{\text{in}}\rangle =$$

$$= \frac{1}{N}\langle 0_{\text{int}}, t = +\infty|\, U(+\infty, 0)\, T(U^{-1}(t_1, 0)\, B^{(\text{int})}(t_1)\, U(t_1, 0)\, U^{-1}(t_2, 0)\cdot \\ \cdot B^{(\text{int})}(t_2)\ldots B^{(\text{int})}(t_n)\, U(t_n, 0))\, U(0, -\infty)|0_{\text{int}}, t = -\infty\rangle = \cdot/. .$$

Using the group property for U we have

$$\cdot/. = \frac{1}{N}\langle 0_{\text{int}}, t = +\infty|\, U(+\infty, 0)\, T(U(0, t_1)\, B^{(\text{int})}(t_1)\, U(t_1, t_2)\cdot \\ \cdot B^{(\text{int})}(t_2)\ldots U(t_n, 0))\, U(0, -\infty)\, |0_{\text{int}}, t = -\infty\rangle =$$

$$= \frac{1}{N}\langle 0_{\text{int}}, t = +\infty|\, T\, B^{(\text{int})}(t_1)\ldots B^{(\text{int})}(t_n)\exp i\int_{-\infty}^{+\infty}\mathfrak{H}^{(\text{int})}\,dt\, |0_{\text{int}}, t = -\infty\rangle.$$

So we have, in compact notation, the following relation between Heisenberg- and interaction picture

$$\langle\text{out}|\, T\, B(t_1)\ldots B(t_n)\, |\text{in}\rangle_{s,i} = \\ = \frac{1}{\langle\text{out}|\text{in}\rangle_i}\langle\text{out}|\, T\, B(t_1)\ldots B(t_n)\exp i\int H_s(x)\, d^4x\, |\text{in}\rangle_i \tag{10}$$

It is easy to do perturbation theory with respect to the external field: One has to expand just the exponential, since the operators are already the source-free ones and the vacua refer to remote times (sources not switched on or already switched off, resp.). For V the expansion would be, for instance

$$V = V|_{J=A^e=0} + \int \Big[J_\mu(x) \left(\frac{\delta V}{\delta J_\mu(x)}\right)_{J=A^e=0} + \\ + A_\mu^e(x) \left(\frac{\delta V}{\delta A_\mu^e(x)}\right)_{J=A^e=0} \Big] d^4x + \ldots \tag{11}$$

By insertion of (8), (9) one may calculate this expression term by term.

3. Green's Functions, Vacuum Current

If we take for B the field operators we get the Green's functions[4], for example

$$G(x, y) = S(x\,y) \cdot V = \frac{i}{N} \langle \text{out}|\, T\, \psi(x)\, \bar{\psi}(y) \exp i \int L_s\, d^4x\, |\text{in}\rangle_i \tag{12}$$

In connection with the Schwinger-Dyson-equations it is more convenient to work with S instead of G. In the absence of sources the functions are identical. Sometimes it is convenient to generate the electron Green's functions by introducing additional "classical" spinor sources $\eta(x)$, $\bar{\eta}(x)$, i.e. adding a term

$$L_{\text{Sp}} = -\int (\eta(x)\, \bar{\psi}(x) + \bar{\eta}(x)\, \psi(x))\, d^4x$$

to the Lagrangian where the η and $\bar{\eta}$ anticommute with themselves and with ψ, $\bar{\psi}$. If we imagine V expanded in powers of η and $\bar{\eta}$, the Green's functions are the coefficients, i.e. $G(x, y)$ is the coefficient of $\bar{\eta}(x)\, \eta(y)$ etc. S may be obtained in the same way from $\ln V$. The same formalism may be obtained by introduction of bilinear, nonlocal sources

$$L_B = \int d^4x\, d^4y\, q_{\alpha\beta}(x, y)\, \psi_\alpha(x)\, \bar{\psi}_\beta(y).$$

This has the advantage, that the rather strange spinor sources are avoided. We will use the spinor sources only in context with gauge

---

[4] It has to be noted, that we have taken a form, in which the closed loop parts are already removed because of the denominator in (12). This differs from the functions used in the literature. In addition we omit a suffix indicating that S and D are the Feynman functions, since no other functions will be considered here. Thus our S corresponds e.g. to $-\, S'_{1R}(x)$ of Jauch-Rohrlich [1], $D_{\mu\nu}$ corresponds to $+\, D'_{1R}$ of that book.

properties. For all other purposes of this lecture these formalisms are not needed.

Similar to G we can define a photon Green's function

$$G_{\mu\nu}(x, y) = i\,\frac{\langle\text{out}|\,T\,A_\mu(x)\,A_\nu(y)\,\exp i\int L_s\,d^4x\,|\text{in}\rangle_i}{\langle\text{out}|\text{in}\rangle_i}. \tag{13}$$

Again the use of quantities like S is more convenient. This time we can generate them readily from $\ln V$. We define

$$\mathfrak{A}_\mu(x) = \frac{1}{i}\,\frac{\delta}{\delta J^\mu(x)}\ln V = \frac{1}{V N}\langle\text{out}|\,T\,A_\mu(x)\exp i\int L_s\,d^4x\,|\text{in}\rangle_i \tag{14}$$

and

$$D_{\mu\nu}(x, y) = \frac{\delta\mathfrak{A}_\mu(x)}{\delta J^\nu(y)} = \frac{1}{i}\,\frac{\delta^2}{\delta J^\mu(x)\,\delta J^\nu(y)}\ln V = \tag{15}$$

$$= \frac{i}{V}\left(\frac{\langle\text{out}|\,T\,A_\mu(x)\,A_\nu(y)\exp i\int L_s\,d^4x\,|\text{in}\rangle_i}{N}\right) - i\,\mathfrak{A}_\mu(x)\,\mathfrak{A}_\nu(y).$$

In the limit of $J \to 0$ the quantity $\mathfrak{A}_\mu$ vanishes. $G_{\mu\nu}$ is then identical with $D_{\mu\nu}$. Note that the Green's functions are the expansion coefficients of some basic functional (V or $\ln V$, resp.) in powers of the external sources, i.e. they are the response to small external perturbations. This is true also for higher Green's functions. The expansion coefficients of $\ln V$ are the truncated functions (η-functions) and are represented by the sum of all connected Feynman diagrams. As an example, how important physical quantities are related to Green's functions we consider the current induced in the vacuum.

By functional differentiation we find

$$\frac{1}{e_0}\,\frac{\delta\ln V}{\delta A_\mu^e(x)} = \frac{i}{V}\,\langle\text{out}|\,j^\mu(x)\,|\text{in}\rangle_{s,i}.$$

If we now vary A_μ^e by a gauge transformation

$$\delta A_\mu^e(x) = \partial_\mu\,\delta\lambda(x), \tag{16}$$

we have

$$\delta\ln V = e_0\,\frac{i}{V}\int\langle\text{out}|\,j_\mu(x)\,|\text{in}\rangle_{s,i}\,\partial_\mu\,\delta\lambda(x) =$$

$$= -\frac{i\,e_0}{V}\int\partial_\mu\,\langle\text{out}|\,j_\mu(x)\,|\text{in}\rangle_{s,i}\,\delta\lambda(x).$$

Since $\delta\lambda$ is arbitrary, the vacuum current should be conserved

$$\partial_\mu\,\langle\text{out}|\,j_\mu(x)\,|\text{in}\rangle_{s,i} = 0. \tag{17}$$

If we now take (3) we see that

$$\frac{i}{V}\,\langle\text{out}|\,j_\mu(x)\,|\text{in}\rangle_{s,i} = \frac{1}{2}\,\frac{i}{V N}\langle\text{out}|\,T\,[\bar\psi_\alpha,(\gamma_\mu)_{\alpha\beta}\,\psi_\beta]\exp i\int L_s\,d^4x\,|\text{in}\rangle_i =$$

$$= -\,\mathrm{Sp}\,\gamma_\mu\,S(x, x).$$

This expression is singular even in the case of a free electron and the divergence is not zero[5]. So we have to be more careful. If we consider $S(x + (\varepsilon/2), x - (\varepsilon/2))$ we note, that this quantity is not gauge-invariant: we know that the whole theory is invariant only against (16), if we change at the same time $\psi(x)$ into $e^{i e_0 \delta\lambda(x)}\psi$. So $S(x, y)$ is transformed into $e^{-i e_0(\delta\lambda(x) - \delta\lambda(y))} S(x, y)$. But since

$$\delta\lambda(x) - \delta\lambda(y) = \int_y^x d\xi_\mu \, \partial^\mu \delta\lambda = \delta \int_y^x d\xi^\mu A_\mu^e(\xi)$$

we notice, that

$$\tilde{S}(x, y) = S(x, y) \exp\left[+ i e_0 \int_x^y d\xi^\mu A_\mu^e(\xi)\right] \tag{18}$$

is gauge invariant: the change due to the variation of A_μ^e is compensated by the corresponding change of the ψ's. Therefore our vacuum current is conserved, if we take[6]

$$\frac{1}{V} \langle \text{out} |\, j_\mu(x) \,| \text{in} \rangle_{s,i} =$$

$$= i \lim_{\varepsilon \to 0} \operatorname{Sp} \gamma_\mu S\left(x + \frac{\varepsilon}{2}, x - \frac{\varepsilon}{2}\right) \exp i e_0 \int_{x-\varepsilon/2}^{x+\varepsilon/2} d\xi^\mu A_\mu^e(\xi) \equiv \tag{19}$$

$$\equiv i \operatorname{Sp} \tilde{S}(x, x)\gamma_\mu = \lim_{\varepsilon \to 0} \operatorname{Sp} \tilde{S}\left(x + \frac{\varepsilon}{2}, x - \frac{\varepsilon}{2}\right)\gamma_\mu .$$

We note that ε_μ has to be a space-like vector if we want to preserve the canonical commutation relations and that the limit has to be perlormed in a symmetrical way (so that odd terms in ε are absent). The fine integral in the exponent is Lorentz-invariant if we take the straight line $x - y$ as path of integration.

We want to stress once more, that the physically interesting case is the one, in which the external sources are absent, so that we are interested in the limit of (19) for $A_\mu^e \to 0$. Since the factor in front of the exponential is singular for $\varepsilon \to 0$, we are, however, not allowed to drop the exponential, but have to expand carefully. This will be done later.

---

[5] That the interaction does not change these facts is clear from Lehmann's results, who showed, that the propagator is in this case at least as singular as for a free field.

[6] The importance of the exponential has been observed already in 1934 [2], [3] and has been discussed since that time frequently [4], [5].

4. Gauge Properties of Propagators

We shall now study very briefly the behaviour of our Green's functions under a special class of gauge transformations. For this, we consider the generating functional with spinor sources present, but without external potential

$$Z(\eta, \bar{\eta}, J_\mu) = \frac{1}{N} \langle \text{out} | \, T \exp i \int (\psi(x)\, \bar{\eta}(x) + \bar{\psi}(x)\, \eta(x) + \\ + J_\mu(x)\, A^\mu(x))\, d^4x \, |\text{in}\rangle_i. \tag{20}$$

If we then make a gauge transformation and change ψ and A according to

$$\psi \to e^{-ie_0 \Lambda(x)}\, \psi(x)$$
$$A_\mu \to A_\mu + \partial_\mu \Lambda$$

it is clear that the same change can be obtained by changing the sources appropriately, so that we have

$$Z_\Lambda(\eta, \bar{\eta}, J_\mu) = Z_{\Lambda=0}(\eta\, e^{-ie_0\Lambda}, \bar{\eta}\, e^{ie_0\Lambda}, J_\mu) \cdot e^{i \int J_\mu \partial^\mu \Lambda\, d^4x} \tag{21}$$

or in differential form

$$i\, \frac{\delta Z_{e_0}}{\delta \Lambda} = \left(\partial^\mu J_\mu + e_0\, \eta\, \frac{\delta}{\delta \eta} - e_0\, \bar{\eta}\, \frac{\delta}{\delta \bar{\eta}} \right) Z_\Lambda. \tag{22}$$

From this, we can, in principle, determine the change of all Green's functions under gauge transformations, since the Green's functions are generated by Z. We restrict the discussion to a special transformation, which relates the manifestly covariant gauges[7] and refer for the general case to a paper by Zumino [6].

We consider a transformation of Z, which produces an infinitesimal change $Z \to Z' = Z + \delta Z$

$$\delta Z = \frac{i}{2} \int \frac{\delta}{\delta \Lambda(x)}\, \delta \hat{M}(x - y)\, \frac{\delta}{\delta \Lambda(y)}\, Z\, d^4x\, d^4y, \tag{23}$$

where $\delta \hat{M}$ is an arbitrary, Lorentz-invariant function.

Since effectively Z is changed via Λ only, any gauge-invariant quantity will remain invariant under the $\hat{M}$ transformation also. If we consider the Z with spinor sources absent we can study the effect of finite transformations. Inserting (22) into (23) and adding up we get

$$Z'(0, 0, J_\mu) = \exp\left\{ -\frac{i}{2} \int \partial^\mu J_\mu(x)\, \delta \hat{M}(x - y)\, \partial^\nu J_\nu(y)\, d^4x\, d^4y \right\} Z(0, 0, J_\mu). \tag{24}$$

---

[7] These gauges are characterized by the property, that they do not introduce external directions. The radiation gauge (Coulomb gauge) is not manifestly covariant, in spite of the fact, that the whole theory is a Lorentz-invariant one also, if the quantisation is done in this gauge.

Now we study the change of the Green's functions[8] studied before. First we observe, that the change of

$$\mathfrak{A}_\mu(x) = \frac{1}{i}\frac{\delta \ln Z(0,0,J_\mu)}{\delta J^\mu}$$

is given by

$$\mathfrak{A}_\mu'(x) = \frac{1}{i}\frac{\delta \ln Z'(0,0,J_\mu)}{\delta J^\mu} = \mathfrak{A}_\mu(x) - \partial_\mu(x)\int \delta\hat{M}(x-y)\,\partial^\nu(y)\,J_\nu(y)\,d^4y, \tag{25}$$

where we have used the symmetry of δM. By functional differentiation we get

$$D_{\mu\nu}'(x,y) = D_{\mu\nu}(x,y) + \partial_\mu(x)\,\partial_\nu(x)\,\delta\hat{M}(x-y). \tag{26}$$

In momentum space this means

$$D_{\mu\nu}'(k) = D_{\mu\nu}(k) - k_\mu\, k_\nu\, \delta\hat{M}(k^2). \tag{27}$$

So the transformation adds a term proportional to $k_\mu\, k_\nu$. We notice the form of $D_{\mu\nu}$ for some special cases: If

$$D_{\mu\nu}(k) = \left(g_{\mu\nu} + \frac{k_\mu\, k_\nu}{k^2}G\right)D(k^2) \tag{28}$$

then we have for

$G = 0$ the Feynman- or Lorentz-gauge,

$G = +2$ the Yennie-Fried-gauge [7] (which is of advantage in dealing with infrared problems),

$G = -1$ the Landau gauge [8] which we will use later.

The connection between these gauges is given by (27). One has sometimes to be careful in doing the transformation because the corresponding δM may be singular.

Now we consider the change in the electron propagator. Since we have

$$S(x,y) = \frac{i}{Z}\frac{\delta^2 Z}{\delta\bar{\eta}(x)\,\delta\eta(y)}\bigg|_{\eta=\bar{\eta}=0}$$

we see that

$$\frac{\delta(Z(0,0,J_\mu)\,S(x\,y))}{\delta\Lambda(z)} =$$

$$= \frac{\delta^2}{\delta\bar{\eta}(x)\,\delta\eta(y)}\left(\partial^\mu J_\mu(z) + e_0\,\eta(z)\,\frac{\delta}{\delta\eta(z)} - e_0\,\bar{\eta}(z)\,\frac{\delta}{\delta\bar{\eta}(z)}\right)Z\big|_{\eta=\bar{\eta}=0} =$$

$$= \frac{1}{i}\left(\partial^\mu J_\mu(z) + e_0\,\delta(y-z) - e_0\,\delta(x-z)\right)Z(0,0,J_\mu)\,S(x,y).$$

[8] Literally the Green's functions considered here agree with our previous definitions only for $A_\mu^e = 0$. Since the interesting case is always the limit $A_\mu^e = J_\mu = 0$, this does not matter.

If we now calculate the change induced by the $\hat{M}$ transformation on $Z(0, 0, J_\mu)\, S(x, y)$ we get by differentiation of (23) with respect to the spinor sources

$$\delta(Z(0, 0, J_\mu)\, S(x, y)) =$$

$$= \frac{i}{2} \int d^4x'\, d^4y'\, \frac{\delta}{\delta \Lambda(x')}\, \delta\hat{M}(x' - y')\, \frac{\delta}{\delta \Lambda(y')}\, (Z(0, 0, J_\mu)\, S(x, y)) =$$

$$= \frac{i}{2\, i^2} \int ((\partial^\mu J_\mu(x') + e_0\, \delta(y - x') - e_0\, \delta(x - x'))\, \delta\hat{M}(x' - y') \cdot$$
$$\cdot (\partial^\mu J_\mu(y') + e_0\, \delta(y - y') - e_0\, \delta(x - y'))\, (Z(0, 0, J_\mu)\, S(x, y))\, d^4x'\, d^4y'.$$

For $J_\mu \to 0$ we get

$$\delta S(x, y) = \frac{e_0{}^2}{i}\, (\delta\hat{M}(0) - \delta\hat{M}(x - y))\, S(x, y).$$

For finite transforms we get therefore, adding up again

$$S'(x, y) = \exp\, [i\, e_0{}^2 (\delta\hat{M}(x - y) - \delta\hat{M}(0))] \cdot S(x, y). \tag{29}$$

5. The Schwinger-Dyson-Equations for the Propagators

We shall now derive exact equations for the Green's functions. As a preliminary step we consider the second time derivative of the operator

$$B_\mu = T\, A_\mu \exp i \int_{-\infty}^{+\infty} L_S\, d^4x' =$$

$$= T\left(\exp i \int_{x_0}^{\infty} L_S\, d^4x'\right) A_\mu(x)\; T\left(\exp i \int_{-\infty}^{x_0} L_S\, d^4x'\right).$$

We have

$$\frac{\partial B_\mu}{\partial t} = T\left(\exp i \int_{x_0}^{\infty} L_S(x)\, d^4x'\right)\Big[- i \int L_S(\vec{x}', x_0)\, d^3x'\, A_\mu(x) +$$

$$+ \dot{A}_\mu(x) + i\, A_\mu(x) \int L_S(\vec{x}', x_0)\, d^3x'\Big]\; T\left(\exp i \int_{-\infty}^{x_0} L_S(x)\, d^4x\right).$$

We will now use the canonical commutation relations for A_μ in the form[9]

---

[9] We use here the Feynman gauge.

$$[A_\mu(\vec{x}, t), \dot{A}_\nu(\vec{y}, t)] = i g_{\mu\nu} \delta(\vec{x} - \vec{y})$$
$$[A_\mu(\vec{x}, t), A_\nu(\vec{y}, t)] = 0.$$

We see that L_S commutes with A_μ at equal times. So we have

$$\frac{\partial B_\mu}{\partial t} = T\left(\dot{A}_\mu \exp i \int L_S \, d^4x'\right).$$

In the same way we find for the second derivative

$$\frac{\partial^2 B_\mu}{\partial t^2} = T\left(\exp i \int_{x_0}^{\infty} L_S \, d^4x'\right)\left(\ddot{A}_\mu(x) - i \int d^3x' \, J^\nu(\vec{x}', x_0) \cdot\right.$$
$$\left. \cdot [A_\nu(\vec{x}', x_0), \dot{A}_\mu(x)]\right) T\left(\exp i \int_{-\infty}^{x_0} L_S(x') \, d^4x'\right) =$$
$$= T \ddot{A}_\mu(x) \exp i \int L_S \, d^4x' + J_\mu(x) \, T\left(\exp i \int L_S \, d^4x'\right).$$

If we use (14) we get therefore the equation (we observe, that A_μ satisfies the sourcefree equation in the bound interaction picture!)

$$\Box \mathfrak{A}_\mu(x) = \frac{-e_0}{V} \langle \text{out} | j_\mu(x) | \text{in} \rangle_{s,i} - J_\mu(x).$$

Use of (19) provides us with an equation between $\mathfrak{A}$ and $\tilde{S}$

$$\Box \mathfrak{A}_\mu(x) = -i e_0 \, \mathrm{Sp} \, \gamma_\mu \tilde{S}(x, x) - J_\mu(x).$$

If we integrate with a Green's function of the free D'Alembert equation

$$\Box D_{\mu\nu}^0(x, y) = -g_{\mu\nu} \delta(x - y) \tag{30}$$

we get[10]

$$\mathfrak{A}_\mu(x) = \int D_{\mu\nu}^0(x, y) \, J^\nu(y) \, d^4y - \frac{e_0}{i} \int D_{\mu\nu}^0(x, y) \, \mathrm{Sp} \, \gamma^\nu \tilde{S}(y, y) \, d^4y. \tag{31}$$

By functional differentiation we find an equation for $D_{\mu\nu}$ [6]:

$$D_{\mu\nu}(x, y) = D_{\mu\nu}^0(x, y) - \frac{e_0}{i} \int D_{\mu\varrho}^0(x, y') \frac{\delta}{\delta J^\nu(y)} \mathrm{Sp} \, \gamma^\varrho \tilde{S}(y', y') \, d^4y'. \tag{32}$$

Proceeding in the same way for S (i.e. observing the source-free field equation and the canonical commutation relation (5) for ψ) we arrive at the equation

$$\left(\frac{1}{i} \gamma_\mu \partial^\mu + m_0\right) S(x, y) = \delta(x - y) +$$
$$+ \frac{i e_0}{V N} \gamma_\mu \langle \text{out} | T \psi(x) \bar{\psi}(y) A^\mu(x) \exp\left(i \int L_S \, d^4x'\right) | \text{in} \rangle_i +$$
$$+ e_0 \gamma^\mu A_\mu^e S(x, y). \tag{32 a}$$

---

[10] This form holds formally in any covariant gauge, if $D_{\mu\nu}^0$ is chosen appropriately: a change by a M-transformation may be absorbed into $D_{\mu\nu}^0$.

From the definition (12) of S we see

$$\frac{\delta S(x,y)}{\delta J_\mu(z)} = \frac{i^2}{VN}\langle \text{out}|\, T\,\psi(x)\,\bar\psi(y)\,A^\mu(z)\exp\left(i\int L_S\,d^4x'\right)|\text{in}\rangle_i - \\ - i\,S(x,y)\,\mathfrak{A}^\mu(z) \tag{33}$$

so that we get the following equation [9] for S

$$\left[\left(\gamma_\mu, \frac{1}{i}\,\partial^\mu - e_0\,\bar{\mathfrak{A}}^\mu + i\,e_0\frac{\delta}{\delta J_\mu(x)}\right) + m_0\right] S(x,y) = \delta(x-y), \tag{34}$$

where we have introduced the abbreviation

$$\bar{\mathfrak{A}}_\mu = \mathfrak{A}_\mu + A_\mu{}^e. \tag{35}$$

We write this equation formally as an integral equation for S by introduction of a mass operator

$$\int M(x,x')\,S(x',y)\,d^4x' = \left(m_0 + i\,e_0\,\gamma_\mu\frac{\delta}{\delta J_\mu(x)}\right) S(x,y) \tag{36}$$

so that (32 a) can be written in the symbolic form

$$S^{-1} = \left(\gamma_\mu, \frac{1}{i}\,\partial^\mu - e_0\,\bar{\mathfrak{A}}^\mu\right) + M. \tag{37}$$

Thereby it has to be understood, that S^{-1} is the inverse of S in the sense

$$\int S_{\alpha\gamma}{}^{-1}(x,x')\,S_{\gamma\beta}(x',y)\,d^4x' = \delta(x-y)\,\delta_{\alpha\beta}.$$

In momentum space this corresponds to the ordinary inverse.

We eliminate now the functional derivative with respect to the external current. We consider S as a functional of $\mathfrak{A}$ instead of J i.e. we do a functional transform of variables, such that

$$\frac{\delta\,\cdot}{\delta J^\varrho(x)} = \int \frac{\delta\,\cdot}{\delta\bar{\mathfrak{A}}_\nu(x')}\,\frac{\delta\bar{\mathfrak{A}}_\nu(x')}{\delta J^\varrho(x)}\,d^4x' = \int \frac{\delta\,\cdot}{\delta\bar{\mathfrak{A}}_\nu(x')}\cdot D_{\nu\varrho}(x,x')\,d^4x'. \tag{38}$$

In addition we write a new symbol for the functional derivative of S^{-1} with respect to $\bar{\mathfrak{A}}$, which we call the vertex function:

$$\frac{\delta S^{-1}(x,y)}{\delta\bar{\mathfrak{A}}_\mu(z)} \underset{\text{Def}}{=} -\,e_0\,\Gamma^\mu(x\,y|z). \tag{39}$$

Then we have

$$\frac{\delta S(x,y)}{\delta J_\varrho(z)} = e_0\int D^{\varrho\nu}(z,z')\,S(x,x')\,\Gamma_\nu(x'\,y'|z')\,S(y',y)\,d^4x'\,d^4y'\,d^4z'. \tag{40}$$

Because of (33) this says, that the vertex is essentially the totally "amputated" three-point function (we have to "amputate" a "S-leg" at each electron-end and a "D-leg" at the photon end to get Γ). For the mass operator we have

$$M(x, y) = m_0\,\delta(x-y) + i\,e_0^2 \int D_{\varrho\nu}(x, z')\,\gamma^\varrho\,S(x, x')\,\Gamma^\nu(x' y|z')\,d^4x'\,d^4z'. \tag{41}$$

Together with (37) this provides us with an equation for S, which is free of functional derivatives, but involves Γ. This form is essentially due to Dyson [10]. In a similar way we can eliminate the derivative from the equation for D. We introduce a polarisation operator

$$\varrho^{\mu\nu}(x, y) = \frac{1}{i\,e_0}\,\frac{\delta}{\delta\bar{\mathfrak{A}}_\nu(y)}\,\mathrm{Sp}\,\gamma^\mu\,\tilde{S}(x, x). \tag{42}$$

Then the equation for D becomes

$$D_{\mu\nu}(x, y) = D_{\mu\nu}{}^0(x, y) - e_0^2 \int D_{\mu\lambda}{}^0(x, x')\,\varrho^{\lambda\sigma}(x', y')\,D_{\sigma\nu}(y', y)\,d^4x'\,d^4y'. \tag{43}$$

It is possible to express the polarisation operator by the vertex by insertion of the definition (19) of $\tilde{S}$. We have, however, to differentiate the exponential too, since $\bar{\mathfrak{A}}$ involves A^e. We get

$$\varrho^{\mu\nu}(x, y) = \lim_{\varepsilon\to 0} \exp\left[i\,e_0 \int\limits_{x+(\varepsilon/2)}^{x-(\varepsilon/2)} d\xi^\varrho\,A_\varrho^e(\xi)\right]\left\{\mathrm{Sp}\,\gamma^\mu\,S\left(x+\frac{\varepsilon}{2}, x-\frac{\varepsilon}{2}\right)\cdot\right.$$
$$\cdot \int\limits_{x+(\varepsilon/2)}^{x-(\varepsilon/2)} d\xi^\nu\,\delta(\xi-y) + \tag{44}$$
$$\left. + \frac{1}{i}\int \mathrm{Sp}\,\gamma^\mu\,S\left(x+\frac{\varepsilon}{2}, x'\right)\Gamma^\nu(x'\,y'|y)\,S\left(y, x-\frac{\varepsilon}{2}\right)d^4x'\,d^4y'\right\}.$$

We shall now derive another equation between S and Γ. For this purpose we calculate the change of S^{-1} induced by an infinitesimal gauge transformation. If we change A_μ^e by $\partial_\mu\,\delta\lambda$ we get from (39)

$$\delta S^{-1}(x, y) = -\,e_0 \int \Gamma^\mu(x\,y|z')\,\delta\bar{\mathfrak{A}}_\mu(z')\,d^4z' = -\,e_0 \int \Gamma^\mu(x\,y|z')\cdot$$
$$\cdot\partial_\mu\,\delta\lambda(z')\,d^4z' = e_0 \int \frac{\partial}{\partial z_\mu'}\,\Gamma_\mu(x\,y|z')\,\delta\lambda(z')\,d^4z'.$$

The same change can be produced by the reciprocal transformation of the ψ's

$$\psi \to \psi' = e^{-i e_0 \delta\lambda(x)}\,\psi(x)$$

so that we have

$$\delta S^{-1}(x, y) = i\, e_0(\delta\lambda(x) - \delta\lambda(y))\, S^{-1}(x, y) = i\, e_0 \int (\delta(x - z') -$$
$$- \delta(y - z'))\, S^{-1}(x, y)\, \delta\lambda(z')\, d^4z'.$$

If we equate the two expressions we get the generalized Ward identity [11]

$$\frac{\partial}{\partial z_\mu} \Gamma_\mu(x\, y|z) = i(\delta(x - z) - \delta(y - z))\, S^{-1}(x, y). \tag{45}$$

We shall now write these equations in momentum space. Literally we cannot use translation invariance as long as the external sources are present. Since we consider these sources to be a mathematical tool without physical reality here, we are interested in the limiting case of $A_\mu{}^e = J_\mu = 0$, which is exclusively of physical interest. Then our Green's functions are translation invariant, as usual. We use this fact in the following formulae, so that they hold literally only in the limit $A_\mu{}^e = J_\mu = 0$ [11]. Our Fourier transform will be

$$S(x, y) = \frac{1}{(2\pi)^4} \int S(p)\, e^{i(p,\, x - y)}\, d^4p$$

similar for D and M, and

$$\Gamma_\mu(x\, y|z) = \frac{1}{(2\pi)^8} \int \Gamma_\mu(p, q)\, e^{i(p, x - z) - i(q,\, y - z)}\, d^4p\, d^4q.$$

The only point in the Fourier inversion which is non trivial is the first term in (44). We get

$$S\left(x + \frac{\varepsilon}{2}, x - \frac{\varepsilon}{2}\right) \int\limits_{x + (\varepsilon/2)}^{x - (\varepsilon/2)} d\xi^\nu\, \delta(\xi - y) =$$

$$= \frac{1}{(2\pi)^8} \int d^4p\, d^4q \int\limits_{x + (\varepsilon/2)}^{x - (\varepsilon/2)} d\xi^\nu\, e^{i(p, \varepsilon)}\, e^{i(q,\, \xi - y)}\, S(p) = \cdot/. .$$

Now we evaluate the line integral along the straight line between the limits and put

$$\xi^\mu = x^\mu + \frac{\lambda}{2}\varepsilon^\mu, \qquad d\xi^\mu = \frac{\varepsilon^\mu}{2}\, d\lambda,$$

---

[11] We shall not introduce different symbols for the Green's functions with and without external sources present. One should, therefore, be careful in differentiating the following expressions functionally. Strictly this should be done only with the corresponding expressions in coordinate space.

where λ varies between -1 and $+1$. So we get

$$\int\limits_{x+(\varepsilon/2)}^{x-(\varepsilon/2)} d\xi^\mu e^{i(q\xi)} = \frac{\varepsilon^\mu}{2} e^{i(qx)} \int\limits_{-1}^{+1} d\lambda\, e^{i(q\varepsilon)(\lambda/2)} = -\varepsilon^\mu e^{i(qx)} \frac{2}{(q\varepsilon)} \sin\frac{(q,\varepsilon)}{2}.$$

Thus

$$\cdot/. = \frac{1}{(2\pi)^8}\int d^4p\, d^4q\, e^{i(q,x-y)} e^{i(p,\varepsilon)} \frac{-2\varepsilon^\nu}{(q\varepsilon)} \sin\frac{(q\varepsilon)}{2} S(p) =$$

$$= \frac{1}{(2\pi)^8}\int d^4p\, d^4q\, e^{i(q,x-y)} \frac{2\sin\frac{(q\varepsilon)}{2}}{(q\varepsilon)} \cdot i \frac{\partial}{\partial p_\nu}(e^{i(p\varepsilon)})\, S(p) =$$

by partial integration

$$= -\frac{i}{(2\pi)^8}\int d^4p\, d^4q\, e^{i(q,x-y)} e^{i(p\varepsilon)} \frac{2\sin\frac{(q\varepsilon)}{2}}{(q\varepsilon)} \frac{\partial S(p)}{\partial p_\nu}.$$

Now we expand for small ε

$$\frac{2\sin\frac{(q\varepsilon)}{2}}{(q\varepsilon)} = 1 - \frac{1}{24}(q\varepsilon)^2 + \ldots$$

and transform, as before, the ε into a derivative $\partial/\partial p$ by partial integration. So we get

$$\cdot/. = -\frac{i}{(2\pi)^8}\int d^4p\, d^4q\, e^{i(q,x-y)} \left\{\frac{\partial}{\partial p_\nu}\left(1 + \frac{1}{24}\left(q\frac{\partial}{\partial p}\right)^2 + \ldots\right) e^{i(p,\varepsilon)}\right\} S(p).$$

The rest of the Fourier inversion is trivial. In the limit $J_\mu = A_\mu^e = 0$ we get

$$S^{-1}(p) = (\gamma p) + M(p) \tag{46}$$

$$M(p) = m_0 + \frac{i e_0^2}{(2\pi)^4}\int d^4q\, D_{\alpha\beta}(q)\, \gamma^\alpha S(p+q)\, \Gamma^\beta(p+q,p) \tag{47}$$

$$D_{\mu\nu}(k) = D_{\mu\nu}^0(k) - e_0^2 D_{\mu\lambda}^{(0)}(k)\, \varrho^{\lambda\sigma}(k)\, D_{\sigma\nu}(k) \tag{48}$$

$$\varrho_{\alpha\beta}(q) = \frac{-i}{(2\pi)^4}\int d^4p\, \mathrm{Sp}\, \gamma_\alpha\left\{S\left(p+\frac{q}{2}\right)\Gamma_\beta\left(p+\frac{q}{2}, p-\frac{q}{2}\right) S\left(p-\frac{q}{2}\right) + \right.$$
$$\left. + \frac{\partial S(p)}{\partial p^\beta} + \frac{1}{24}\left[\frac{\partial}{\partial p^\beta}\left(q\frac{\partial}{\partial p}\right)^2\right] S(p) + \ldots\right\} \tag{49}$$

$$q^\mu \Gamma_\mu(p+q,p) = S^{-1}(p+q) - S^{-1}(p). \tag{50}$$

The first equation is the Dyson equation for the electron propagator, (48) is the corresponding one for the photon propagator. (47) and (49) express the mass- resp. polarization operator by the vertex function. The appearence of the contributions from the line integral in the current expression has to be noted. Formally the first term on the right hand side is quadratically divergent at the upper limit in perturbation theory, the divergence is however cancelled exactly by the second term. This can be seen easily by use of the last equation, the generalized Ward identity. Putting $q \to 0$ we get

$$\Gamma_\mu(p, p) = \frac{\partial S^{-1}(p)}{\partial p^\mu}, \tag{51}$$

which is the form originally obtained by Ward [12].

We add a brief remark on the structure of $\varrho_{\mu\nu}$. Since it has the physical notion of a polarization, it should be gauge invariant. Indeed we see from (42) that we have

$$\frac{\partial}{\partial x^\mu} \varrho^{\mu\nu}(x, y) = 0 \tag{52}$$

because of current conservation (formally this is fulfilled only, if we include the terms from the exponential! This can be checked e.g. by making the perturbation expansion in (49)). Since there are no external vectors (except possibly from the gauge) in the problem, if the external sources are zero, the most general form for $\varrho_{\mu\nu}$ is in this case

$$\varrho_{\mu\nu}(k) = (g_{\mu\nu}\, k^2 - k_\mu\, k_\nu)\, \varrho(k^2), \tag{53}$$

so that one has to consider only the invariant $\varrho(k^2)$.

In the Landau gauge the equation for $D_{\mu\nu}$ can be turned into a scalar one. Inserting (28) with $G = -1$ and the same form for $D_{\mu\nu}{}^0$ with $D^0(k^2) = (1/k^2)$ we get

$$D(k^2) = \frac{1}{k^2(1 + e_0{}^2\, \varrho(k^2))}. \tag{54}$$

5. Equations for Higher Functions

The equations written down so far are not sufficient, since they serve only to define all interesting quantities in terms of Γ, for which we have no equation. In principle one might think of an iterative procedure to solve the equations. For this purpose it is necessary to derive equations, which relate Γ to the higher Green's functions.

We start from eq. (37). By functional differentiation we have, using (39)

$$\Gamma_\mu(x\, y|z) = \gamma_\mu\, \delta(x - y)\, \delta(x - z) - \frac{1}{e_0} \frac{\delta M(x, y)}{\delta \bar{\mathfrak{A}}^\mu(z)}. \tag{55}$$

From this equation we can derive either a functional differential eq. for M or one for Γ. If we fold with S and D and use (41) we get

$$M(x, y) = m_0\,\delta(x-y) + i\,e_0^2\,\gamma_\mu\,S(x, y)\,\gamma_\nu\,D^{\mu\nu}(x, y) -$$

$$- i\,e_0\,\gamma_\mu \int S(x, x')\,\frac{\delta M(x', y)}{\delta\bar{\mathfrak{A}}^\nu(z')}\,D^{\mu\nu}(z', x)\,d^4x'\,d^4z'. \tag{56}$$

On the other hand, if we insert (41) into (55) and carry out the differentiation we get

$$\Gamma_\lambda(x\,y|z) = \gamma_\lambda\,\delta(x-y)\,\delta(x-z) - i\,e_0 \int D^{\varrho\nu}(x, z')\,\gamma_\varrho \Bigl[\iint S(x, x'')$$

$$\Gamma_\lambda(x''\,y''|z)\,S(y'', x')\,\Gamma_\nu(x'\,y|z')\,d^4x''\,d^4y'' + S(x, x')\,\frac{\delta\Gamma_\nu(x'\,y|z')}{\delta\bar{\mathfrak{A}}^\lambda(z)}\Bigr]\,d^4x'\,d^4z'. \tag{57}$$

We have made use of the fact, that $\delta D/\delta\bar{\mathfrak{A}}$ is zero in the physical limit $J_\mu = A_\mu{}^e = 0$ by Furry's theorem, and have therefore omitted this term. Eq. (56) or (57) may serve as basis for an iterative procedure.

Similar to the introduction of the mass operator we can write the eq. for Γ as an integral equation. By iteration of (56) we see, that M depends on $\mathfrak{A}$ only via S and D. Since $\delta D/\delta\mathfrak{A}$ is zero in the physical limit, we can neglect the latter dependence and put

$$\frac{\delta M(x, y)}{\delta\bar{\mathfrak{A}}_\nu(z)} = \int K(x\,x', y\,y')\,\frac{\delta S(x', y')}{\delta\bar{\mathfrak{A}}_\nu(z)}\,d^4x'\,d^4y', \tag{58}$$

where we have

$$K(x\,x', y\,y') = \frac{\delta M(x\,y)}{\delta S(x', y')} \tag{59}$$

if the iteration of (56) is possible. This turns (55) into the integral equation

$$\Gamma_\mu(x\,y|z) = \gamma_\mu\,\delta(x-y)\,\delta(x-z) - \tag{60}$$

$$- \int K(x\,x', y\,y')\,S(x', x'')\,\Gamma_\mu(x''\,y''|z)\,S(y'', y')\,d^4x'\,d^4y'\,d^4x''\,d^4y''.$$

To any given M (which one may find e.g. by iteration) we can construct a corresponding kernel K for the vertex equation.

In principle also the derivative with respect to $\mathfrak{A}$ could be eliminated: if we calculate K by differentiation of (41) with respect to S and insert the result in (60), we get an equation, which contains only Γ, D, S and their derivatives with respect to S. For the high-energy-approximation which will be discussed in the second part of this lecture, it is however not necessary to study this complicated equation (cf. Appendix 2).

Renormalization Constants

We shall now discuss briefly the various renormalization constants and their definition. It is well known, that we can write a spectral representation for the propagator [13], which can be deduced only from Lorentz invariance and the property, that the eigenstates of the operator of total energy-momentum form a complete set

$$S(p) = \int_m^\infty \frac{\varrho_1(x)}{(\gamma p) + x} dx + \int_m^\infty \frac{\varrho_2(x)}{(\gamma p) - x} dx \tag{61}$$

with positive spectral functions ϱ_i and m the physical mass of the electron. In the same way we can write a spectral representation for S^{-1}, i.e. for M

$$M \equiv M(\gamma p) = m_0 - \int_\pm \frac{r(x)\, dx}{(\gamma p) + x}, \tag{62}$$

where $\int_\pm$ is an abbreviation for $\int_{-\infty}^{-m} + \int_{+m}^{+\infty}$.

If a state with mass m exists, it contributes a δ-function to ϱ and S has a pole[12] for $(\gamma p) = -m$. So we have

$$S^{-1}|_{\gamma p = -m} = ((\gamma p) + M(\gamma p))|_{\gamma p = -m} = 0$$

and therefore

$$m = m_0 - \int dx \frac{r(x)}{x - m}. \tag{63}$$

In perturbation theory the integral diverges logarithmically (because of $r(+\infty) = r(-\infty)$ the linear divergence drops out). Therefore we make subtractions to isolate the divergence:

$$\gamma p + M = \gamma p + M + m - m = \gamma p + m + m_0 - \int \frac{r}{(\gamma p) + x} dx - m_0 +$$

$$+ \int \frac{r}{x - m} dx = (\gamma p) + m + \int \frac{((\gamma p) + m)\, r\, dx}{(x - m)\,(x + (\gamma \cdot p))} =$$

$$= ((\gamma p) + m) \left(1 + \int \frac{r\, dx}{(x + (\gamma p))\,(x - m)}\right) =$$

---

[12] There is a clean pole only, if one cuts off the infrared frequencies by giving the photon a small mass μ. Then we get a gap in the spectrum between m and $m + \mu$. The cuts corresponding to photon emission start at μ, 3μ etc. (an even number of photons cannot be emitted because of Furry's theorem). If $\mu \to 0$ there is no gap, all cuts start at m and we have a singularity of the type pole times logarithms. By renormalization-group arguments it can be shown, that all the logarithms pile up to a non-integer exponent. See Bogoliubov-Shirkov, Introduction to the Theory of Quantized Fields, Chap. **44**, p. 531.

$$= ((\gamma p) + m)\left(1 + \int dx \frac{r(x)}{(x-m)^2}\right) -$$

$$- ((\gamma p) + m)^2 \int dx \frac{r(x)}{(x-m)^2 (x + (\gamma p))}.$$

The last integral is convergent. Thus we have

$$S^{-1}(p) = \frac{1}{Z_2} ((\gamma p) + m) - ((\gamma p) + m)^2 \int dx \frac{r(x)}{(x-m)^2 (x + (\gamma p))} \tag{64}$$

with

$$Z_2^{-1} = 1 + \int_{\pm} dx \frac{r(x)}{(x-m)^2}. \tag{65}$$

In the neighbourhood of the pole the last term can be omitted and we see, that Z_2 is the residue of S on the pole. The integral in (65) diverges again logarithmically in perturbation theory.

Another renormalization constant is introduced by the observation, that the vertex function for equal momenta "on shell" $(\gamma p) = (\gamma q) = -m$ has to be proportional to γ_μ:

$$\Gamma_\mu(p, q)\big|_{(\gamma p) = (\gamma q) = -m} = \frac{1}{Z_1} \gamma_\mu. \tag{66}$$

Because of Ward's identity (49) we see, that we must have

$$Z_1 = Z_2. \tag{67}$$

Next we consider the vacuum polarization. Since ϱ is closely related to D^{-1} (see (54)) we can write a similar spectral representation (dispersion relation) as for M

$$\varrho(k^2) = \int_0^\infty dx^2 \frac{\sigma(x^2)}{x^2 + k^2}. \tag{68}$$

It can be proved by similar methods as those used in Lehmann's paper that $\sigma(x^2)$ is positive definite, if the metric in Hilbert space has this property. We define now a renormalization constant Z_3 by

$$\frac{1}{Z_3} = 1 + \int dx^2 \frac{\sigma(x^2)}{x^2} = 1 + \varrho(0), \tag{69}$$

which obviously has the property

$$0 \leqslant Z_3 < 1 \tag{70}$$

because σ is positive. The notion is that of charge renormalization for the following reason: consider the scattering of two electrons from each other. The matrix element for the process will then be

$$M \sim e_0^2 D(k^2),$$

where k^2 is the momentum transfer[13]. The physical charge will then be the one which is felt by the particles, if they are very far apart (k^2 very small) and almost free. Thus we *define* the physical charge e by postulating[14]

$$\lim_{k^2 \to 0} M = e^2 D_0(k^2),$$

where $D_0(k^2) = 1/k^2$ is the free propagator. Comparing the expressions for M we see

$$e^2 = \frac{1}{\frac{1}{e_0{}^2} + \varrho(0)} = e_0{}^2 Z_3. \tag{71}$$

In perturbation theory the spectral integral diverges and one again isolates the divergence into Z_3 by subtraction

$$\frac{1}{e_0{}^2} + \varrho(k^2) = \frac{1}{e_0{}^2} + \varrho(0) - k^2 \int dx^2 \frac{\sigma(x^2)}{x^2(x^2 + k^2)}.$$

Passing to renormalized quantities by splitting off the Z-factors from S, Γ and $e_0{}^2 D$ it is possible to show, that the bare quantities can be avoided. Expansion in powers of e gives a solution of the basic equations, which is finite term by term, the well-known renormalized perturbation series. We shall not go into details here.

We consider now the gauge properties of the renormalization constants. Z_3 is clearly gauge invariant, since it is related to the charge and derived from the gauge-invariant quantity ϱ. The same is certainly not true for Z_2, which is related to the gauge dependent function S. The gauge behaviour of Z_2 is easily obtained [14] from our previous results on the $\hat{M}$-transformation. Let us assume, that the infrared frequencies are cut off by introduction of a small photon mass μ. The behaviour of the function $S(x, y)$ for large distances $(x - y)^2 \to \infty$ is given by the contribution of the state with the lowest invariant mass in (61) (the lowest mass gives the longest range in coordinate space), which is then the pole:

$$\lim_{(x-y)^2 \to \infty} S(x, y) = Z_2 S^0(x - y, m) + \dots, \tag{72}$$

---

[13] Higher graphs containing more exchanged photons contain short range forces which are not interesting in this context.

[14] Here we have taken Ward's identity already into account, which ensures $Z_1 = Z_2$ and have omitted the vertex and electron wave-function renormalization.

where S^0 is the free propagator with mass m. If the $\hat{M}$-transformation is not singular, we must have

$$\delta\hat{M}(x-y)\to 0 \qquad \text{for} \qquad (x-y)^2\to\infty$$

(otherwise the Fourier integral does not exist). Looking at (29) we see that

$$Z_2' = \exp\left[-i\,e_0^2\,\delta\hat{M}(0)\right]Z_2. \tag{73}$$

We will consider a useful example, which sheds some light on the high-energy-behaviour of our theory. Consider the free propagator in a theory with infrared cutoff μ and ultraviolet cutoff Λ in the Feynman gauge as a starting point

$$D_{\mu\nu}{}^0(k) = g_{\mu\nu}\left(\frac{1}{k^2+\mu^2}-\frac{1}{k^2+\Lambda^2}\right) \tag{74}$$

and take

$$\delta\hat{M}(k^2) = -\frac{G}{k^2}\left(\frac{1}{k^2+\mu^2}-\frac{1}{k^2+\Lambda^2}\right),$$

so that we arrive at one of the covariant gauges studied in (28) after the transformation. Transforming to coordinate space we get

$$\delta\hat{M}(0) = -\frac{iG}{16\pi^2}\ln\frac{\Lambda}{\mu},$$

so that we get

$$Z_2' = \left(\frac{\Lambda^2}{\mu^2}\right)^{-(G/4\pi)\alpha_0} Z_2 \tag{75}$$

with

$$\alpha_0 = \frac{e_0{}^2}{4\pi}.$$

Speculations on the High-Energy-Behaviour

We add now some speculations on the high-energy-behaviour of the propagators, based on the assumption, that the equations and the spectral representations exist as they are written down (without renormalization) and that all divergences are only due to perturbation theory, which would then be invalid at high energies. "High energies" denotes the limit of the theory, in which the momentum vectors under consideration are large and spacelike (we should rather use the terminus "small distances" in coordinate space), so we are considering the behaviour at rather "unphysical" momenta (far "off-shell").

If the spectral integral (62) exists at the upper limit, $r(x)$ has to decrease for large argument and we must have

$$M\to m_0 \qquad \text{for} \qquad (\gamma p)\to\infty. \tag{76}$$

Physically: if we look into the interior of the electron dressed by its interaction with the field coupled to it (electromagnetic field, virtual pairs) we see the *bare* mass, whereas we observe the physical mass m, if we look at it from the distance. (76) means for the propagator

$$S^{-1}(p) \to (\gamma p) \tag{77}$$

and because of Ward's identity (50)

$$\Gamma_\mu \to \gamma_\mu \tag{78}$$

for both arguments large.

In the same way we can argue for the vacuum polarization: If the spectral integral (68) exists, $\sigma(x)$ has to go to zero for large argument. This means, that

$$\varrho(k^2) \to 0 \qquad \text{and} \qquad D^{-1}(k^2) \to k^2 \qquad \text{for} \qquad k^2 \to \infty. \tag{79}$$

Looking at our previous consideration of scattering and the notion of the charge, this means, that we should observe the *bare* charge e_0 concentrated at a point, if we make a measurement at high transfer k^2 (small distances), i.e. if we look through the polarization cloud into the electron.

If we consider now the equations for the corresponding quantities, namely (47) and (49), in perturbation theory, none of the above asymptotic properties results. So we have to rely on an essentially non perturbative treatment, if we want to check our speculations. From the experimental point of view the attack of the corresponding problems is hopeless presently, since the characteristic high-energy-behaviour of perturbation theory is $\sim \alpha \ln ((\hbar/m\,c)\,(1/r))$. If we are to observe deviations (i.e. if the whole series is needed) we have to go to distances of order $r \sim e^{-1/\alpha}$ Compton wave lengths. We see therefore, that we have to recur to theoretical investigations.

To conclude this chapter, we present now an argument due to Gell-Mann and Low [15] on the behaviour of the electron propagator at high momenta in perturbation theory of arbitrary order. For this purpose we *assume* the finiteness of Z_3, so that we have $D(k^2) \to 1/k^2$ at high momenta. Then we are able to neglect the photon self-energy-parts in this limit. We consider again a theory containing an ultraviolet cutoff, i.e. we take

$$D(k^2) = \frac{1}{k^2} - \frac{1}{k^2 + \Lambda^2}. \tag{80}$$

We split off a factor $(\gamma p)^{-1}$ from S. The remainder is dimensionless and we have therefore

$$S(p) = \frac{1}{(\gamma p)}\, S_1\left(\frac{\Lambda^2}{p^2}, \frac{p^2}{m^2}\right). \tag{81}$$

At first we note, that it is possible to consider the limiting case $\Lambda^2 \gg p^2 \gg m^2$, because there are no divergences present for $m \to 0$:

if one computes S term by term in perturbation theory, one gets a finite answer in any order, since Λ prevents ultraviolet divergences and μ^2 acts as an "infrared" cutoff even if we put $m = 0$. From the renormalizability we know, that we can write

$$S_1\left(\frac{\Lambda^2}{p^2}, \frac{p^2}{m^2}\right) = Z_2\left(\frac{\Lambda^2}{m^2}\right) S_f\left(\frac{\Lambda^2}{p^2}, \frac{p^2}{m^2}\right), \tag{82}$$

where for very high Λ^2 the renormalized propagator is cutoff-independent

$$S_f\left(\frac{\Lambda^2}{p^2}, \frac{p^2}{m^2}\right) \to S_f\left(\frac{p^2}{m^2}\right) \qquad \text{for} \qquad \Lambda^2 \to \infty. \tag{83}$$

If we neglect the mass terms in S (which is allowed, as was argued above) we get therefore a functional equation, valid in the limit $\Lambda^2 \gg p^2 \gg m^2$

$$S_f\left(\frac{\Lambda^2}{p^2}\right) = Z_2\left(\frac{\Lambda^2}{m^2}\right) S_f\left(\frac{p^2}{m^2}\right). \tag{84}$$

The general solution of this equation is

$$Z_2 = A\left(\frac{\Lambda^2}{m^2}\right)^{\varphi}, \qquad S_f = B\left(\frac{p^2}{m^2}\right)^{-\varphi}, \qquad S_1 = A\,B\left(\frac{\Lambda^2}{p^2}\right)^{\varphi}. \tag{85}$$

The quantities A, B and φ are functions of the coupling constant and the gauge constant G. The lowest perturbation approximation is (in Feynman gauge)

$$\varphi = -\frac{\alpha_0}{4\pi} + \ldots, \qquad A = 1 + \ldots, \qquad B = 1 + \ldots. \tag{86}$$

If we start from the Feynman gauge and pass to a covariant gauge of one of the types considered before we get because of (75)

$$Z_2' = \left(\frac{\Lambda^2}{\mu^2}\right)^{-\alpha_0(G/4\pi)} \left(\frac{\Lambda^2}{m^2}\right)^{\varphi(\alpha_0)} \cdot A(\alpha_0) = A_1\left(\alpha_0, \frac{m^2}{\mu^2}\right)\left(\frac{\Lambda^2}{m^2}\right)^{\varphi(\alpha_0) - \alpha_0(G/4\pi)}$$

$$A_1 = \left(\frac{m^2}{\mu^2}\right)^{-\alpha_0(G/4\pi)}. \tag{87}$$

Thus we see, that there is one and only one gauge, in which Z_2' is cutoff-independent and finite, namely the one in which

$$-G = -\frac{4\pi}{\alpha_0}\varphi(\alpha_0). \tag{88}$$

In lowest order perturbation theory this is the Landau gauge, so that we have

$$-G = 1 + \ldots. \tag{89}$$

Indeed it turns out, that the divergences in the perturbation expansion of Z_2 start with α_0^2 in this gauge, as was noticed by LANDAU [8].

References

1. J. M. Jauch, F. Rohrlich, The Theory of Photons and Electrons, Addison-Wesley, Cambridge **1954**.
2. P. A. M. Dirac, Proc. Camb. Phil. Soc. **30**, 150 (1934).
3. W. Heisenberg, Zs. f. Phys. **90**, 209 and **92**, 692 (1934).
4. J. G. Valatin, Proc. Roy. Soc. A **222**, 93 and 228 (1954).
5. J. Schwinger, Phys. Rev. Lett. **3**, 296 (1959).
6. B. Zumino, Journ. Math. Phys. **1**, 1 (1960).
7. H. M. Fried, D. R. Yennie, Phys. Rev. **112**, 1391 (1958).
8. L. D. Landau, A. A. Abrikosov, I. M. Khalatnikov, Dokl. Akad. Nauk SSSR **95**, 773 (1954).
9. J. Schwinger, Proc. Nat. Acad. Sci. **37**, 452 and 455 (1951).
10. F. J. Dyson, Phys. Rev. **75**, 1736 (1949).
11. Y. Takahashi, N. Cim. **6**, 370 (1957).
12. J. C. Ward, Phys. Rev. **78**, 182 (1950).
13. H. Lehmann, N. Cim. **11**, 342 (1954).
14. K. Johnson, B. Zumino, Phys. Rev. Lett. **3**, 351 (1959).
15. M. Gell-Mann, F. Low, Phys. Rev. **95**, 1300 (1954).

Appendix 1: Perturbation Theory of Arbitrary Order

In order to establish the connection with the formal solution given in Wess' lecture we consider briefly the perturbation expansion of V with regard to the interaction, i.e. we expand V in powers of e_0. This expansion can be obtained in the following way: perform a canonical transformation similar to (6) but now to the interaction picture (with sources present!) so that we have now

$$U_i(t_1, t_2) = T \exp i \int_{t_2}^{t_1} L_i^{(\mathrm{int})}(x)\, d^4x.$$

Then we get, working as before

$$V = \frac{1}{N'} \langle \mathrm{out}|\, U_i(+\infty, -\infty)\, |\mathrm{in}\rangle_S.$$

N' is the transformation functional without interaction, but with sources and does therefore not contain e_0. Expanding in powers of e_0 we get

$$V = (V)_{e_0=0} + e_0 \left.\frac{\partial V}{\partial e_0}\right|_{e_0=0} + \cdots.$$

The lowest term is clearly

$$\left.\frac{\partial V}{\partial e_0}\right|_{e_0=0} = \frac{1}{N'} i \int d^4x\, \langle \mathrm{out}|\, T\, j^\mu(x)\, A_\mu(x)\, |\mathrm{in}\rangle_S|_{e_0=0}.$$

If we now take our previous form (9) for V we see, that we can produce this expression by functional derivation:

$$+ \frac{\delta^2}{\delta J^\mu(x)\, \delta A_\mu^e(x)} V = \frac{-e_0}{N} \langle \mathrm{out}|\, T \left(j^\mu(x)\, A_\mu(x) \exp i \int_{-\infty}^{+\infty} L_S(x)\, d^4x \right) |\mathrm{in}\rangle_i = \cdot/. .$$

Now this can be transformed back to the Heisenberg picture, using (10)

$$\cdot/. = -e_0 \langle \text{out} | \, T \, j^{\mu}(x) \, A_{\mu}(x) \, | \text{in} \rangle_{s,i}.$$

For $e_0 = 0$ the Heisenberg picture is identical with the interaction picture (apart from N') so that we have

$$e_0 \frac{\partial V}{\partial e_0}\bigg|_{e_0=0} = +i \int d^4x \, \frac{\delta^2}{\delta J^{\mu}(x) \, \delta A_{\mu}^{e}(x)} \, V \bigg|_{e_0=0}.$$

For the higher derivatives we can proceed in the same way. We see, that the whole perturbation expansion can be summed: we only have to pay attention to the vacuum diagrams in the denominator. The result is

$$V = \frac{-1}{N} \exp\left(-i e_0 \int d^4x \, \frac{\delta}{\delta J^{\mu}(x)} \frac{\delta}{\delta A_{\mu}^{e}(x)}\right) \cdot \langle \text{out} | \text{in} \rangle_s. \tag{A1}$$

Since all Green's functions can be calculated if V is given as a functional of the sources, the whole problem is reduced to the theory of the electron — and electromagnetic field in interaction with external sources J_{μ}, A_{μ}^{e} but *without* interaction between each other: we have to calculate $\langle \text{out}|\text{in}\rangle_s$. In spite of the fact, that this functional can be determined exactly (see Wess' lecture) the formula is of no use for explicit calculations apart from perturbation theory, since for practical purposes the exponential has to be expanded. The form (A1) is however very useful for the discussion of the formal structure of the whole perturbation series, especially in connection with gauge transformations. We must not forget, however, that the convergence of the perturbation series has never been proved. The formal expansion in powers of the external fields of Section 2 is less dangerous in this respect, since we left the special form of the external sources entirely open: we could use the exponential as a formal series, without ever touching the question of convergence.

Appendix 2: Elimination of K and $(\delta/\delta\mathfrak{A})$

We consider equation (41) for the mass operator and differentiate with respect to S:

$$\begin{aligned} \frac{\delta M_{\alpha\beta}(x, y)}{\delta S_{\gamma\delta}(z, w)} &= i e_0^2 (\gamma^{\varrho})_{\alpha\gamma} \int D_{\varrho\nu}(x, z') \, \Gamma_{\delta\beta}{}^{\nu}(w\, y|z') \, \delta(x - z) \, d^4z' + \\ &+ i e_0^2 \int D_{\varrho\nu}(x, z') \, (\gamma^{\varrho} S(x, x'))_{\alpha\lambda} \frac{\delta \Gamma_{\lambda\beta}{}^{\nu}(x'\, y|z')}{\delta S_{\gamma\delta}(z\, w)} \, d^4x' \, d^4z' + \\ &+ i e_0^2 \int (\gamma^{\varrho} S(x\, x') \, \Gamma^{\nu}(x'\, y|z'))_{\alpha\beta} \frac{\delta D_{\varrho\nu}(x, z')}{\delta S_{\gamma\delta}(z, w)} \cdot d^4x' \, d^4z' = \\ &= \underset{\alpha\,\gamma\,\beta\,\delta}{K(x\, z, y\, w)} \end{aligned} \tag{A2}$$

According to (59) this is the Bethe-Salpeter-Kernel K. Therefore we can insert this expression in the integral equation (60) for Γ and get an equation, in which neither K nor derivations with respect to $\mathfrak{A}$ occur:

$$\begin{aligned} \Gamma_{\mu}(x\, y|z)_{\alpha\beta} = (\gamma_{\mu})_{\alpha\beta} \, \delta(x - y) \, \delta(x - z) - i e_0^2 \int (\gamma_{\varrho} S(x\, x'')\, \cdot \\ \cdot \Gamma_{\mu}(x''\, y''|z) \, S(y''\, y') \, D^{\varrho\nu}(x, z') \, \Gamma_{\nu}(y'\, y|z'))_{\alpha\beta} \, d' \, d'' - \end{aligned}$$

$$- i e_0^2 \int D_{\varrho\nu}(x z') (\gamma^\varrho S(x w'))_{\alpha\lambda} \frac{\delta \Gamma^\nu(w' y|z')_{\lambda\beta}}{\delta S_{\varrho\sigma}(x' y')} \cdot \tag{A3}$$

$$\cdot (S(x' x'') \Gamma_\mu(x'' y''|z) S(y'' y'))_{\varrho\sigma} d' d'' - i e_0^2 \int (\gamma^\varrho S(x w') \cdot$$

$$\cdot \Gamma^\nu(w' y|z'))_{\alpha\beta} \frac{\delta D_{\varrho\nu}(x, z')}{\delta S_{\varrho\sigma}(x' y')} (S(x' x'') \Gamma_\mu(x'' y''|z) S(y'' y'))_{\varrho\sigma} d' d''.$$

Together with (46)—(49) this equation could again provide a basis for an iterative procedure, which is not identical with perturbation theory.

Appendix 3: Graphical Representation

Sometimes it is convenient to represent the equations for Green's functions in graphical form. We will do this in the following way: An electron propagator S is represented by a solid line |—◂—|. A photon propagator D is represented by a wavy line |∼∼∼|. It has to be noted, that we always consider *exact* Green's functions, so our propagator corresponds to the "primed" functions S', D' of the literature, which usually are represented by

Since we will deal always with the exact functions, we will not draw the circles and will denote the free propagator by a zero, for instance

Kernels (amputated functions) are represented by circles with endpoints at the periphery, e.g. (one has to indicate whether a photon line or an electron line can be attached to the point)

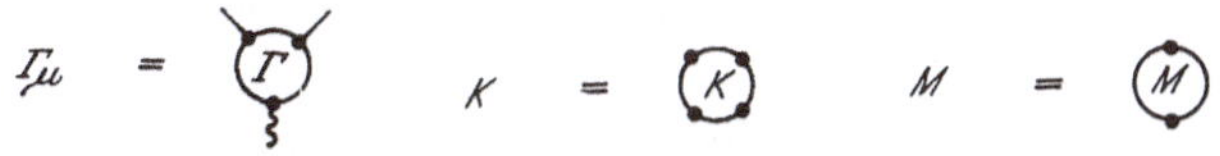

We give some examples: equations (48) and (49) are

(we do not indicate the "subtraction terms" graphically). Equations (46) and (47) are

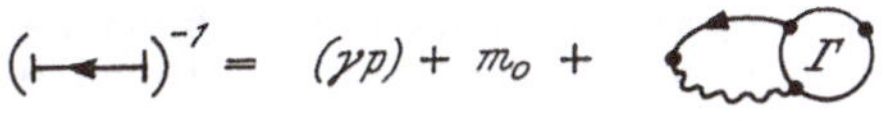

Equation (57) would read

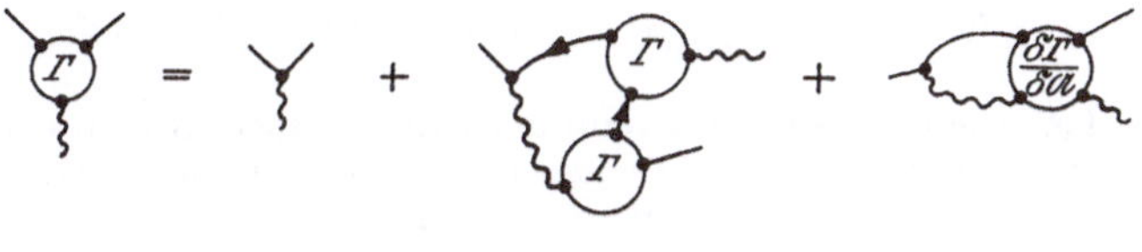

For equation (60) we have

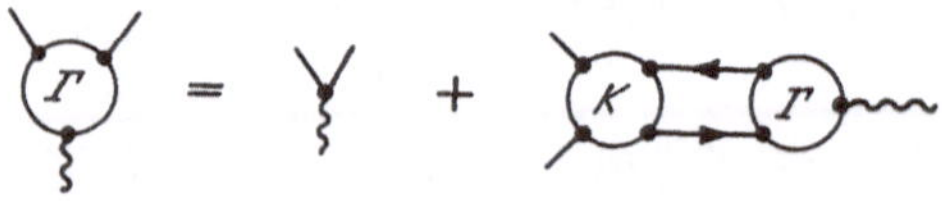

and so on. The graphs are more useful for iterative solutions than for the exact equations.

Note that functional differentiation of an S-line with respect to $\bar{\mathfrak{A}}$ corresponds to insertion of $S\,\Gamma\,S$:

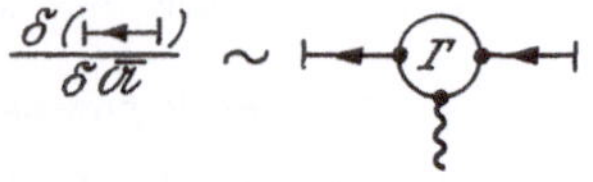

Functional differentiation of D with respect to $\bar{\mathfrak{A}}$ gives zero in the limit $J_\mu = A_\mu^e = 0$ since we would have

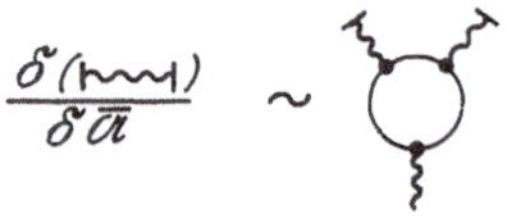

which is zero by Furry's theorem. Functional differentiation of ⊢◄⊣ with respect to S corresponds to cutting off (removing) the corresponding line.

By these results it is easy to "derive" the higher equations graphically.

Appendix 4: Vacuum Polarization

As an example we consider the polarization of the vacuum induced by an external electromagnetic field A_μ^e in lowest order perturbation theory, i.e. in lowest order with respect to A_μ^e *and* e_0. For simplicity we put $m_0 = 0$. We start from the current induced in the vacuum as given by (19) and expand

$$\frac{1}{V}\langle\text{out}|\, j_\mu\, |\text{in}\rangle_{S,i} = i\,\mathrm{Sp}\,\tilde{S}(x, x)\,\gamma_\mu = i\,\mathrm{Sp}\,\tilde{S}(x, x)\,\gamma_\mu|_{e_0 = A_\mu^e = 0} +$$

$$+\, e_0 \int d^4x'\,\bar{\mathfrak{A}}^\varrho(x')\,\frac{\delta}{\delta\bar{\mathfrak{A}}^\varrho(x')}\, i\,\mathrm{Sp}\,\tilde{S}(x, x)\,\gamma_\mu|_{e_0 = A_\mu^e = 0} + \cdots =$$

$$= i\,\mathrm{Sp}\,\tilde{S}(x, x)\,\gamma_\mu|_0 + e_0^2 \int d^4x'\,\bar{\mathfrak{A}}^\varrho(x')\,\varrho_{\mu\varrho}(x, x')|_0 + \cdots.$$

This exhibits the notion of ϱ as a polarization! In addition, we can use our previous result (49) for the expansion of ϱ in powers of A_μ^e. To lowest order in e_0 we have to replace S and Γ by the free functions

$$S(p) \to S^0(p) = \frac{1}{(\gamma\, p)}, \qquad \Gamma_\mu(p, q) \to \Gamma_\mu^0(p, q) = \gamma_\mu$$

and $\bar{\mathfrak{A}}_\mu$ by A_μ^e. The first term gives zero: in coordinate space $S^0(\varepsilon)$ is odd in ε and we have to perform the limit in a symmetrical way. The next term gives the polarization current. In momentum space we get

$$\varrho_{\alpha\beta}{}^{0}(q) = -\frac{i}{(2\pi)^4}\int d^4p\,\mathrm{Sp}\,\gamma_\alpha\left[\frac{1}{\left(\gamma, p+\frac{q}{2}\right)}\gamma_\beta\frac{1}{\left(\gamma, p-\frac{q}{2}\right)}+\frac{\partial}{\partial p^\beta}\frac{1}{(\gamma p)}+\right.$$

$$\left.+\frac{1}{24}\frac{\partial}{\partial p^\beta}\left(q\frac{\partial}{\partial p}\right)^2\frac{1}{(\gamma p)}+\ldots\right].$$

First we want to show, that ϱ is indeed conserved. This can be done in the following way: putting $p_\pm = p \pm (q/2)$ we have

$$q^\beta\,\varrho_{\alpha\beta}{}^{0}(q) = \frac{-i}{(2\pi)^4}\int d^4p\,\mathrm{Sp}\,\gamma_\alpha\left[\frac{1}{(\gamma p_+)}(\gamma q)\frac{1}{(\gamma p_-)}+\right.$$

$$\left.+\left(q\frac{\partial}{\partial p}\right)\left(1+\frac{1}{24}\left(q\frac{\partial}{\partial p}\right)^2\right)\frac{1}{(\gamma p)}+\ldots\right].$$

In the first term we observe

$$(\gamma q) = (\gamma p_+) - (\gamma p_-),$$

so that

$$\cdot/. = \frac{1}{(\gamma p_+)}(\gamma q)\frac{1}{(\gamma p_-)} = \frac{1}{(\gamma p_-)} - \frac{1}{(\gamma p_+)}.$$

We expand in powers of q. The terms even in q drop out and we get

$$\cdot/. = -\sum_{n=0}^{\infty}\frac{2}{(2n+1)!\,2^{2n+1}}\left(q\frac{\partial}{\partial p}\right)^{2n+1}\frac{1}{(\gamma p)}.$$

The first term is quadratically divergent, the second diverges logarithmically. These terms are exactly cancelled by the contributions from the line integral. The rest of the sum is zero, as is easily shown by integration, since each term falls off rapidly enough at large p.

From this argument we can already see qualitatively, that we have to expect a logarithmic divergence for $\varrho_{\alpha\beta}$: if we do not multiply with q the quadratic divergence drops out as well as one can see e.g. by putting $q = 0$ in the first term. For the logarithmically divergent term the cancellation does however not occur.

In order to see this explicitly we use the fact, that $\varrho_{\alpha\beta}$ has the form

$$\varrho_{\alpha\beta} = (g_{\alpha\beta}\,q^2 - q_\alpha\,q_\beta)\,\varrho(q^2)$$

from which we see that

$$q^2\,\varrho(q^2) = \frac{1}{3}\,\varrho_\mu{}^\mu.$$

Computation of the trace gives

$$q^2\,\varrho(q^2) = \frac{-8i}{3(2\pi)^4}\int d^4p\left[\frac{-(p_+\cdot p_-)}{p_+{}^2\,p_-{}^2}+\frac{1}{p^2}-\frac{q^2}{12\,p^4}+\frac{(p\,q)^2}{3\,p^6}\right].$$

The integral is most conveniently evaluated by the following method, which is used in part II of this lecture very frequently: we pass to Euclidean space (this is allowed for Green's functions) and use spherical coordinates in four dimensions

$$i\, d^4 p = p^3\, dp\, d\Omega$$

$$\int d\Omega = \int_0^{\pi} \sin^2\theta\, d\theta \int_0^{\pi} \sin\lambda\, d\lambda \int_0^{2\pi} d\varphi.$$

For θ we take the angle between p and q, so that $(p\, q) = p\, q \cos\theta$. We write

$$\frac{1}{p_+^{\,2}} = \frac{1}{p^2 + 2\,p\,\frac{q}{2}\cos\theta + \frac{q^2}{4}} = \frac{1}{p_>^{\,2}(1 + 2\,x\cos\theta + x^2)},$$

where $p_>$ is the greater of q and $q/2$

$$p_> = \begin{cases} p & \text{for} \quad p > \frac{q}{2} \\ \frac{q}{2} & \text{for} \quad p < \frac{q}{2} \end{cases}$$

and x is the ratio of the smaller and the greater of the two vectors

$$x = \begin{cases} \frac{q}{2\,p} & \text{for} \quad p > \frac{q}{2} \\ \frac{2\,p}{q} & \text{for} \quad p < \frac{q}{2} \end{cases} \qquad (x < 1).$$

We expand into spherical harmonics (Gegenbauer polynomials) in four dimensions

$$\frac{1}{1 + 2\,x\cos\theta + x^2} = \sum_{l=0}^{\infty} x^l C_l^{\,1}(\cos\theta).$$

The C_l^1 are elementary and much simpler than the three-dimensional analogs. They fulfill the orthogonality relation

$$\int_0^{\pi} d\theta\, C_l^{\,1}\, C_n^{\,1} \sin^2\theta = \frac{\pi}{2}\,\delta_{ln}.$$

If we expand also $1/p_-^{\,2}$ in the way just indicated we get

$$\int d\Omega\, \frac{1}{p_+^{\,2}\, p_-^{\,2}} = \frac{4\,\pi}{p_>^{\,4}} \sum_{l,m=0}^{\infty} x^{l+m}(-)^m \int_0^{\pi} \sin^2\theta\, C_l^{\,1}(\cos\theta)\, C_m^{\,1}(\cos\theta)\, d\theta =$$

$$= \frac{2\,\pi^2}{p_>^{\,4}} \sum_{l,m} x^{l+m}(-)^m\, \delta_{lm} = \frac{2\,\pi^2}{p_>^{\,4}} \sum_l x^{2l}(-1)^l = \frac{2\,\pi^2}{p_>^{\,4}(1 + x^2)}.$$

The rest of the angular integration is trivial. If all terms are taken together we get q^2 times a logarithmic divergence from the upper limit of the p-integration. Thus we see that also $\varrho(0)$ is logarithmically divergent.

If S is computed in higher orders with respect to A^e

$$S \sim S_0 + \int S^0 \gamma\, S^0 A^e + \int S^0 \gamma\, S^0 \gamma\, S^0 A^e A^e + \int S^0 \gamma\, S^0 \gamma\, S^0 \gamma\, S^0 A^e A^e A^e + \ldots$$

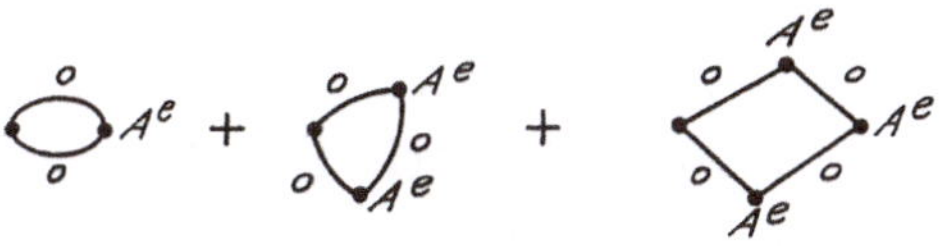

(zero by Furry's theorem)

one has to take also the higher order terms from the exponential, e.g. for the next approximation

$$S^0 \left(-ie_0 \int d\xi^\mu A^e_\mu\right)^3 +$$

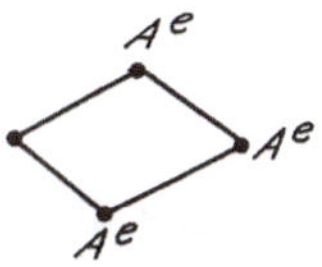

Though the first term is $\sim \varepsilon^3$ it gives contributions, since $S_0 \sim 1/\varepsilon^3$. (There are also important cancellations of divergences.) In the next approximation the corresponding term can be omitted, since it is $\varepsilon^5/\varepsilon^3 \sim 0$.

Quantumelectrodynamics with Functional Calculus*

By

J. Wess[1]

Universität Wien

Lecture I

The purpose of this talk is to give an introduction to functional methods and how these methods can be applied to Quantumelectrodynamics [1]. As we want to learn the formulism, I shall go through the derivations in quite some detail.

Field equations:

$$\{\gamma^\nu(i\,\partial_\nu - e\,A_\nu) - m\}\,\psi = 0 \qquad (\Box + \mu_0{}^2)\,A_\nu = j_\nu \tag{1}$$

j_ν is the current, conserved as a consequence of the invariance under the phase transformation $\psi \to e^{i\alpha}\,\psi$. By Noethers Theorem $j_\nu = e\,\bar{\psi}(x)\gamma_\nu\,\psi(x)$, but in a quantized theory such expressions, where the product of the fieldoperators are taken at the same space time point have to be handled with special care and we shall have enough occasions to comment on that.

Commutation relations:

$$\{\psi^+(x), \psi(x')\}_{x_0 = x_0'} = \delta(\vec{x} - \vec{x}'),\ [A_\nu(x), \dot{A}_\mu(x')]_{x_0 = x_0'} = -\,i\,g_{\mu\nu}\,\delta(\vec{x} - \vec{x}'). \tag{2}$$

Generating functional:

$$G(J, \eta, \bar{\eta}) = \langle 0|(e^{i\int dy\,\{\bar{\eta}(y)\,\psi(y) + \bar{\psi}\eta + A^\nu J_\nu\}})_t\,|0\rangle \tag{3}$$

$(\ )_t$ means time ordering, operators at an earlier time are to the right of operators at a later time. Under the time ordering symbol the operators can be looked upon as if they commute or anticommute respectively.

J_μ is an arbitrary c-number function, η, $\bar{\eta}$ anticommute with ψ, ψ^+, η and η^+ and commute with A.

---

* Lecture given at the IV. Internationale Universitätswochen für Kernphysik, Schladming, 25 February—10 March 1965.

[1] This article is dedicated to Prof. Urban on the occation of his 60th birthday.

Functional derivatives lead to the following expressions:

$$\frac{1}{i}\frac{\delta}{\delta J_\mu(x)} G(J,\eta,\eta^+) = \langle 0|(e^{i\int\{\}} A^\mu(x))_t|0\rangle =$$

$$= \langle 0|\left(e^{i\int_{x_0}^{\infty}\{\}}\right)_t A^\mu(x)\left(e^{i\int_{-\infty}^{x_0}\{\}}\right)_t|0\rangle. \tag{4}$$

Thus there is the correspondence

$$\frac{1}{i}\frac{\delta}{\delta J_\mu(x)} \to A^\mu(x), \qquad \frac{1}{i}\frac{\delta}{\delta\bar\eta(x)} \to \psi(x), \qquad \frac{1}{i}\frac{\delta}{\delta\eta(x)} \to -\bar\psi(x).$$

By taking an arbitrary number of functional derivatives and then setting $\eta = \bar\eta = J = 0$ we can derive all the Greens functions from our functional. Therefore G is called the generating functional.

The knowledge of G therefore would imply the knowledge of all Greens functions and we know that the Greens functions contain all the information of our quantized theory.

In order to calculate G we set up a number of differential equations.

Spatial derivatives can be taken through the time ordering. For time derivatives we calculate:

$$\frac{\partial}{\partial x_0}\frac{1}{i}\frac{\delta}{\delta\bar\eta(x)} G = \frac{\partial}{\partial x_0}\langle 0|\left(e^{i\int_t^{\infty}\{\}}\right)_{t=x_0}\psi(x)\left(e^{i\int_{-\infty}^{t=x_0}\{\}}\right)_t|0\rangle =$$

$$= \langle 0|(\ldots)_t \frac{\partial}{\partial x_0}\psi(x)(\ )_t|0\rangle +$$

$$+ i\langle 0|(\ )_t\left[\psi(x), \int_{x_0=y_0} d\vec y\,\{\bar\eta\psi + \bar\psi\eta + J_\nu A^\nu\}\right]_-(\ )_t|0\rangle =$$

$$= \langle 0|(\ldots)_t \frac{\partial}{\partial x_0}\psi(x)(\ )_t|0\rangle + i\gamma_0\eta(x)\,G$$

$$(i\gamma^\nu\partial_\nu - m)\frac{1}{i}\frac{\delta}{\delta\bar\eta(x)} G = \langle 0|(e^{i\int\{\}} e\gamma^\nu A_\nu(x)\,\psi(x))_t|0\rangle - \eta(x)\,G$$

$$\left\{i\gamma^\nu\left[\frac{\partial}{\partial x_\nu} + ie\frac{1}{i}\frac{\delta}{\delta J^\nu(x)}\right] - m\right\}\frac{1}{i}\frac{\delta}{\delta\bar\eta(x)} G = -\eta(x)\,G. \tag{5}$$

Similarly

$$(\Box + \mu_0{}^2)\frac{\delta G}{i\,\delta J_\mu(x)} = \left(-J^\mu(x) + e\frac{\delta}{\delta\eta(x)}\gamma^\mu\frac{\delta}{\delta\bar\eta(x)}\right)G. \tag{6}$$

Any matrixelement of the operator functional has to satisfy these eqs. In (3) we have written the matrixelement with respect to the vacuum of the interacting fields. Clearly (5) and (6) are not sufficient to determine this particular matrixelement. We have to add boundary conditions to make sure that the solution corresponds to the vacuum expectation value.

For

$$t \to +\infty: \lim_{t\to+\infty} \frac{1}{i}\frac{\delta}{\delta\bar\eta(x)} G = \lim_{t\to+\infty} \langle 0|\psi(x)(e^{i}\textstyle\int)_t|0\rangle$$

$$t \to -\infty: \lim_{t\to-\infty} \frac{1}{i}\frac{\delta}{\delta\bar\eta(x)} G = \lim_{t\to-\infty} \langle 0|(e^{i}\textstyle\int)_t\,\psi(x)|0\rangle.$$

By the asymptotic condition we demand that the fields approach free outgoing or incoming fields, multiplied by a constant factor. $\psi \to z_2^{1/2}\,\psi_{\text{in}}^{\text{out}}$. For a free field we know that the positive frequency parts annihilate the vacuum, therefore only negative frequency parts will contribute:

$$\psi^{\text{in}}|0\rangle = \psi^{-\text{in}}|0\rangle$$
$$\langle 0|\,\psi^{\text{out}} = \langle 0|\,\psi^{+\text{out}}.$$

The boundary conditions for $1/i\,(\delta/\delta\bar\eta)\,G$ therefore are that it has only positive (negative) frequency parts as the time goes to plus (minus) infinity.

Moreover

$$G(\eta\,\bar\eta\,J)|_{\eta=\bar\eta=J=0} = 1.$$

The first step to solve eq. (5) is to find the electron propagator in an external field $\mathscr{A}_\mu(x)$:

$$\{\gamma^\nu[i\,\partial_\nu - e\,\mathscr{A}_\nu(x)] - m\}\,S(x, x', \mathscr{A}) = \delta(x - x'). \tag{7}$$

The boundary conditions imposed on $S(x\,x'\,\mathscr{A})$ are the ones discussed before.

However, (7) is already an equation for which a solution in a closed form has not been found. To proceed we shall do as if $S(x\,x'\,\mathscr{A})$ is known and we shall find a formal solution, which of course contains $S(x\,x'\,\mathscr{A})$.

First we list some properties of $S(x\,x'\,\mathscr{A})$. A functional derivative of (7) gives:

$$\{\gamma^\nu[i\,\partial_\nu - e\,\mathscr{A}_\nu] - m\}\frac{\delta}{\delta\mathscr{A}_\varrho(y)} S(x\,x'\,\mathscr{A}) = e\,\gamma^\varrho\,\delta(x-y)\,S(x\,x'\,\mathscr{A})$$

$$\frac{\delta}{\delta\mathscr{A}_\varrho(y)} S(x\,x'\,\mathscr{A}) = e\,S(x\,y\,\mathscr{A})\,\gamma^\varrho\,S(y\,x'\,\mathscr{A}) \tag{8}$$

(8) is known as *Ward identity*.

We shall also use:

$$\left[J^\mu(y), F\left(\frac{1}{i}\frac{\delta}{\delta J}\right)\right]_- = i\frac{\delta F}{\delta\frac{1}{i}\frac{\delta}{\delta J_\mu(x)}}$$

for arbitrary F.

Eq. (5) is easily converted into an integral equation:

$$\frac{1}{i}\frac{\delta}{\delta\bar{\eta}(x)}G = -\int dz\, S\left(x\, z \frac{1}{i}\frac{\delta}{\delta J}\right)\eta(z)\, G$$

$$G = e^{-i\int dz\, dz'\, \bar{\eta}(z')\, S\left(z'\, z \frac{1}{i}\frac{\delta}{\delta J}\right)\eta(z)}\, G_J \tag{9}$$

G_J is a functional of J only and we have to use eq. (6) to determine it. First we calculate the term:

$$e\frac{\delta}{\delta\eta(x)}\gamma^\mu\frac{\delta}{\delta\bar{\eta}(x)}G = -i\, e\frac{\delta}{\delta\eta(x)}\gamma^\mu\int dz'\, S\left(x\, z'\,\frac{1}{i}\frac{\delta}{\delta J}\right)\eta(z')\, G =$$

$$= -i\, e\,\mathrm{tr}\,\gamma^\mu S\left(x\, x\frac{1}{i}\frac{\delta}{\delta J}\right)G +$$

$$+ e\int dz\, dz'\,\bar{\eta}(z)\, S\left(z\, x\frac{1}{i}\frac{\delta}{\delta J}\right)\gamma^\mu S\left(x\, z'\,\frac{1}{i}\frac{\delta}{\delta J}\right)\eta(z')\, G =$$

$$= -i\, e\,\mathrm{tr}\,\gamma^\mu S\left(x\, x\frac{1}{i}\frac{\delta}{\delta J}\right)G +$$

$$+ \int dz\, dz'\,\bar{\eta}(z)\frac{\delta}{\delta\frac{1}{i}\frac{\delta}{\delta J_\mu(x)}}S\left(z\, z'\,\frac{1}{i}\frac{\delta}{\delta J}\right)\eta(z')\, G.$$

Next we commute $J^\mu(x)$ according to:

$$e^{i\phi}A\, e^{-i\phi} = A + i[\phi, A] + \frac{i^2}{2!}[\phi,[\phi, A]] + \dots$$

$$e^{i\int dz\, dz'\,\bar{\eta}(z)\, S\left(z\, z'\frac{1}{i}\frac{\delta}{\delta J}\right)\eta(z')}\, J^\mu(x)\, e^{-i\int dz\, dz'\,\bar{\eta}(z)\, S\left(z\, z'\frac{1}{i}\frac{\delta}{\delta J}\right)\eta(z')} =$$

$$= J^\mu(x) + i\int dz\, dz'\,\bar{\eta}(z)\left[S\left(z\, z'\,\frac{1}{i}\frac{\delta}{\delta J}\right), J^\mu(x)\right]\eta(z') =$$

$$= J^\mu(x) + \int dz\, dz'\,\bar{\eta}(z)\frac{\delta}{\delta\frac{1}{i}\frac{\delta}{\delta J_\mu(x)}}S\left(z\, z'\,\frac{1}{i}\frac{\delta}{\delta J}\right)\eta(z').$$

We find:

$$\{\Box + \mu_0{}^2\}\frac{1}{i}\frac{\delta}{\delta J_\mu(x)}G_J = \left\{-J^\mu(x) - i\, e\,\mathrm{tr}\,\gamma^\mu S\left(x,\, x\frac{1}{i}\frac{\delta}{\delta J}\right)\right\}G_J. \tag{6'}$$

To evaluate it further we first concentrate on the term:

$$e\,\mathrm{tr}\,\gamma^\mu S\left(x\, x\frac{1}{i}\frac{\delta}{\delta J}\right).$$

We define $F(1/i \cdot \delta/\delta J)$ through:

$$e \operatorname{tr} \gamma^\mu S\left(x\, x\, \frac{1}{i} \frac{\delta}{\delta J}\right) = \frac{\delta F\left(\frac{1}{i} \frac{\delta}{\delta J}\right)}{\delta \frac{1}{i} \frac{\delta}{\delta J_\mu(x)}}.$$

As before:

$$e^{-F} J_\mu(x)\, e^F = J_\mu(x) + i \frac{\delta}{\delta \frac{1}{i} \frac{\delta}{\delta J_\mu(x)}} F.$$

Thus eq. (6′) becomes for:

$$G_J = e^{-F} \tilde{G}_J$$

$$\{\square + \mu_0^2\} \frac{1}{i} \frac{\delta}{\delta J_\mu(x)} e^{-F} \tilde{G}_J = e^{-F} J_\mu(x)\, \tilde{G}_J$$

or

$$(\square + \mu_0^2) \frac{1}{i} \frac{\delta}{\delta J_\mu(x)} \tilde{G}_J = - J_\mu(x)\, \tilde{G}_J.$$

This equation is easily integrated with the help of the Greens function $\Delta(x-y)$:

$$(\square + \mu_0^2)\, \Delta(x-y) = \delta(x-y)$$

$$\frac{\delta}{\delta J_\mu(x)} \tilde{G}_J = - i \int dz\, \Delta(x-z)\, J_\mu(z)\, \tilde{G}_J.$$

$$\tilde{G}_J = e^{-\frac{i}{2} \int dz\, dz'\, J^\mu(z')\, \Delta(z'-z)\, J_\mu(z)} \cdot \text{Const}$$

The final result is:

$$\begin{aligned} G(\eta\, \bar{\eta}\, J) = {} & e^{-i \int dz\, dz'\, \bar{\eta}(z')\, S\left(z'\, z\, \frac{1}{i} \frac{\delta}{\delta J}\right) \eta(z)}\, e^{-F\left(\frac{1}{i} \frac{\delta}{\delta J}\right)} \times \\ & \times e^{-\frac{i}{2} \int dz\, dz'\, J^\mu(z')\, \Delta(z'-z)\, J_\mu(z)} \times \\ & \times \left[e^{-F\left(\frac{1}{i} \frac{\delta}{\delta J}\right)}\, e^{-\frac{i}{2} \int J' \Delta J'} \right]^{-1}_{J'=0}. \end{aligned} \tag{10}$$

A short discussion of F:

$$e \operatorname{tr} \gamma^\mu S(x\, x\, \mathscr{A}) = \frac{\delta}{\delta \mathscr{A}} F(\mathscr{A})$$

can be formely integrated:

$$F(\mathscr{A}(y)) = \operatorname{tr} \int dz\, dz'\, \delta(z-z') \log S(z\, z'\, \mathscr{A})\, \delta(y-z)$$

Multiplication to the Greensfunction is to be understood in the sense of multiplication of integral kernels.

Thus we have given a formal solution of our field theory involving $S(x\, y\, \mathscr{A})$. But $S(x\, y\, \mathscr{A})$ is not known in a closed form. Also the whole expression (10) is very complicated and can only be handled in terms of power series expansions. This leads to nothing else but perturbation theory with all its difficulties. Thus the formal solution (10) does not even guarantee the existence of a solution as the whole expression (10) might not exist.

Lecture II

To exhibit the difficulties we meet, when dealing with expressions like (10), I should like to discuss what is known as Schwingers Paradoxon [2]. It arises from the fact that the product of two field operators, taken at the same space time point, is a not to well defined object. Ignoring this fact (as we have done frequently by using the current in the form $\bar{\psi}(x)\, \gamma^\mu\, \psi(x)$) can lead to contradictions: $j^0(x)$ is the generator density of a gauge transformation. If the current is defined in an gauge invariant way we expect:

$$[j^0(x), \vec{j}(y)]_{x_0=y_0} = 0. \qquad (11)$$

The current conservation law tells us: $j_0{}^0 = -\vec{\nabla}\vec{j}$ or

$$[j_0{}^0(x), j_0(y)]_{x_0=y_0} = 0.$$

Now

$$j_0{}^0(x) = i[H, j_0(x)]$$

and therefore

$$[[H, j_0(x)], j_0(y)] = 0$$

we take the vacuumexpectation value

$$\langle 0\,|\,[[H, j_0(x)], j_0(y)]\,|0\rangle = -2\,\langle 0|\, j_0(x)\, H\, j_0(x)\,|0\rangle = 0.$$

As the energy should be positive this expression can only be zero if the vacuum is an eigenstate of the current density, this is clearly not the case. Positive energy and the commutator (11) are therefore contradictory and we have to modify the definition of the current. To illustrate this problem I should like to discuss it in a two dimensional model, i.e. a model with one space and one time dimension. The equations are the same as before (1), (2), (5), (6), with the only difference that there is only one time and one space dimension. The γ-algebra, i.e. the matrices satisfying $\gamma^\mu\gamma^\nu + \gamma^\nu\gamma^\mu = 2\, g^{\mu\nu}$ can be represented by:

$$\gamma^0 = \sigma_x, \qquad \gamma^1 = -i\,\sigma_y, \qquad \gamma_5 \equiv \gamma = \sigma_z.$$

The complete antisymmetric tensor has two indices:

$$\varepsilon_{\mu\nu} \qquad \varepsilon_{01} = 1 \qquad \varepsilon_{10} = -1 \qquad \varepsilon_{00} = \varepsilon_{11} = 0.$$

It is easy to calculate the free field Green's functions and, we find that the causal Green's function has the following singularity:

$$S(\varepsilon) = -\frac{i}{2\pi}\frac{\gamma^\nu \varepsilon_\nu}{\varepsilon^2} \qquad \text{for} \qquad \varepsilon^2 \neq 0. \tag{12}$$

Thus, even in the case of free fields the product

$$\bar{\psi}_\alpha(x)\, \psi_\beta(x)$$

is not defined.

$$\langle 0|\, (\bar{\psi}_\alpha(x+\varepsilon)\, \psi_\beta(x))_t\, |0\rangle = -i\, S_{\alpha\beta}(\varepsilon).$$

To define a meaningful current, even for a free field, we have to prescribe some kind of limiting procedure, such as to cancel the singular terms.

To investigate the singularity of $\psi_\alpha{}^*(x+\varepsilon)\,\psi_\beta(x)$ we observe:

$$\left[\lim_{\varepsilon\to 0} \psi_\alpha{}^*(x+\varepsilon)\, \psi_\beta(x)\, \varepsilon^\varrho, \int dy\, f(y)\, \psi_\varrho(y)\right]_{x_0 = y_0} =$$
$$= -\lim_{\varepsilon\to 0} \varepsilon^\varrho\, \psi_\beta(x)\, f(\boldsymbol{x}+\boldsymbol{\varepsilon})\, \delta_{\alpha\varrho} = 0.$$

Schur's Lemma, therefore, tells us that the singular term of $\psi_\alpha{}^*(x+\varepsilon)\cdot\psi_\beta(x)$ is proportional to the unit matrix in Hilbertspace. Therefore, for the purpose of studying the singularity we can take the vacuum expectation value. It is easy to check that the following definition of the current is reasonable:

$$j_\varepsilon{}^\mu(x) = (\bar{\psi}(x+\varepsilon)\, \gamma^\mu\, \psi(x))^t \qquad j^\mu(x) = \lim_{\varepsilon\to\pm 0} j_\varepsilon{}^\mu(x) \tag{13}$$

$$\varepsilon \text{ spacelike.}$$

1) The singularity cancels by virtue of the symmetric limit.

2) The current is conserved as already $\partial_\mu\, j\, \varepsilon^\mu(x) = 0$ (using the free-field equations).

3) $j^0(x)$ is the generatordensity of the gauge transformation:

$$[j_\varepsilon{}^0(x), \psi_\varrho(y)]_{x_0 = y_0, \varepsilon_0 = 0} = -\delta(x+\varepsilon-y)\, \psi_\varrho(x).$$

To derive the current commutator we calculate

$$\left[\int j^0(x)\, f(\boldsymbol{x})\, d\boldsymbol{x}, j_\varepsilon{}^1(y)\right]_{x_0 = y_0\ \varepsilon_0 = 0} = \int d\boldsymbol{x}\, f(\boldsymbol{x})\, \{\delta(\boldsymbol{x}-\varepsilon-\boldsymbol{y}) - \delta(\boldsymbol{x}-\boldsymbol{y})\}\cdot$$
$$\cdot\, \bar{\psi}(y+\varepsilon)\, \gamma^1\, \psi(y) = \tag{14}$$
$$= \varepsilon^\sigma\, \partial_\sigma f(\boldsymbol{y})\, \bar{\psi}(y+\varepsilon)\, \gamma^1\, \psi(y) =$$
$$= \varepsilon^\sigma\, \partial_\sigma f(\boldsymbol{y})\, \langle 0|\, \bar{\psi}(y+\varepsilon)\, \gamma^1\, \psi(y)\, |0\rangle =$$
$$= -\frac{i}{2\pi}\, \partial_\sigma\, f(\boldsymbol{y})\, \frac{\varepsilon^\sigma\, \varepsilon^\varrho}{\varepsilon^2}\, \underbrace{\operatorname{tr} \gamma_\varrho \gamma^1}_{2 g_\varrho{}^1} = -\frac{i}{\pi}\frac{\partial}{\partial y_1}\, f(\boldsymbol{y})$$

or:

$$[j^0(x), j^1(y)]_{x_0=y_0} = \frac{i}{\pi} \frac{\partial}{\partial x_1} \delta(\mathbf{x}-\mathbf{y}).$$

Exactly this form was predicted by SCHWINGER for the 4-dimensional case also.

There is now no contradiction with gauge invariance as $j^0(x)$ generates a local gauge transformation, but j^1 is only invariant under a phase transformation which is generated by

$$Q = \int d\mathbf{x}\, j^0(x) \qquad \text{and} \qquad [Q, j^1(y)] = 0.$$

Next we consider the case, that an external electromagnetic field is present. If the mass of the electron is zero, the two-dimensional model can be solved explicitly.

$$i\gamma^\nu \partial_\nu \psi = e\gamma^\nu \mathcal{A}_\nu \psi. \tag{15}$$

Multiplication by γ_0 decouples the two components in this equation.

Solution:

$$\psi = e^{-ie\int dx' S_0(x-x')\gamma^\nu \mathcal{A}_\nu(x')} \cdot \psi_0(x) \tag{16}$$

$$\gamma^\nu \partial_\nu S_0(x-y) = \delta(x-y)$$

$$\psi_0: \qquad \text{free field.}$$

Green's function:

$$S(x\,x'\,\mathcal{A}) = e^{-ie\int dy \{S_0(x-y) - S_0(x'-y)\}\gamma^\nu \mathcal{A}_\nu(y)} \cdot S_0(x-x). \tag{17}$$

It satisfies

$$\gamma^\nu(i\,\partial_\nu - e\,\mathcal{A}_\nu)\, S(x\,x'\,\mathcal{A}) = \delta(x-x')$$

and as we know, it is the key to the general solution.

But before going into this, let us define the current properly. Eq. (15) is invariant under the following gauge transformation:

$$\psi \to e^{-ie\lambda(x)}\psi(x) \qquad \mathcal{A}^\nu \to \mathcal{A}^\nu + \partial^\nu \lambda. \tag{18}$$

With the gauge invariant definition

$$j_\varepsilon^\mu(x) = \bar{\psi}(x+\varepsilon)\,\gamma^\mu\, e^{-ie\int_x^{x+\varepsilon} \mathcal{A}_\nu d\xi^\nu} \cdot \psi(x). \tag{19}$$

We guarantee that also the limit $\varepsilon \to 0$ will be gauge invariant. To find out what kind of limiting procedure we have to employ we take advantage of knowing the solution of (15).

Therefore:

$$j_\varepsilon^\mu(x) = \bar{\psi}_0(x+\varepsilon)\,\gamma^\mu\, e^{ie\int dx' S_0(x+\varepsilon-x')\mathcal{A}^\nu\gamma_\nu} \cdot$$

$$\cdot\, e^{-ie\int_x^{x+\varepsilon} d\xi^\mu \mathcal{A}_\mu}\, e^{-ie\int dx' S_0(x-x')} \cdot \psi_0(x)$$

we develope in ε and write only the terms which will contribute to the limit $\varepsilon \to 0$.

$$j_\varepsilon^\mu(x) = \bar{\psi}_0(x+\varepsilon)\,\gamma^\mu\,\{1 + i\,e\,\varepsilon^\varrho\,B_\varrho(x)\}\,\psi_0(x)$$

$$B_\varrho(x) = \partial_\varrho \int dx'\,S_0(x-x')\,\gamma^\nu\,\mathscr{A}_\nu(x') - \mathscr{A}_\varrho(x).$$

This gives further

$$j_\varepsilon^\mu(x) = j_{0\,\varepsilon}{}^\mu(x) + i\,e\,\varepsilon^\varrho\,\langle 0|\,\bar{\psi}_0(x+\varepsilon)\,\gamma^\mu\,B_\varrho\,\psi_0(x)\,|0\rangle =$$

$$= j_{0\,\varepsilon}{}^\mu(x) + \frac{i\,e}{2\,\pi}\operatorname{tr}\gamma^\sigma\gamma^\mu\,B_\varrho\,\frac{\varepsilon_\sigma\,\varepsilon_\varrho}{\varepsilon^2}.$$

This expression seems to depend on the path $\varepsilon \to 0$. But when we examine it more carefully we find: $\gamma\,B_\varrho = \varepsilon_{\varrho\nu}\,B^\nu$ and also

$$\gamma^\sigma\,\gamma = \varepsilon_\nu{}^\sigma\,\gamma^\nu$$

With $\tilde{\varepsilon}^\alpha = \varepsilon^{\alpha\beta}\,\varepsilon_\beta$ we can write

$$\frac{i\,e}{2\,\pi}\operatorname{tr}\gamma^\sigma\gamma^\mu\,B_\varrho\,\frac{\varepsilon_\sigma\,\varepsilon_\varrho}{\varepsilon^2} = -\frac{e\,i}{2\,\pi}\operatorname{tr}\gamma^\sigma\,\gamma^\mu\,B^\nu\,\frac{\tilde{\varepsilon}_\sigma\,\tilde{\varepsilon}_\nu}{\varepsilon^2} =$$

$$= \frac{i\,e}{2\cdot 2\,\pi}\operatorname{tr}\gamma^\sigma\,\gamma^\mu\,B^\nu\left\{\frac{\varepsilon_\nu\,\varepsilon_\sigma}{\varepsilon^2} - \frac{\tilde{\varepsilon}_\sigma\,\tilde{\varepsilon}_\nu}{\varepsilon^2}\right\}.$$

This shows that the limit is actually path independent as

$$\frac{\varepsilon_\nu\,\varepsilon_\sigma}{\varepsilon^2} - \frac{\tilde{\varepsilon}_\sigma\,\tilde{\varepsilon}_\nu}{\varepsilon^2} = g_{\sigma\nu}.$$

Thus we find

$$j_\mu(x) = \lim_{\varepsilon \to \pm 0} j_\mu{}^\varepsilon(x) = j_{0\,\mu}(x) + \frac{e\,i}{4\,\pi}\operatorname{tr}\gamma^\sigma\,\gamma_\mu\,B_\sigma$$

$$\varepsilon \text{ spacelike.}$$

The trace can easily be evaluated and finally we get

$$j_\mu(x) = j_{0\,\mu}(x) - \frac{e}{\pi}\,\mathscr{A}_\mu{}^{\mathrm{tr}}(x)$$

when

$$\mathscr{A}_\mu{}^{\mathrm{tr}}(x) = P_\mu{}^\varrho\,\mathscr{A}_\varrho(x) = \int dx'\,\{g^{\mu\varrho}\,\delta(x-x') - \partial^\mu\,\partial^\varrho D(x-x')\}\,\mathscr{A}_\varrho(x').$$

We see that

1) j_μ is conserved

2) j_μ is gauge invariant, the factor $e^{-ie\int_x^{x+\varepsilon}\mathscr{A}_\nu d\xi^\nu}$ we had to introduce in order to achieve it is also necessary to make the limit $\varepsilon \to 0$ covariant, using spacelike ε's only.

3) As before, $j^0(x)$ generates the gauge transformation

$$\left[e\int j^0(x)\,\lambda(x)\,dx,\,\psi_\varrho(x)\right] = -\,e\,\lambda(x)\,\psi_\varrho(x).$$

But $[j^0(x), \mathscr{A}] = 0$ and therefore the gauge invariance argument, used to demonstrate Schwingers Paradoxon does not apply. Now we are ready to give the solution of eqs (5) and (6) in the 2-dimensional case ($m = 0$). We redefine the current term in (6)

$$\frac{\delta}{\delta\eta(x)}\gamma^\mu\frac{\delta}{\delta\bar\eta(x)} \to \frac{\delta}{\delta\eta(x+\varepsilon)}\gamma^\mu\left(e^{-ie\int_x^{x+\varepsilon}d\xi_\mu\frac{1}{i}\frac{\delta}{\delta J_\mu(\xi)}}\right)\cdot\frac{\delta}{\delta\bar\eta(x)}.$$

Consequently:

$$\operatorname{tr}\gamma^\mu S(x\,x\,\mathscr{A}) \to \operatorname{tr}\gamma^\mu S(x, x+\varepsilon, \mathscr{A})\, e^{-ie\int_x^{x+\varepsilon}\mathscr{A}_\mu d\xi^\mu} =$$

$$= \operatorname{tr}\gamma^\mu\{1 + i\,e\,\varepsilon^\sigma B_\sigma\}\,S_0(\varepsilon).$$

The limit $\varepsilon \to 0$ can be carried out as before and yields:

$$-\frac{i\,e}{\pi}\,P^{\mu\varrho}\,\mathscr{A}_\varrho.$$

The eq. which defines F becomes

$$\frac{\delta F(\mathscr{A})}{\delta\mathscr{A}_\mu(x)} = -\frac{i\,e^2}{\pi}\int dx'\,P^{\mu\varrho}(x-x')\,\mathscr{A}_\varrho(x')$$

and can be solved:

$$F(\mathscr{A}) = -\frac{i\,e^2}{2\,\pi}\int dx\,dx'\,\mathscr{A}^\mu(x)\,P_{\mu\varrho}(x-x')\,\mathscr{A}^\varrho(x').$$

The generating functional:

$$G(\eta\,\bar\eta\,J) =$$

$$= \exp\left[-i\int dz\,dz'\,\bar\eta(z')\,e^{-ie\int dy\,\{S_0(z-y)-S_0(z'-y)\}\gamma^\mu\frac{\delta}{i\delta J_\mu(y)}}\cdot S_0(z-z')\,\eta(z')\right]\cdot$$

$$\cdot\exp\left[-\frac{i\,e^2}{2\,\pi}\int dx\,dx'\,\frac{\delta}{i\,\delta J_\mu(x)}\,P_{\mu\varrho}(x-x')\,\frac{\delta}{i\,\delta J_\varrho(x')}\right]\cdot$$

$$\cdot\exp\left[-\frac{i}{2}\int dz\,dz'\,J^\mu(z)\,\Delta(z-z')\,J_\mu(z')\right]\cdot[\;]_{J'=0}{}^{-1}.$$

We are left to carry out the functional differentiations involved in $G(\eta\,\bar\eta\,J)$. This can be done formally by relaying on the similarity with ordinary differentiations.

The formula:

$$e^{iC\frac{\partial^2}{\partial x^2}} e^{iBx^2} = \frac{1}{(1+4\,B\,C)^{1/2}} e^{i\frac{B}{1+4\,BC}x^2}$$

can be used to evaluate

$$e^{-F} e^{-\frac{i}{2}\int J\Delta J},$$

when we identify

$$C = -\frac{e^2}{2\,\pi} P \qquad \text{and} \qquad B = -\frac{1}{2}\frac{1}{\Box + \mu_0{}^2 - i\,\varepsilon}.$$

These expressions are diagonal in momentum space.

We find:

$$\frac{B}{1+4\,B\,C} = -\frac{1}{2}\mathcal{G} = -\frac{1}{2}\left[\Box + \mu_0{}^2 + \frac{e^2}{\pi} P\right]^{-1} =$$

$$= -\frac{1}{2}\left[(\Box + \mu^2)\,P + (\Box + \mu_0{}^2)\,(1-P)\right]^{-1}$$

with $\mu^2 - \mu_0{}^2 = e^2/\pi$.

P and $(1 - P)$ are projection operators which project into orthogonal subspaces. In momentum space:

$$\tilde{P}_{\mu\varrho} = \left(g_{\mu\varrho} - \frac{k_\mu k_\varrho}{k^2}\right) \qquad (1-\tilde{P})_{\mu\varrho} = \frac{k_\mu k_\varrho}{k^2}.$$

Therefore:

$$\mathcal{G} = P\frac{1}{\mu^2 - k^2} + (1-P)\frac{1}{\mu_0{}^2 - k^2} =$$

$$= \left(g_{\mu\varrho} - \frac{k_\mu k_\varrho}{k^2}\right)\frac{1}{\mu^2 - k^2} + \frac{k_\mu k_\varrho}{k^2}\,\frac{1}{\mu_0{}^2 - k^2}.$$

The term $(1 + 4\,B\,C)^{-1/2}$ is of no interest according to the normalization $G(\eta, \bar{\eta}, J)|_{\eta=\bar{\eta}=J=0} \equiv 1$

$$e^{-F} e^{-\frac{i}{2}\int J\Delta J}\left[e^{-F} e^{-\frac{i}{2}\int J'\Delta J'}\right]^{-1}_{J'=0} =$$

$$= e^{-\frac{i}{2}\int dz\,dz'\,J^\mu(z)\,\mathcal{G}_{\mu\varrho}(z-z')\,J^\varrho(z')},$$

where the Fourier transform of $\mathcal{G}$ is given above.

The functional derivatives involved in

$$e^{-i\int \bar{\eta}(z') S\left(z', z \frac{1}{i}\frac{\delta}{\delta J}\right)\eta(z)}$$

can be evaluated in a power series expansion, using the fact that $e^{\partial/\partial x}$ is the translation operator. The result is rather complicated [3] and not too instructive. Instead of writing it down, I would like to give the expressions for the two point functions only:

$$\langle 0|(A_\mu(x)\, A_\nu(x'))_t|0\rangle = i\, \mathcal{G}_{\mu\nu}(x - x').$$

The momentum transform of $\mathcal{G}_{\mu\nu}$ shows the mass renormalization $\mu^2 = \mu_0{}^2 + (e^2/\pi)$. One also finds that it is consistent with the canonical commutation rules and the current conservation.

$$(\Box + \mu_0{}^2)\, \partial_\nu A^\nu = 0$$

$$\langle 0|(\psi(x)\psi^+(x'))_t|0\rangle = i\, S_0(x - x')\, e^{ie^2\{\mathcal{D}(x-x') - \mathcal{D}(0)\}},$$

where

$$\mathcal{D}(x) = \frac{1}{\mu_0{}^2}(\Delta(x, \mu_0) - \Delta(x, 0)) - \frac{1}{\mu^2}(\Delta(x, \mu) - \Delta(x, 0)).$$

It should be noted that $\mathcal{D}(x)$ together with its first three derivatives is finite and continuous for all values of x.

Lecture III

The two-dimensional model, which we solved by functional methods, can be solved in operator form also. Although we might not learn much for real Quantum Electrodynamics, it seems to be worthwhile to discuss the operator solution as it exhibits some features which are interesting by themselves. There are essentially two ways to solve equations (1) and (2) with electronmass $m = 0$, we shall discuss both.

1) We shall see that it is possible to express the interacting fields ψ and A_μ in terms of free canonical fields, such that (1) holds in a subspace of the whole Hilbert space, spanned by the free fields. As the operators A and ψ do not lead out of this subspace we have a solution of eqs. (1) and (2) in this subspace.

We have to introduce four fields:

$$a_\nu: (\Box + \mu^2)\, a_\nu = 0 \qquad [a_\nu(x), a_\mu(y)] = i\left(g_{\mu\nu} - \frac{\partial_\nu\, \partial_\mu}{\mu^2}\right)\Delta(x - y, \mu)$$

$$\partial_\nu a^\nu = 0$$

$$b: \qquad \Box\, b = 0 \qquad [b(x), b(y)] = -\, i\, D(x - y)$$

$$C: \qquad \Box\, C = 0 \qquad [C(x), C(y)] = i\, D(x - y)$$

$$B: (\Box + \mu_0{}^2)\, B = 0 \qquad [B(x), B(y)] = i\Delta(x - y, \mu_0).$$

Notice that B and C satisfy commutation rules corresponding to negative energies.

We make the following Ansatz:

$$A_\mu = a_\mu + \frac{1}{\mu}\varepsilon_{\mu\nu}\, \partial^\nu C + \frac{1}{\mu_0}\partial_\mu(B + b) \tag{21}$$

$$\psi = e^{i\phi}\chi,$$

where

$$\phi = -\frac{e}{\mu_0}(b + B) + \frac{e}{\mu}\gamma\left(C + \frac{1}{\mu}\varepsilon^{\sigma\nu}\,\partial_\nu\, a_\sigma\right)$$

and χ is a free canonical spinor field.

It is easy to verify that ψ satisfies the Dirac equation:

$$i\gamma^\nu\,\partial_\nu\psi = e\gamma^\nu A_\nu\,\psi$$

and that $\partial_\nu A^\nu = 0$.

To show that the fields A_μ and ψ obey the equal-time canonical commutation rules, we calculate:

$$[A_\mu(x), A_\nu(y)] = i\left(g_{\mu\nu} - \frac{\partial_\mu\,\partial_\nu}{\mu^2}\right)\Delta(x-y,\mu) - \frac{i}{\mu^2}\partial_\mu\,\partial_\nu D(x-y) -$$

$$- \frac{i}{\mu_0{}^2}\partial_\mu\,\partial_\nu\Delta(x-y,\mu_0) + \frac{i}{\mu_0{}^2}\partial_\mu\,\partial_\nu D(x-y)$$

$$[A_\nu(x), \phi(y)] = -\frac{i\,e}{\mu_0{}^2}[\partial_\nu\Delta(x-y,\mu_0) - \partial_\nu D(x-y)] -$$

$$- \frac{i\,e\,\gamma}{\mu^2}\varepsilon_{\nu\sigma}\,\partial^\sigma\,[\Delta(x-y,\mu) - D(x-y)]$$

$$[\phi(x), \phi(y)] = \frac{i\,e^2}{\mu_0{}^2}[\Delta(x-y,\mu_0) - D(x-y)] -$$

$$- \frac{i\,e^2}{\mu}\gamma_x\gamma_y\,[\Delta(x-y,\mu) - D(x-y)].$$

The equal-time commutation rules therefore become:

$$[A_\mu(x), A_\nu(y)]_{x_0=y_0} = 0 \qquad [\dot{A}_\mu(x), A_\nu(y)]_{x_0=y_0} = i\,g_{\mu\nu}\,\delta(\vec{x}-\vec{y})$$

$$[A_\mu(x), \psi(y)]_{x_0=y_0} = [\dot{A}_\mu(x), \psi(y)]_{x_0=y_0} = 0$$

$$\{\psi^+(x), \psi(y)\}_{x_0=y_0} = \delta(\vec{x}-\vec{y}).$$

To evaluate Maxwell's equation we first have to calculate the current according to (19):

$$j_\varepsilon{}^\mu(x) = e\,\bar{\psi}(x+\varepsilon)\,\gamma^\mu\, e^{-ie\int_x^{x+\varepsilon} d\xi^\mu A_\mu}\,\psi(x) =$$

$$= j_{0\varepsilon}{}^\mu(x) - \frac{e}{2\pi}\operatorname{tr}\gamma^\sigma\gamma^\mu(e\,A_\varrho + \partial_\varrho\,\phi)\,\frac{\varepsilon_\sigma\,\varepsilon^\varrho}{\varepsilon^2}.$$

As $\gamma(e\,A_\varrho + \partial_\varrho\phi) = \varepsilon_{\varrho\sigma}(e\,A^\sigma + \partial^\sigma\phi)$ holds, we find by a manipulation described already in detail:

$$j_\mu(x) = \lim_{\varepsilon\to\pm 0\,\text{spacelike}} j_{\mu\varepsilon}(x) = j_{0\,\mu}(x) + \frac{e^2}{\pi}\left(a_\mu + \frac{1}{\mu}\varepsilon_{\mu\nu}\,\partial^\nu C\right).$$

Maxwell's equation, therefore, would lead to the following equation:

$$\mu_0 \, \partial_\mu b + \mu \, \varepsilon_{\mu\nu} \, \partial^\nu C - e \, \bar{\chi} \, \gamma_\mu \, \chi = \mathfrak{K}_\mu = 0. \tag{24}$$

This equation cannot hold as an operator equation as the free fields entering it are supposed to be independent. The best we can hope is that all matrixelements of "physical" interest will be zero. We first observe that $\Box \, \mathfrak{K}_\mu = 0$ and that $\mathfrak{K}_\mu$ can be split into positive and negative frequencies $\mathfrak{K}_\mu^+$ and $\mathfrak{K}_\mu^-$. For the vacuum we demand:

$$\mathfrak{K}_\mu^+ \, |0\rangle = \langle 0| \, \mathfrak{K}_\mu^- = 0 \tag{25}$$

The fact that

$$[\mathfrak{K}_\mu^\pm(x), A_\varrho(y)] = [\mathfrak{K}_\mu^\pm(x), \psi(y)] = [\mathfrak{K}_\mu^\pm(x), \psi^+(y)] = 0. \tag{26}$$

show that the matrix elements of $\mathfrak{K}_\mu$ will be zero for all states which can be created from the vacuum by applying A's and ψ's only. But these are exactly the matrix elements of "physical interest". Therefore, we can assert that Maxwell's equation holds for all states which can be physically realized. It can also be shown that the Green's functions calculated from the operator solution correspond to the Green's function calculated before.

2) There is another way to solve the model in operator form [4]. This solution does not require a larger Hilbert space, but it has other features which are intimately connected with the two-dimensionality. In order to discuss this solution it is necessary to re-examine the free spinor field once more:

$$\gamma^\nu \, \partial_\nu \phi = 0 \qquad \{\phi(x), \phi^*(y)\}_{x_0 = y_0} = \delta(\vec{x} - \vec{y}).$$

ϕ is a free canonical field. We have seen (13) that

$$j_0^\mu(x) = \lim_{\varepsilon \to \pm 0, \text{ spacelike}} \bar{\phi}(x + \varepsilon) \, \gamma^\mu \, \phi(x)$$

is a reasonable definition of the current. From γ-invariance follows that

$$\varepsilon^{\nu\mu} \, j_{0\,\mu}(x) = \lim_{\varepsilon \to \pm 0} \bar{\phi}(x + \varepsilon) \, \gamma^\mu \gamma \, \phi(x)$$

is conserved too. This two conservation laws imply:

$$j_0^\mu = \partial^\mu \Lambda = \varepsilon^{\mu\nu} \, \partial_\nu \Sigma \qquad \Box \, j_0^\mu = 0 \qquad \Box \, \Lambda = \Box \, \Sigma = 0. \tag{27}$$

In a quantized theory Λ and Σ are operators. In order that the current commutation relations can be satisfied

$$[j_0^\mu(x), j_0^\nu(y)] = \frac{i}{\pi} \, \partial^\mu \, \partial^\nu D(x - y), \tag{28}$$

we have to demand that:

$$\begin{aligned} [\Lambda(x), \Lambda(y)] &= [\Sigma(x), \Sigma(y)] = -\frac{i}{\pi} D(x - y) \\ [\Lambda(x), \Sigma(y)] &= [\Sigma(x), \Lambda(y)] = -\frac{i}{\pi} \tilde{D}(x - y). \end{aligned} \tag{29}$$

$\tilde{D}(x)$ is defined through $\partial^\mu \tilde{D} = \varepsilon^{\mu\nu}\partial_\nu D$ and it is a simple function as is $D(x)$:

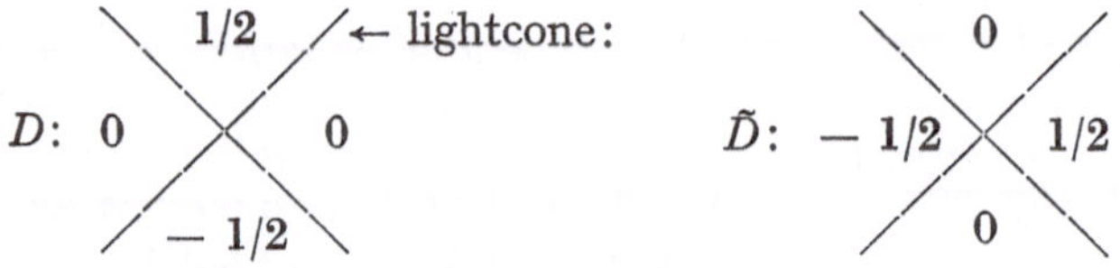

Since in two dimensions the right spacelike cone is separated from the left spacelike cone, $\tilde{D}(x)$ is an invariant function, too.

Moreover, j_0^μ is the generator of the phase transformation:

$$[\phi(x), j_0^\mu(y)] = (g^{\mu\nu} + \gamma\,\varepsilon^{\mu\nu})\,\partial_\nu D(y-x)\,\phi(x) \tag{30}$$

and therefore

$$\begin{aligned}[][\phi(x), \Lambda(y)] &= (D(y-x) + \gamma\,\tilde{D}(y-x))\,\phi(x)\\ [\phi(x), \Sigma(y)] &= (\tilde{D}(y-x) + \gamma\,D(y-x))\,\phi(x).\end{aligned} \tag{31}$$

The commutation rules (29) and (31) define the algebra of the ϕ, Λ and Σ fields, and a representation of this algebra in a Hilbert space can be constructed completely analogous to the free field case. In particular all the matrix elements of the fields can be constructed.

With the help of the fields ϕ, Λ, Σ, we can construct another field:

$$\chi_c = e^{ic\{\Lambda + \gamma\Sigma\}}\,\phi. \tag{32}$$

It is easy to check that $\gamma^\nu\,\partial_\nu\chi_c = 0$ and that

$$j_c^\mu(x) = \lim_{\varepsilon \to \pm 0 \text{ spacelike}} \bar{\chi}_c(x+\varepsilon)\,\gamma^\mu\,\chi_c(x) = \left(1 - \frac{c}{\pi}\right) j_0^\mu(x). \tag{33}$$

The most interesting case is $c = \pi$. The corresponding field we call χ and from (33) follows

$$j_\chi^\mu = \lim_{\varepsilon \to \pm 0} \bar{\chi}(x+\varepsilon)\,\gamma^\mu\,\chi(x) = 0 \tag{34}$$

and that the χ field does not carry charge. Eq. (34) shows that the finite part of the limit tends to zero after the singularity of the product has been eliminated via the symmetrization of the limit.

χ is not a canonical field. We can work out the following commutation rules:

$$\chi(x)\,\chi^*(y) + \chi^*(y)\,\chi(x)\,e^{i\pi(1+\gamma_x\gamma_y)D(x-y) + (\gamma_x+\gamma_y)\tilde{D}(x-y)} = -\,i\,S(x-y)\,\gamma^0. \tag{35}$$

For equal time:

$$x_0 = y_0 \begin{cases} \chi(x)\,\chi^*(y) - \gamma_x\gamma_y\,\chi^*(y)\,\chi(x) = 0 & \text{for} \quad \vec{x} \neq \vec{y} \\ \chi(x)\,\chi^*(y) + e^{i\pi(\gamma_x+\gamma_y)\tilde{D}(x-y)}\,\chi^*(y)\,\chi(x) = \delta(\vec{x}-\vec{y}) \end{cases}.$$

Moreover, we find:

$$[\chi(x), \Lambda(y)] = [\chi(x), \Sigma(y)] = 0. \tag{36}$$

This shows that the Hilbert space of the χ field is smaller than the Hilbert space of the ϕ field. Actually we expect that $\mathscr{H}_\phi$ be the direct product of $\mathscr{H}_\chi$ and $\mathscr{H}_{A,\Sigma}$. Again all matrix elements of the χ field can be constructed from (35). For example:

$$\langle 0|\, \chi(x)\, \chi^*(y)\, |0\rangle = -i\, S^-(x-y)\, \gamma^0\, e^{2i\pi\{D^-(y-x)+\gamma\tilde{D}^-(y-x)\}}.$$

It is now possible to give a solution of the following eqs.:

$$i\gamma^\nu\, \partial_\nu \psi = e\, A_\nu \gamma^\nu \psi \qquad j_\nu = \lim_{\varepsilon\to\pm 0} \bar{\psi}(x+\varepsilon)\, e^{-ie\int_x^{x+\varepsilon} A_\nu\, d\xi^\nu}\, \psi(x)$$

$$(\Box + \mu_0^2)\, A_\nu = e\, j_\nu \qquad \partial_\nu A^\nu = 0. \tag{37}$$

Canonical commutation rules for $x_0 = y_0$

$$\begin{aligned}
&\{\psi(x), \psi^*(y)\}_{x_0=y_0} = \delta(\vec{x}-\vec{y})\\
&[A_1(x), A_1(y)]_{x_0=y_0} = [\pi(x), \pi(y)]_{x_0=y_0} = 0 \qquad \pi = \partial_0 A_1 - \partial_1 A_0\\
&[A_1(x), \pi(y)]_{x_0=y_0} = i\,\delta(\vec{x}-\vec{y})\\
&[A_1(x), \psi(y)]_{x_0=y_0} = [\pi(x), \psi(y)]_{x_0=y_0} = 0.
\end{aligned} \tag{38}$$

A_1 and ψ are the only canonically independent fields as

$$A_0 = \frac{1}{\mu_0^2}\{e\, j_0 - \partial_1 \pi\}, \qquad \dot{A}_0 = \partial_1 A_1.$$

The solution will be based on the χ field and on the fields a, $\tilde{a}$, b_μ defined as follows:

$$\begin{aligned}
&\Box\, a = \Box\, \tilde{a} = 0 \qquad \partial_\mu a = \varepsilon_{\mu\nu}\, \partial^\nu \tilde{a}\\
&[a(x), a(y)] = [\tilde{a}(x), \tilde{a}(y)] = \frac{1}{i} D(x-y)\\
&[a(x), \tilde{a}(y)] = \frac{1}{i} \tilde{D}(x-y)\\
&(\Box + \mu^2)\, b_\mu = 0 \qquad \partial_\mu b^\mu = 0\\
&[b^\mu(x), b^\nu(y)] = i\left(g^{\mu\nu} + \frac{1}{\mu^2}\, \partial^\mu \partial^\nu\right) \Delta(x-y, \mu)\\
&[\chi, a] = [\chi, b] = [a, b] = 0.
\end{aligned} \tag{39}$$

We make the following Ansatz:

$$\psi = e^{-ie\{c\gamma^\nu \partial_\nu \gamma^\mu b_\mu + c_1 a + c_2 \gamma \tilde{a}\}}\, \chi. \tag{40}$$

and we have to determine c, c_1 and c_2.

The Dirac equation yields:

$$A_\nu = -c\,\mu^2\, b_\nu + (c_1 - c_2)\, \partial_\nu a.$$

Maxwell's equation:

$$e\, j_\nu = \mu^2(\mu^2 - \mu_0^2)\, c\, b_\nu + \mu_0^2 (c_1 - c_2)\, \partial_\nu a.$$

This reduces eq. (37) to:

$$e \lim_{\varepsilon \to \pm 0} \bar{\psi}(x+\varepsilon)\, \gamma^{\nu} e^{-ie \int_{x}^{x+\varepsilon} A_{\nu}\, d\xi^{\nu}} \psi(x) = \{\mu^2(\mu^2 - \mu_0{}^2)\, c\, b^{\nu} + \mu_0{}^2(c_1 - c_2)\, \partial^{\nu} a\}.$$

This equation is satisfied and independent of the direction $\varepsilon \to 0$ if:

$$\mu^2 - \mu_0{}^2 = \frac{e^2}{\pi} \qquad \text{and} \qquad \mu^2 c_2 = \mu_0{}^2 c_1.$$

We are left to show that the canonical commutation rules can be satisfied. We calculate for $x_0 = y_0$:

$$\psi(x)\, \psi^*(y) = \exp[-i e \{-c \gamma_x \partial_{\nu} \varepsilon^{\nu\mu} b_{\mu} + c_1 a + c_2 \tilde{a} \gamma_x\}] \cdot$$

$$\cdot \exp[i e \{-c \gamma_y \partial_{\sigma} \varepsilon^{\sigma\varrho} b_{\varrho} + c_1 a + c_2 \tilde{a} \gamma_y\}]\, \chi(x)\, \chi^*(y) =$$

$$= -\exp[i(\gamma_x + \gamma_y)(\pi - e^2 c_1 c_2)\, \tilde{D}(x-y)]\, \psi^*(y)\, \psi(x) + \delta(\vec{x} - \vec{y}),$$

Therefore $c_1 c_2 = (\pi/e^2)$ and consequently:

$$c_1 = \frac{\mu}{\mu_0} \frac{1}{e} \sqrt{\pi} \qquad c_2 = \frac{\mu_0}{\mu} \frac{1}{e} \sqrt{\pi}.$$

We also find:

$$[A_{\mu}(x), A_{\nu}(y)] =$$

$$= i \left\{ c^2 \mu^2 \left(g_{\mu\nu} + \frac{1}{\mu^2} \partial_{\mu} \partial_{\nu} \right) \Delta(x-y) + (c_1 - c_2)^2 \partial_{\mu} \partial_{\nu} D(x-y) \right\}.$$

This determines the last coefficient:

$$c = -\frac{1}{\mu^2}.$$

The relations:

$$[\psi(x), A_1(y)]_{x_0 = y_0} = [\psi(x), \pi(y)]_{x_0 = y_0} = 0$$

can be checked by a direct calculation.

Our Ansatz (40) has led to the following solution:

$$\psi = e^{i \left\{ \frac{e}{\mu^2} \gamma\, \partial_{\nu} \varepsilon^{\nu\mu} b_{\mu} - \frac{\mu}{\mu_0} \sqrt{\pi} a - \frac{\mu_0}{\mu} \sqrt{\pi} \tilde{a} \gamma \right\}} \cdot \chi \tag{41}$$

$$A_{\mu} = b_{\mu} + \frac{e}{\sqrt{\pi}} \frac{1}{\mu \mu_0} \partial_{\mu} a.$$

It might be interesting to notice that for $e \to 0$ we have:

$$A_{\mu} \to b_{\mu} \qquad \psi \to \phi$$

if we identify

$$a = \sqrt{\pi}\, \Lambda \qquad \tilde{a} = \sqrt{\pi}\, \Sigma.$$

References

1. J. Schwinger, Proc. Natl. Acad. Sci. U.S. **37**, 452 (1951).
K. Symanzik, Z. Naturforsch. **9a**, 809 (1954).
B. Zumino, Journal of Math. Physics I, 1 (1960).
2. J. Schwinger, Phys. Rev. Lett. **3**, 296 (1959).
3. W. Thirring and J. Wess, Ann. of Phys. **27**, 331 (1964).
4. This solution has been worked out by B. Zumino and the author during the 1962 "Summer Institute of Physics" held at the University of Washington, Seattle.

Theory of Laser Radiation*

By

W. Thirring

Universität Wien

With 2 Figures

1. Introduction

I was asked to start this sekond week nonrelativistically with a theory of lasers. By this I mean the description of radiation in interaction with matter of negative (absolute) temperature. It is clear that for a treatment satisfactory for theorists we have to idealize the situation. Firstly we replace the atoms by two-level systems (spins) and ignore the other levels. Secondly we start with matter of $T = -0\infty$ (all spins up, e.g. in the state of higher energy), which means we consider flashing ruby-lasers rather than continous Ne-He-lasers. Finally we shall ignore the atomistic structure of matter and treat it as a continuum.

It turns out [1] that many spins up coupled to the radiation field at the same point behave like an oscillator with negative energy. This means that the relevant matrix elements agree to the order (Number of spins down/Number of spins) with the corresponding oscillator matrix elements. Similarly a field of spins is equivalent to a field of oscillators with negative energy as long as most of the spins are up. Thus we arrive at a theory similar to the usual treatment of a dielectric where the atoms are replaced by a continuum of oscillators. The difference is the negative energy of the matter field which creates a dynamical instability of the system. This instability describes the development of the light avalanche which in our model would go on infinitely long, since we have infinitely many spins.

Our model has the virtue that it yields a complete theoretical treatment since it leads to linear equations. As is to be expected, the quantum effects are not dominating in this system, nevertheless it is useful for studying questions like coherence, correlation function, induced emission etc.

---

* Lecture given at the IV. Internationale Universitätswochen für Kernphysik, Schladming, 25 February—10 March 1965.

2. The Infinite Laser

First we shall study the case where matter is homogeneously distributed over all space. To stress the correspondence with dielectric theory, matter will be described by a polarization field $\vec{P}(x)$ proportional to the oscillator coordinate at the point x. All oscillators have the same frequency ω_0 and their density is related to the susceptibility $\varkappa$. With this notation we write for the Lagrangian of the matter field

$$L_m = -\frac{1}{2\varkappa\omega_0^2}\int d^4x(\dot{\vec{P}}^2(x) - \omega_0^2\,\vec{P}^2(x)). \tag{1}$$

For the radiation field we shall employ a gauge where the time component of the four-potential is zero so that $\vec{E} = -\dot{\vec{A}}$, $\vec{H} = \operatorname{curl}\vec{A}$ [2]. Hence the radiation Lagrangian is

$$L_r = \frac{1}{2}\int d^4x(\dot{\vec{A}}^2 - (\operatorname{curl}\vec{A})^2). \tag{2}$$

The interaction is simply $\vec{E}\cdot\vec{P}$ or in a symmetrized form

$$L_i = \frac{1}{2}\int d^4x(\dot{\vec{P}}\cdot\vec{A} - \dot{\vec{A}}\cdot\vec{P}). \tag{3}$$

The field equations deduced from $L = L_m + L_r + L_i$ are

$$\ddot{A} + \operatorname{curl}\operatorname{curl} A = \dot{P}$$
$$\left(\frac{\partial^2}{\partial t^2} + \omega_0^2\right)P = \omega_0^2\varkappa\dot{A} \tag{4}$$

e.g. Maxwell's equations with $\dot{P}$ as current and an oscillator driven by the electric field.

The total energy becomes

$$H = \frac{1}{2}\int d^3x\,\{\dot{A}^2 + (\operatorname{curl} A)^2 - \varkappa^{-1}\omega_0^{-2}(\dot{P}^2 + \omega_0^2\,P^2)\}. \tag{5}$$

Inserting a plane wave ansatz $A, P \sim \exp i(\vec{k}\,\vec{x} - \omega\,t)$ and separating into longitudinal and transversal modes we find for the former the frequencies

$$\omega_l^2 = \omega_0^2(1-\varkappa) \qquad \text{and} \qquad 0 \tag{6}$$

and for the latter

$$\omega_{T\pm}^2 = \frac{\omega_0^2}{2}\left(1 - \varkappa + \frac{k^2}{\omega_0^2} \pm \sqrt{\left(1 + \varkappa - \frac{k^2}{\omega_0^2}\right)^2 - 4\varkappa}\right). \tag{7}$$

The longitudinal mode with $\omega = 0$ is just a gauge field with $P = 0$, $A = \operatorname{grad}\lambda$ and without physical interest. The others are longitudinal oscillator waves with a slightly renormalized frequency but without dispersion since the P-oscillators are not coupled to their neighbours. More interesting are the transversal modes where the connection between k^2 and ω^2 is plotted in Fig. 1.

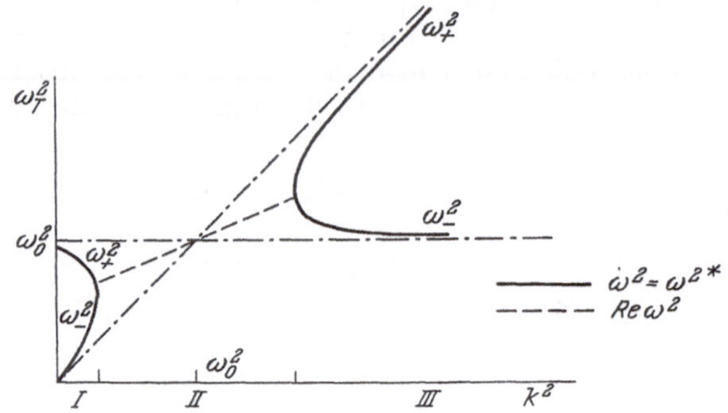

Fig. 1

We see that in regions I and III for each k^2 there are two real ω^2 whereas in region II near $k^2 = \omega_0{}^2$ there is no real but two complex conjugate ω^2. The dielectric constant

$$\varepsilon = \frac{k^2}{\omega_T{}^2} = 1 - \frac{\omega_0{}^2 \varkappa}{\omega_0{}^2 - \omega_T{}^2} \tag{8}$$

becomes complex in this region. This is to be contrasted with an ordinary dielectric with positive static polarisability $-\varkappa > 0$ where we have two ω^2 for each k^2 (Fig. 2).

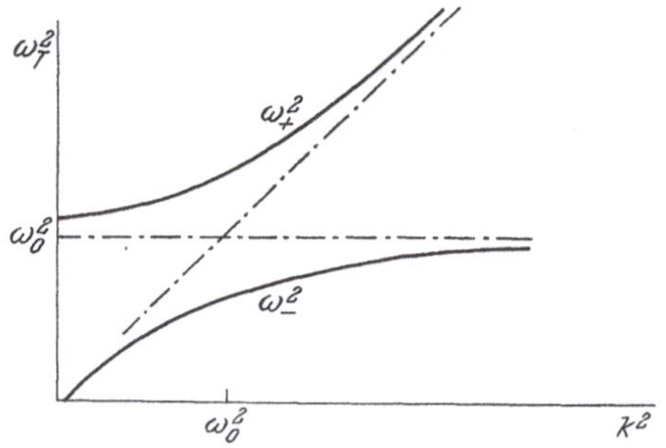

Fig. 2

In quantum theory we have to observe that the Lagrangian (1) to (3) leads to somewhat unusual commutation rules.

$$[A_i(k), \dot{A}_j{}^+(k')] = -\varkappa^{-1}\,\omega_0{}^{-2}\,[P_i(k), \dot{P}_j{}^+(k')] = -$$
$$-\varkappa^{-1}\,\omega_0{}^{-2}\,[\dot{A}_i(k'), P_j{}^+(k')] = i\,\delta_{ij}\,\delta(k-k'). \tag{9}$$

To satisfy these we first remark that the fields and the Hamiltonian separate into a longitudinal part

$$A_l, \qquad \operatorname{curl} A_l = 0, \qquad H_l(A_l)$$

and a transversal part

$$A_T, \qquad \operatorname{div} A_T = 0, \qquad H_T(A_T).$$

The former is not very interesting and will not be discussed further. For the latter we have to distinguish between the regions I and III where the frequencies are real and II where they are complex. In the first region the commutation relations can be satisfied by the ansatz

$$\text{I.} \qquad k^2 < \omega_-^2 < \omega_+^2 < \omega_0^2$$

$$A_T(k) = (\alpha_k + \alpha_{-k}^+) \cdot$$

$$\sqrt{\frac{\omega_0^2 - \omega_+^2}{2\,\omega_+(\omega_+^2 - \omega_-^2)}} + (\beta_k + \beta_{-k}^+) \sqrt{\frac{\omega_-^2 - \omega_0^2}{2\,\omega_-(\omega_-^2 - \omega_+^2)}}$$

$$\dot{A}_T(k) = -i(\alpha_k - \alpha_{-k}^+) \sqrt{\frac{\omega_+(\omega_0^2 - \omega_+^2)}{2(\omega_+^2 - \omega_-^2)}} - i(\beta_k - \beta_{-k}^+) \sqrt{\frac{\omega_-(\omega_-^2 - \omega_0^2)}{2(\omega_-^2 - \omega_+^2)}}$$

$$P_T(k) = -i(\alpha_k - \alpha_{-k}^+) \cdot$$

$$\cdot \sqrt{\frac{(\omega_+^2 - k^2)\,\varkappa\,\omega_0^2}{2\,\omega_+(\omega_+^2 - \omega_-^2)}} - i(\beta_k - \beta_{-k}^+) \sqrt{\frac{(k^2 - \omega_-^2)\,\varkappa\,\omega_0^2}{2\,\omega_-(\omega^2 - \omega_+^2)}} \qquad (10)$$

$$\dot{P}_T(k) = -(\alpha_k + \alpha_{-k}^+) \cdot$$

$$\cdot \sqrt{\frac{\omega_+(\omega_+^2 - k^2)\varkappa\,\omega_0^2}{2(\omega_+^2 - \omega_-^2)}} - (\beta_k + \beta_{-k}^+) \sqrt{\frac{\omega_-(-\omega^2 + k^2)\,\varkappa\,\omega_0^2}{2(\omega_-^2 - \omega_+^2)}}$$

where the commutators between the α and β which are different from zero are

$$-[\alpha_k, \alpha_{k'}^+] = P_T\,\delta(k - k') \qquad (11)$$

$$+[\beta_k, \beta_{k'}^+] = P_T\,\delta(k - k')$$

here P_T is the transversal projection operator

$$(P_T)_{lm} = \delta_{lm} - \frac{k_l\,k_m}{k^2}.$$

Inserting into the Hamiltonian we find

$$H_T^I = \int_I d^3k\,\{\beta_k^+\,\beta_k\,\omega_-(k) - \alpha_k^+\,\alpha_k\,\omega_+(k)\} \qquad (12)$$

as was to be expected since $\alpha(t) = \alpha(0)\,e^{-i\omega_+ t}$, $\beta(t) = \beta(0)\,e^{-i\omega_- t}$. Thus in I the transversal part has a discrete energy spectrum corresponding to oscillators with positive and negative energy. Since they are uncoupled they do not cause any dynamic instability. The same situation occurs in region III.

The laser action happens only in region II with the complex frequencies $\omega_+, \omega_- = (\omega_+)^*$. Here the commutation relations (9) are satisfied by

$$A_T(k) = (\alpha_k + \beta_{-k}^+) \cdot$$

$$\cdot \sqrt{\frac{\omega_+^2 - \omega_0^2}{2\,\omega_+(\omega_+^2 - \omega_-^2)}} + (\beta_k + \alpha_{-k}^+) \sqrt{\frac{\omega_-^2 - \omega_0^2}{2\,\omega_-(\omega_-^2 - \omega_+^2)}} \qquad (13)$$

$$\dot{A}_T(k) = -i(\alpha_k - \beta_{-k}^+)\cdot$$
$$\cdot\sqrt{\frac{\omega_+(\omega_+^2 - \omega_0^2)}{2(\omega_+^2 - \omega_-^2)}} - i(\beta_k - \alpha_{-k}^+)\sqrt{\frac{\omega_-(\omega_-^2 - \omega_0^2)}{2(\omega_-^2 - \omega_+^2)}}$$
$$P_T(k) = i(\alpha_k - \beta_{-k}^+)\cdot$$
$$\cdot\sqrt{\frac{(k^2 - \omega_+^2)\,\varkappa\,\omega_0^2}{2\,\omega_+(\omega_+^2 - \omega_-^2)}} + i(\beta_k - \alpha_{-k}^+)\sqrt{\frac{(k^2 - \omega_-^2)\,\varkappa\,\omega_0^2}{2\,\omega_-(\omega_-^2 - \omega_+^2)}}$$
$$\dot{P}_T(k) = (\alpha_k + \beta_{-k}^+)\cdot$$
$$\cdot\sqrt{\frac{\omega_+(k^2 - \omega_+^2)\,\varkappa\,\omega_0^2}{2(\omega_+^2 - \omega_-^2)}} + (\beta_k + \alpha_{-k}^+)\sqrt{\frac{\omega_-}{2}\,\frac{(k^2 - \omega_-^2)\,\varkappa\,\omega_0^2}{\omega_+^2 - \omega_-^2}}$$

if α and β obey

$$[\alpha_k, \beta_{k'}^+] = P_T\,\delta(k - k') = [\beta_k, \alpha_{k'}^+],$$
$$[\alpha, \alpha^+] = [\alpha, \beta] = [\beta, \beta^+] = 0. \tag{14}$$

The transversal energy becomes

$$H_T^{\mathrm{II}} = \int d^3k\,\{\alpha_k\,\beta_k^+\,\omega_+ + \beta_k\,\alpha_k^+\,\omega_-\} \tag{15}$$

in accordance with the time dependence

$$\alpha \sim e^{-i\omega_+ t}, \qquad \beta^+ \sim e^{i\omega_+ t}, \qquad \alpha^+ \sim e^{i\omega_- t}, \qquad \beta \sim e^{-i\omega_- t}. \tag{16}$$

Since $\omega_\pm$ are complex quantities $\omega \pm i\,\Gamma$, α and α^+ increase exponentially with time. Correspondingly the Hamiltonian H_T^{II} has no discrete eigenstates. Before starting the discussion we still have to solve the initial value problem. Using the time dependence (16) we find

$$A(k, t = 0) = \mathrm{A}, \qquad \text{e.t.c.} \tag{17}$$

$$A(t) = \Big\{A\,[(k^2 - \omega_-^2)\cos\omega_+ t - (k^2 - \omega_+^2)\cos\omega_- t] +$$
$$+ \dot{A}\left[\frac{\omega_+}{k}(k^2 - \omega_-^2)\sin\omega_+ t - \frac{\omega_-}{k^2}(k^2 - \omega_+^2)\sin\omega_- t\right] +$$
$$+ P\left[\frac{\omega_0^2}{\omega_t}\sin\omega_+ t - \frac{\omega_0^2}{\omega_-}\sin\omega_- t\right] - \dot{P}[\cos\omega_+ t - \cos\omega_- t]\Big\}\cdot$$
$$\cdot\frac{1}{\omega_+^2 - \omega_-^2}$$

$$\dot{A}(t) = \{A\,[\omega_-(k^2 - \omega_+^2)\sin\omega_- t - \omega_+(k^2 - \omega_-^2)\sin\omega_+ t] -$$
$$- \dot{A}\,[(\omega_0^2 - \omega_+^2)\cos\omega_+ t - (\omega_0^2 - \omega_-^2)\cos\omega_- t] +$$
$$+ P\,[\cos\omega_+ t - \cos\omega_- t]\,\omega_0^2 +$$
$$+ \dot{P}\,[\omega_+\sin\omega_+ t - \omega_-\sin\omega_- t]\}\,\frac{1}{\omega_+^2 - \omega_-^2}$$

$$P(t) = \Bigg\{ A\,\varkappa\,\omega_0^2\,k^2 \left[\frac{1}{\omega_+} \sin \omega_+ t - \frac{1}{\omega_+} (\sin \omega_- t) \right] -$$

$$- \dot{A}\,\varkappa\,\omega_0^2 [\cos \omega_+ t - \cos \omega_- t] -$$

$$- P \left[\frac{k^2 - \omega_+^2}{\omega_+^2} \cos \omega_+ t - \frac{k^2 - \omega_-^2}{\omega_-^2} \cos \omega_+ t \right] \omega_0^2 -$$

$$- \dot{P} \left[(k^2 - \omega_+^2) \frac{\sin \omega_+ t}{\omega_+} - (k^2 - \omega_-^2) \frac{(\sin \omega_{-0} t)}{\omega_{-0}} \right] \Bigg\} \frac{1}{\omega_+^2 - \omega_-^2}$$

$$\dot{P}(t) = \Bigg\{ A\,\varkappa\,\omega_0^2\,k^2 [\cos \omega_+ t - \sin \omega_- t] +$$

$$+ \dot{A}\,\varkappa\,\omega_0^2 [\omega_+ \sin \omega_+ t - \omega_- \sin \omega_- t] +$$

$$+ P \left[\frac{k^2 - \omega_+^2}{\omega_+} \sin \omega_+ t - \frac{k^2 - \omega_-^2}{\omega_-} \sin \omega_- t \right] \omega_0^2 -$$

$$- \dot{P}\,[(k^2 - \omega_+^2) \cos \omega_+ t - (k^2 - \omega_-^2) \cos \omega_- t] \Bigg\} \frac{1}{\omega_+^2 - \omega_-^2}.$$

After having shown how the quantum theoretic formalism works if the energy is not positive definite, we are in a position to answer physical questions. We shall restrict ourselves to modes near ω_0 (region II) where the excitations are amplified.

a) If we are initially in the groundstate $|0)$ of radiation and matter

$$\alpha_k^+|0) = \beta_k^+|0) = 0 \tag{18}$$

the expectation value of the electric field remains zero for all times

$$(0|\vec{A}(k, t)|0) = 0. \tag{19}$$

This means that the light comes out with no definite phase. This is clear intuitively because when the laser is self-exciting there is nothing which fixes the phase. The total energy remains, of course, zero, but when one calculates the energy in the radiation field

$$\left(0 \left| \frac{1}{2} \int d^3x (\dot{\vec{A}}(x)^2 + (\text{curl}\,\vec{A})^2) \right| 0\right)$$

one sees immediately that this increases with $e^{+2\Gamma t}$.

b) Next we study the excitation of the laser by a wave with a definite phase. This is done in our model by starting at $t = 0$ with a coherent state $|c)$ such that

$$\begin{aligned} (c|\vec{A}_T(k)|c) &= \vec{l}[\delta(\vec{k} - \vec{k}_0) + \delta(\vec{k} + \vec{k}_0)] \cdot \frac{1}{2} \\ (c|\dot{\vec{A}}(k)|c) &= (c|P|c) = (c|\dot{P}|c) = 0. \end{aligned} \tag{20}$$

$\vec{l}$ being a polarization vector, $\vec{l}\cdot\vec{k}=0$. This represents a standing wave $\sim\vec{l}\cos\vec{k}_0\,\vec{x}\cdot\cos|\vec{k}_0|\,t$. According to (17) this develops for times $\gg\Gamma$ into

$$(c|\vec{A}_T(x,t)|c)=\vec{l}\cos\vec{k}_0\,\vec{x}\cdot e^{\Gamma t}\left(\frac{1}{2}\cos\bar{\omega}\,t-\frac{k_0{}^2-\bar{\omega}^2+\Gamma^2}{4\,\bar{\omega}\,\Gamma}\sin\bar{\omega}\,t\right). \quad (21)$$

Thus we get an exponentially increasing wave with the same direction and polarization but with a slightly modified frequency (remember that now $\bar{\omega}$ and Γ depend on k for which here k_0 has to be inserted).

However the Laser radiation $\sim e^{\Gamma t}$ does not correspond to a coherent state for $t\gg\Gamma^{-1}$. This can easily be seen by calculating the square fluctuation

$$(c|A^2(t)|c)-(c|A(t)|c)^2=(\Delta A)^2$$

which also increases exponentially with the time whereas it is constant for a coherent state. Thus both, the amplitude and the square fluctuation are amplified, the relative square fluctuation $\Delta A/(c|A|c)$ remaining constant. Thus this laser amplification cannot be used for measuring the phase of a wave more accurately.

3. Other Geometries

Finally we shall see what happens if the laser material is distributed nonhomogeneously in space. These cases can be formulated by multiplying the integrand in L_m and L_i with the density of laser atoms. There are some situations where even this more complicated problem can be mathematically analyzed. I shall not be able to reproduce the arithmetic here and have to refer for that to ref. [1]. I shall discuss only the relevant points.

a) The Point Laser

In this case we assume that the laser material is concentrated at one point corresponding to a density $\delta(\vec{x})$. For simplicity let us ignore polarization effects and work with a scalar field. This problem is similar to the soluble problem of the interaction of a dipole with the radiation field, except that now the dipole energy E_d is negative. In both cases the initial value problem can be solved explicitly, i.e. the fields at any time t can be expressed in terms of the fields and their time derivatives at any other time t_0. In the limit $t_0\to\pm\infty$ one gets for $E_d>0$ the fields expressed in terms of the asymptotic fields. This is not possible for $E_d<0$ where we have a time dependence $e^{\pm\Gamma(t-t_0)}$ corresponding to a continous energy spectrum. What corresponds to region II in the previous case are now all spherically symmetric modes (s-states). Thus there are no asymptotic states but we have to discuss the situation in terms of bare states at, say $t_0=0$. If the electromagnetic field is in its ground state at $t=0$ the laser radiation develops exponentially and spherically symmetric as one would expect. If we start with N photons with wave-vector $\vec{q}$ the electromagnetic intensity consists of three contributions: the incident intensity, the laser radiation and the interference between the two. The laser radiation comes out with a frequency around ω_0 with a width Γ corresponding to the exponential increase $e^{2\Gamma t}$. It is proportional to $1 + 2\,N(q_0)$ and thus has a contribution stimulated by photons of the resonance frequency $\omega_0 = |\vec{q}_0|$. The laser radiation is spherically symmetric, a preference for the foreward direction being contained in the interference term. This is analogous to the case of $E_d > 0$ if the oscillator is in an excited state at $t = 0$ and incident photons are present. Again the angular distribution of the dipole radiation is unchanged and only the interference term distinguishes the incident direction. This shows that a quick argument using Einstein's induced emission, according to which the stimulated radiation should go in the forward direction may be misleading. The factors $\sqrt{N}$, $\sqrt{N+1}$ describe only the natural clustering of Bosons and thus the enhancement of the forward direction by the interference term. This is simply illustrated by a state of two independent (but identical) Bosons, one with momentum $\vec{k}_0$ and one with a spherically symmetric wave-function $f(\vec{k})$, $\Sigma_k |f(k)|^2 = 1$.

The properly normalized 2 particle state is

$$|2) = [1 + |f(k_0)|^2]^{-1/2} \sum_k f(k)\, a_k^+\, a_{k_0}^+ |0).$$

The probability for finding one particle with momentum k_0 and one with q in this state is

$$\frac{|f(q)|^2}{1 + |f(k_0)|^2}$$

whereas the probability for finding 2 particles with momentum k_0 is

$$2\,\frac{|f(k_0)|^2}{1 + |f(k_0)|^2}.$$

This factor two expresses the boson clustering according to which they prefer to travel together in spite of the product form of the 2 particle wave function. Correspondingly the expectation value of the intensity is

$$(2|a_q^+ a_q|2) = [1 + |f(k_0)|^2]^{-1} \Big(|f(q)|^2 + \delta_{q k_0} \sum_k |f(k)|^2 + $$
$$+ \delta_{q k_0}\, 2 \cdot \mathrm{Re}\, f^*(k_0)\, f(q) \Big) =$$
$$= [|f(q)|^2 + \delta_{q k_0} (1 + 2|f(k_0)|^2)]\; [1 + |f(k_0)|^2]^{-1}.$$

Thus the enhancement in the forward direction according to the factor 2 comes in through the interference term. Exactly this situation appears in the case of an oscillator with an initial photon.

If we have initially a definite number of photons, all expectation values of expressions linear in the fields vanish. This means that the absolute phase is not fixed. Nevertheless the radiation has coherence properties since in the usual coherence experiments only the relative phases at different space-time point are relevant. This is expressed by the range of the correlation function $(|E(x)\, E(x')|)$.

For incoherent radiation (e.g. a thermal ensemble) this is a rapidly decaying function of $|x - x'|$ whereas for coherent radiation (e.g. a plane wave photon) it has an infinite range. In this sense the laser radiation is as coherent as photons whose frequency is defined within a width Γ. If the laser is excited by a wave of definite phase, then the absolute phase of the laser radiation becomes determined. But again it is not a so-called coherent state which comes out. The zero point fluctuations of the field are amplified to the same degree as the amplitude, whereas for the coherent state they equal the vacuum fluctuations.

b) The Laser Rod.

After the previous discussion one might wonder why in actual lasers the radiation comes out so well collimated. This collimation occurs if the geometry is such that modes with frequencies near ω_0 have a fourier transform concentrated around a certain $\vec{k}_0$. This happens if the laser material forms a rod. Mathematically, this problem can be handled like that of the classical exercise of a dielectric cylinder. It turns out that in this case the laser radiation comes out in the axial direction.

General References

SINGER, J. R., Masers, New York: Wiley and Sons **1959**.

DUNSMUIR, R., Electronics and Control **10**, **453** (**1961**).

GÜRS, K., Z. Naturforschung **18** a, 510 (1963); **17** a, 990 (1962).

HOUTERMANS, F. G., Helv. Phys. Acta **33**, **933** (1960).

SCHWINGER, J., J. Math. Phys. **2**, 407 (1961).

LAMB, W. E. Jr., Lectures in Theoretical Physics (BOULDER) Vol. II, p. 435.

GORDON, J. P., L. R. WALKER and W. H. LOUISELL, Phys. Rev. **130**, 806 (1963).

MCCUMBER, D. E., Phys. Rev. **130**, 675 (1963).

SERBER, R. and C. TOWNES, Quantum Electronics New York: Columbia University Press 1960, p. 233.

HAKEN, H. and H. SAUERMANN, Z. Physik **173**, 261 (1963).

---

[1] A more complete demonstration of the various points can be found in F. SCHWABL, W. THIRRING: Quantum Theory of Laser Radiation. Ergebnisse der Exakten Naturw. Springer 1964.

[2] We use natural units $\hbar = c = 1$.

Non-Local Quantumelectrodynamics*

By

M. Lévy

Laboratoire de Physique Theorique et Hautes Energies,
Universite de Paris, France

1. Introduction

In these notes we shall omit many details which can be found in a paper which was published in Nuclear Physics a few months ago [1]. We are concerned, through a non-local treatment of quantum electrodynamics (QED), with two problems:

1. Gauge invariance and the photon mass: it has been thought for a long time that gauge invariance leads automatically to a vanishing physical mass of the photon. Recently [2], examples have been given which clearly indicate that this is not necessarily so. One can also show [3] that a consistent formulation of QED can be constructed which allows for a non-zero mass from the beginning.

Even in the conventional perturbation expansion of QED, the calculation of the photon self-energy has always been an ambiguous problem [4]. In the presence of a weak external field $A_\mu^{\text{ext}}(x)$ the vacuum expectation value of the induced current can be written

$$\langle j_\mu(x)\rangle_0 = \int \Pi_{\mu\nu}(x-x')\, A_{\text{ext}}{}^{\nu}(x')\, d^4x'. \tag{1.1}$$

To order e^2 the Fourier transform of $\Pi_{\mu\nu}$ can be expressed as follows:

$$\hat{\Pi}_{\mu\nu}(p) = g_{\mu\nu} A(0) + (g_{\mu\nu} p^2 - p_\mu p_\nu)\, B(p^2). \tag{1.2}$$

The first term on the right hand side is a quadratically divergent constant which violates current conservation and gauge invariance since these laws impose the conditions

$$p^\mu \hat{\Pi}_{\mu\nu} = \hat{\Pi}_{\mu\nu} p^\nu = 0. \tag{1.3}$$

A possible explanation for this result is that the current operator, when expressed in terms of the Dirac field, is not well defined because

---

* Lecture given at the IV. Internationale Universitätswochen für Kernphysik, Schladming, 25 February—10 March 1965.

it involves a product of distributions computed at the same point. A better procedure should be to express $j_\mu(x)$ as a gauge invariant limit which we can write using the distribution theory:

$$j_\mu(x) = -\frac{e}{2}\int g(x - x_1)\, g^*(x - x_2)\, [\tilde{\psi}(x_1), \gamma_\mu \psi(x_2)] \times$$
$$\times \exp\,[i\, e\, U(x_2) - i\, e\, U(x_1)]. \tag{1.4}$$

Here g is an infinitely differentiable function of compact support and $U(x)$ is a functional of A_μ such that, if we make a gauge transformation $A_\mu \to A_\mu + \partial_\mu \Lambda$, $\psi \to e^{ie\Lambda}\, \psi$, etc., we have $U \to U + \Lambda$. There are many equivalent forms for U as we shall see later. We assume that U is a linear functional of A_μ:

$$U = \int h_\mu(x - x')\, A^\mu(x')\, d^4x'. \tag{1.5}$$

Then the only requirement coming from gauge invariance is:

$$-\,i(p \cdot \hat{h}) = 1. \tag{1.6}$$

because in momentum space $\hat{U} = \hat{h} \cdot \hat{A}$ and if we make $\hat{A}_\mu \to \hat{A}_\mu -$ $-\, i\, p_\mu \hat{\Lambda}$ we want $\hat{U} \to \hat{U} + \hat{\Lambda}$.

With eq. (1.4) we can compute $\langle j_\mu(x)\rangle_0$ to order e^2. We find

$$\hat{\Pi}_{\mu\nu} = \Pi_{\mu\nu}{}^{(a)} + \Pi_{\mu\nu}{}^{(b)}, \tag{1.7}$$

where $\hat{\Pi}_{\mu\nu}{}^{(a)}$ is the value one gets in the usual theory, except for the form factors $\hat{g}$ (which we assume to be real for simplicity):

$$\hat{\Pi}_{\mu\nu}{}^{(a)}(p) = -\frac{e^2}{2\,\pi^3}\int \hat{g}(k)\, \hat{g}(p + k)\, T_{\mu\nu}(k, p + k) \times$$
$$\times\, \delta(k^2 - m^2)\, \frac{1}{p^2 + 2\, p\, k}\, d^4k, \tag{1.8}$$

where

$$T_{\mu\nu}(p, q) = p_\mu q_\nu + p_\nu q_\mu - g_{\mu\nu}(p \cdot q - m^2). \tag{1.9}$$

If we compute $\hat{\Pi}_{\mu\nu}{}^{(a)}\, p^\nu$:

$$\hat{\Pi}_{\mu\nu}{}^{(a)}\, p^\nu = -\frac{e^2}{2\,\pi^3}\int \hat{g}(k)\, \hat{g}(p + k)\, k_\mu\, \delta(k^2 - m^2)\, d^4k, \tag{1.10}$$

we see that it does not vanish by symmetry if we make it convergent through the form factors. On the other hand, we have

$$\hat{\Pi}_{\mu\nu}{}^{(b)} = -\frac{i\, e^2}{2\,\pi^3}\int \hat{g}(k)\, \hat{g}(p + k)\, k_\mu\, \delta(k^2 - m^2)\, \hat{h}_\nu(p)\, d^4k \tag{1.11}$$

Because of eq. (1.6) we see clearly that:

$$\hat{\Pi}_{\mu\nu}{}^{(b)}\, p^\nu = -\,\hat{\Pi}_{\mu\nu}{}^{(a)}\, p^\nu. \tag{1.12}$$

independently of the form of $\hat{h}_\nu$.

However, the above argument is not, in our opinion, very convincing, because we must take into account the change in the equations of motion due to the new non-local character of the current. We shall see, that, independently of any limit, the change in the equations of motion exactly cancels the contribution from $\Pi_{\mu\nu}^{(b)}$ so that the original result is restored.

2. Convergence of S-matrix elements in a non-local theory

This problem has been extensively studied in the past [5, 6, 7] and it has been shown that Lorentz invariant form factors always lead to a divergent S-matrix. However, we shall see that, if the fields are averaged not over 4-dimensional space-time volumes, but over 1-dimensional time like paths, the resulting equations have finite solutions to every order of perturbation theory.

For a given field $A(x)$, we define:

$$A[x; f] = \int_{-\infty}^{+\infty} f(s)\, A(x + \zeta(s))\, ds, \tag{1.13}$$

where $f(s)$ is a one-dimensional infinitely differentiable function of compact support and $x + \zeta(s)$ the coordinates of a time like path going through the point x ($\zeta(0) = 0$). It is convenient to reexpress formally $A[x; f]$ as an integral over a 4-dimensional volume:

$$A[x; f] = \int G[x - x'; f]\, A(x')\, d^4x', \tag{1.14}$$

where, in momentum space:

$$\hat{G}(p; f] = \int_{-\infty}^{+\infty} e^{-ip \cdot \zeta(s)}\, f(s)\, ds \tag{1.15}$$

is the generalized Fourier transform of $f(s)$. If we take straight lines $\zeta_\mu(s) = \lambda_\mu \cdot s$, then we simply have:

$$\hat{G}[p; f] = \hat{f}(\lambda \cdot p). \tag{1.16}$$

Borchers [8] has shown that, if the field operator $A(x)$ acts on a Hilbert space with reasonable spectrum, an average like (1.13) defines a distribution on the 4-dimensional space, provided $\zeta(s)$ is time-like.

II. Formulation of the non-local theory

1. Gauge invariant fields

In order to build a non-local theory, it is first convenient to introduce gauge-invariant fields.

$$\Psi = e^{-ieU(x)}\, \psi, \; \tilde{\Psi} = \tilde{\psi}\, e^{ieU(x)} \tag{2.1}$$

and

$$\Phi_\mu(x) = A_\mu(x) - \partial_\mu U, \tag{2.2}$$

where U is the functional defined in the previous section. In addition to the condition that $U \to U + \Lambda$ in a gauge transformation we also require that

$$\lim_{t \to -\infty} U(\vec{x}, t) = 0, \tag{2.3}$$

so that $\Psi \to \psi_{\text{in}}$ etc.... at $t \to -\infty$.

U can be evaluated as a line integral [9]. A more convenient form is given by:

$$U = \int D_R(x - x')\, \partial_\nu A^\nu(x')\, d^4x', \tag{2.4}$$

where

$$\Box D_R(x - x') = \delta(x - x').$$

In this case we can write

$$\Phi_\mu(x) = \int \tau_{\mu\nu}(x - x')\, A^\nu(x')\, d^4x' \tag{2.5}$$

with

$$\hat{\tau}_{\mu\nu}(p) = g_{\mu\nu} - \frac{p_\mu p_\nu}{p^2 + i\varepsilon}. \tag{2.6}$$

Equations (2.1) and (2.2) are very similar to those of the Stückelberg formalism [10], when the photon has a mass μ. There, one writes:

$$A_\mu = \phi_\mu + \partial_\mu B \tag{2.7}$$

with the supplementary condition:

$$\partial_\mu A^\mu + \mu^2 B = 0. \tag{2.8}$$

It is then found that B obeys the free equation of motion:

$$(\Box + \mu^2)\, B = 0, \tag{2.9}$$

so that we can write:

$$\Box B = \partial_\mu A^\mu \tag{2.10}$$

and consequently:

$$B = \int D_R(x - x')\, \partial_\mu A^\mu(x')\, d^4x'. \tag{2.11}$$

In addition we can eliminate the B field by a canonical transformation [11]: if we define

$$S = e \int \psi^*(x)\, B(x)\, \psi(x)\, d^3x \tag{2.12}$$

then

$$e^{iS}\, \psi(x)\, e^{-iS} = e^{-ieB}\, \psi = \Psi. \tag{2.13}$$

2. Construction of the non-local theory

Putting

$$F_{\mu\nu} = \partial_\mu \Phi_\nu - \partial_\nu \Phi_\mu \tag{2.14}$$

we write the Lagrangian of the non-local theory exactly as in the usual one:

$$\mathscr{L} = -\frac{1}{4} F_{\mu\nu} F^{\mu\nu} - \tilde{\Psi}(x) \left[-i\gamma^\mu \partial_\mu + m\right] \Psi(x) -$$

$$- j_\mu(x) \Phi^\mu(x) + \delta m \tilde{\Psi}(x) \Psi(x) + \frac{1}{2} \partial\mu^2 \Phi_\mu(x) \Phi^\mu(x) \tag{2.15}$$

except that we have added, without spoiling gauge invariance, a mass renormalization for the photon, and that j_μ is now expressed in terms of the averaged Dirac field:

$$j_\mu(x) = -\frac{e}{2} [\tilde{\Psi}[x; f^*], \gamma_\mu \Psi[x; f]]. \tag{2.16}$$

This expression is very similar to the one of eq. (1.4), except that the form factor g is now replaced by the path averaging function f.

The equations of motion can be written as follows:

$$(-i\gamma^\mu \partial_\mu + m) \Psi(x) = \Theta(x). \tag{2.17a}$$

$$i \partial_\mu \tilde{\Psi} \gamma^\mu + m \tilde{\Psi} = \tilde{\Theta}(x). \tag{2.17b}$$

$$\Box \Phi_\mu = J_\mu(x), \tag{2.17c}$$

where

$$\Theta(x) = \delta_m \Psi(x) + \frac{1}{2} e \int G[x - x'; f^*] [\Phi_\mu(x'), \gamma^\mu \Psi[x'; f]]_+ \, d^4x'. \tag{2.18}$$

$$J_\mu(x) = \int \tau_{\mu\nu}(y - x) \left[\delta\mu^2 \Phi^\nu(y) + j^\nu(y)\right] d^4y \tag{2.19}$$

($\tilde{\Theta}$ is the adjoint of Θ). In order to prepare for the quantization of the theory, we have expressed Θ and J_μ in terms of symmetrical and anti-symmetrical products of the field operators resp. We can now replace eqs. (2.17) by integral equations:

$$\Psi(x) = \psi_{\text{in}} - \int S_R(x - x') \Theta(x') \, d^4x'. \tag{2.20a}$$

$$\Phi_\mu(x) = \Phi_\mu^{(\text{in})} + \int D_R(x - x') J_\mu(x') \, d^4x' \tag{2.20b}$$

and solve these equations by successive iterations [12]. The only thing that is needed for the quantization is the commutation relation for the in-fields, which are:

$$[\psi_{\text{in}}(x), \tilde{\psi}_{\text{in}}(x')]_+ = -i S(x - x') \tag{2.21}$$

$$[\Phi_\mu^{\text{in}}(x), \Phi_\nu^{\text{in}}(x')] = -i \int \tau_{\mu\nu}(x-y)\, D(y-x')\, d^4y. \qquad (2.22)$$

(The expression on the right hand side can be defined mathematically by putting a small mass for the photon).

3. Conservation laws

As is well known, it is possible to deduce a certain number of conservation laws from the fact that the Lagrangian is stationary when the fields obey the equations of motion. This can be done even for a non-local theory, as was shown by Pauli [13]. Pauli's method to construct a conserved quantity, for example the current I_μ, is to write:

$$I_\mu = j_\mu + j_\mu', \qquad (2.23)$$

where j_μ' is defined by:

$$\partial_\mu j^{\mu\prime} = -\partial_\mu j^\mu. \qquad (2.24)$$

Integrating over a time-like path $\zeta(s)$, this gives:

$$j_\mu' = -\int d^4x' \int_{-\infty}^{0} ds\, \delta(x - x' + \zeta(s))\, \partial_\nu j^\nu(x'), \qquad (2.25)$$

so that I_μ is nothing but:

$$J_\mu = \int \tau_{\nu\mu}(y-x)\, j^\nu(y)\, d^4y, \qquad (2.26)$$

where $\tau_{\mu\nu}$ is defined with a line integral. But, as we have seen, there is a great deal of arbitrariness in the definition of $\tau_{\mu\nu}$.

III. S-Matrix

1. Definition

The S-matrix is defined, in the non-local theory, according to the method of Källén or Yang-Feldman [12]. From the eqs. of motion (2.20), we can define the out-fields:

$$\begin{aligned} \Psi_{\text{out}}(x) &= \psi_{\text{in}}(x) + \int S(x-x')\, \Theta(x')\, d^4x' \\ \Phi_{\text{out}}^\mu(x) &= \Phi_{\text{in}}^\mu(x) - \int D(x-x')\, J_\mu(x')\, d^4x' \end{aligned} \qquad (3.1)$$

Then the S-matrix is the one which transforms the in-fields into the out-fields

$$\Psi_{\text{out}} = S^{-1} \psi_{\text{in}} S, \quad \Phi_{\text{out}}^\mu = S^{-1} \Phi_{\text{in}}^\mu S \quad \text{etc.} \qquad (3.2)$$

To prove that the S-matrix is unitary, one has to prove that the out-fields also satisfy the free commutation relations. When symmetrical or antisymmetrical products are used in the definition of $\Theta(x)$ and $J_\mu(x)$, it has been shown by Hayashi [14] that this is so up to the 4th order in e. No general proof exists at present that this is so to all orders.

2. Convergence

Using the reduction formulae of Lehmann, Symanzik and Zimmermann [15] one can express the elements of the S-matrix in terms of matrix elements of the fields themselves. On the other hand, since we know that an operator P_μ corresponding to the total energy-momentum can be constructed, we can use the displacement properties of these matrix elements. For example, $\langle p|\Phi_\mu(x)|p'\rangle$ can be written:

$$\langle p|\Phi_\mu(x)|p'\rangle = \exp\{-i\,(p'-p)\,x\}\,\langle p|\Phi_\mu(0)|p'\rangle \tag{3.3}$$

Putting $q = p' - p$, and using the equations of motion, we have:

$$\langle p|\Phi_\mu(0)|p'\rangle = -\frac{1}{q^2+\delta\mu^2}\,\hat{t}_{\nu\mu}(q)\,\langle p|j^\nu(0)|p'\rangle \tag{3.4}$$

This expression can be transformed, using the explicit expression for $j_\nu(0)$, by standard techniques:

$$\langle p|j_\nu(0)|p'\rangle = -\frac{e}{(2\pi)^4}\int d^4k\,\hat{f}(\lambda\cdot(p-k))\,\hat{f}(\lambda\cdot(p'-k))\,\times$$
$$\times\,\theta(k^2)\,\theta(k_0)\,F_\nu(k,p,p';\lambda), \tag{3.5}$$

where F_ν is a sum over a finite number of states:

$$F_\nu = \frac{1}{2}\sum_\varrho \gamma_\nu^{\beta\alpha}\,[\langle p|\tilde{\Psi}_\beta(0)|k,\varrho\rangle\,\langle k,\varrho|\Psi_\alpha(0)|p'\rangle -$$
$$-\langle p|\Psi_\alpha(0)|k,\varrho\rangle\,\langle k,\varrho|\tilde{\psi}_\beta(0)|p'\rangle] \tag{3.6}$$

Similar expressions can be found for matrix elements of $\Psi(x)$ or $\tilde{\Psi}(x)$.

That the integral (3.5) can always be made convergent by a suitable choice of $\hat{f}$ can be seen from the following remark due to Bloch [6]: If a time-like vector k is such that

$$|\lambda\cdot k| < A$$

where λ is a fixed time-like vector, then the space components $\vec{k}$ are also bounded. Putting $k^2 = M^2$ and $\lambda^2 = L^2$, we find that the integral over $\vec{k}$ vanishes if $M^2 > (A^2/L^2)$, and is limited by

$$|\vec{k}| < \frac{1}{L^2}\,[|\vec{\lambda}|\,A + \lambda_0(A^2 - M^2L^2)^{1/2}]$$

if $M^2 < (A^2/L^2)$.

Actually, it is not necessary to assume that $\hat{f}$ vanishes outside a fixed domain. It is possible to choose a rapidly decreasing smooth function.

Another way to prove the convergence of the S-matrix elements is to use a graphical method which has been already given by Dyson [16] and Bloch [8]. (See reference [1]).

IV. The Averaging Process

So far we have only considered calculations of observable quantities over a fixed family of paths. However, since we do not have a dynamical principle to determine the dependence of these quantities on the path variables, we must perform an average over all possible paths. We shall limit our study to the case of a family of straight lines $\zeta_\mu = \lambda_\mu s$, such that $\lambda_\mu^2 = L^2$.

Let us consider a *renormalized* matrix element $M(p_i; \lambda \cdot p_i)$, where the p_i's are fixed physical momenta.

We want to calculate

$$M_{av.}(p_i) = \frac{\int M(p_i; \lambda \cdot p_i)\, d\mu(\lambda)}{\int d\mu(\lambda)} \tag{4.1}$$

where $d\mu(\lambda)$ is a measure over the family of paths. In order to avoid the difficulty of the lack of compactness of the real Lorentz group, we shall define instead of (4.1)

$$M_{av.}(p_i) = \left\{ \frac{\int M(p_i; (\Lambda z) \cdot p_i)\, d\mu(\Lambda)}{\int d\mu(\Lambda)} \right\}_{z^2 = L^2} \tag{4.2}$$

where z is now a complex vector such that $z^2 = L^2$, Λ a matrix of the complex Lorentz group and $d\mu(\Lambda)$ the invariant measure on the group. Actually, it is sufficient to average over the four-dimensional rotation group, which is a subgroup of the complex Lorentz group.

Averages like (4.2) can be written explicitly if we assume that M has a Fourier transform in λ_μ:

$$M = \frac{1}{(2\pi)^4} \int \hat{M}(p_i; x)\, e^{-i x \cdot \lambda}\, d^4 x \tag{4.3}$$

Then

$$M_{av.}(p_i) = \frac{1}{(2\pi)^4} \int \hat{M}(p_i; x)\, \langle e^{-i x \cdot \lambda} \rangle_{av}\, d^4 x \tag{4.4}$$

The average over the exponential can be evaluated:

$$\langle \exp\{-i x \cdot \lambda\} \rangle_{av} = U_1(x)\, \theta(x^2) + U_2(x)\, \theta(-x^2), \tag{4.5}$$

where:

$$\begin{aligned} U_1 &= \frac{2}{L\sqrt{x^2}} [J_1(L\sqrt{x^2}) - i\,\varepsilon(x_0)\, \mathbf{H}_1(L\sqrt{x^2})] \\ U_2 &= \frac{2}{L\sqrt{-x^2}} [I_1(L\sqrt{-x^2}) - \mathbf{L}_1(L\sqrt{-x^2})] \end{aligned} \tag{4.6}$$

Here, J_1 and I_1 are Bessel functions, $\mathbf{H}_1$ is a Struve function and $\mathbf{L}_1$ a modified Struve (or Nicholson) function.

One can verify that the average (4.4) involves only time-like values of λ by defining $\varrho(\lambda)$ such that:

$$M_{av.}(p_i) = \int M(p_i; \lambda \cdot p_i)\, \varrho(\lambda)\, d^4\lambda \tag{4.7}$$

From eqs. (4.3), (4.4) and (4.7) we have:

$$\langle \exp\{-i\, x \cdot \lambda\}\rangle_{av.} = \int e^{-ix\lambda}\, \varrho(\lambda)\, d^4\lambda$$

The Fourier inversion of this equation gives [17], for $\lambda^2 < 0$:

$$\varrho(\lambda) = \frac{1}{4\pi^2 L \sqrt{-\lambda^2}} \int\limits_0^\infty dm^2 \left\{ \frac{2}{\pi} J_1(m\,L)\, K_1(m\sqrt{-\lambda^2}) - \right.$$
$$\left. - [I_1(m\,L) - \mathbf{L}_1(m\,L)]\; Y_1(m\sqrt{-\lambda^2}) \right\} = 0 \tag{4.8}$$

Similarly, we can verify that $M_{av}(p_i)$ is an invariant function of p_i^2 and $p_i \cdot p_j$ by writing:

$$M(p_i; \lambda \cdot p_i) = \frac{1}{(2\pi)^n} \int\limits_{-\infty}^{+\infty} \cdots \int\limits_{-\infty}^{+\infty} \hat{M}(p_i; s_1 \ldots s_n)\, e^{-s_i(\lambda_i \cdot p_i)} \times$$
$$\times\, ds_1 \ldots ds_n \tag{4.9}$$

Putting $P = \sum_i s^i p^i$, we see that the average of (4.9) can be obtained by calculating $\exp\{-i\,\lambda \cdot P\}$ and we have seen that this average depends only on P^2.

Finally, it should be noted that, in order to preserve the unitarity of the S-matrix elements $M(p_i; \lambda \cdot p_i)$ should actually be elements of the R-matrix defined by

$$S = e^{iR} \tag{4.10}$$

with the prescription that

$$S_{av.} = e^{i R_{av.}} \tag{4.11}$$

V. Photon Mass Renormalization

Now since we have a prescription to calculate finite matrix elements of the field operators in a gauge invariant way, we can return to the problem with which we started, namely the problem of the photon mass.

1. Vacuum expectation value of the current

First of all, we can calculate the lowest order expectation value of $\langle j_\mu(x)\rangle_0$, namely

$$\langle j_\mu^{\text{in}}(x)\rangle_0 = \frac{e}{4\pi^3}\int |f(\lambda\cdot p)|^2 p_\mu\, \delta(p^2 - m^2)\, d^4p. \tag{5.1}$$

Whereas in the usual theory, or in non-local theories with spherically symmetrical form factors, this expression is completely ambiguous, here we can see that all four components of $\langle j_\mu^{\text{in}}(x)\rangle_0$ vanish identically in a truly convergent manner. (To see this, it is sufficient to take p_0 along λ_μ.)

The next order can be written:

$$\langle j_\mu(x)\rangle_0 = \int \Pi_{\mu\nu}(x-y)\, \Phi_{\text{ext}}^\nu(y)\, d^4y, \tag{5.2}$$

where

$$\Phi_\nu^{\text{ext}}(y) = \int \tau_{\nu\varrho}(y - x')\, A_{\text{ext}}^\varrho(x')\, d^4x' \tag{5.3}$$

and:

$$\hat{\Pi}_{\mu\nu}(p) = \frac{e^2}{4\pi^3}\int |f(\lambda\cdot k)|^2 |f(\lambda\cdot(p+k))|^2\, T_{\mu\nu}(k, p+k)\, \times \tag{5.4}$$
$$\times\, [\Delta_R(p+k)\,\delta(k^2-m^2) + \Delta_A(k)\,\delta[(p+k)^2 - m^2]]\, d^4k,$$

where $T_{\mu\nu}$ is the same here as in (1.9). After averaging over λ one finds that $\langle J_\mu(x)\rangle_0$ defined by (2.26) contains a singular term proportional to Φ_μ^{ext}:

$$[\langle J_\mu(x)\rangle_0]_{A\nu} \simeq \left(\delta\mu^2 + \frac{e^2}{4\pi^2 L^2}\right)\Phi_\mu^{\text{ext}} \tag{5.5}$$

Since, experimentally, we know that the physical mass of the photon is zero, we must choose for $\delta\mu^2$ a finite value, which, to the lowest order, is given by:

$$\delta\mu^2 \simeq -\frac{e^2}{4\pi^2 L^2} \tag{5.6}$$

so that there is no term directly proportional to the external field in the induced current.

2. The photon propagator

In order to formulate in a general way the condition to be imposed to the theory if the photon physical mass vanishes, we shall investigate the properties of the photon propagator.

The definition of the photon propagator cannot be taken as the expectation value of a T product of the fields Φ_μ, since this is not a covariant quantity in a non-local theory, because the fields do not commute for space-like distances. However, we can choose

$$D_{\mu\nu}'(x-y) = \left\{ \frac{\delta\langle\Phi_\mu(x)\rangle}{\delta J_\nu^{\text{ext}}(4)} \right\}_{J_\nu^{\text{ext}}\to 0} \tag{5.7}$$

which is the general definition of Green's functions in QED.

In this way, we find that $D_{\mu\nu}'$ is related to the polarization tensor $\Pi_{\mu\nu}$ through the relation:

$$\hat{D}_{\mu\nu}'(p) = \hat{D}_{\mu\nu}^F(p) + \hat{D}_{\mu\varrho}^F(p)\,\hat{\Pi}_{\varrho\sigma}(p)\,\hat{D}_{\sigma\nu}'(p), \tag{5.8}$$

where we here define:

$$\hat{D}_{\mu\nu}^F(p) = \hat{t}_{\mu\nu}(p)\,\hat{D}_F(p) \tag{5.9}$$

We can decompose $\hat{\Pi}_{\mu\nu}$ in terms of two invariant functions Π_1 and Π_2 and calculate its average over all λ_μ's:

$$[\hat{\Pi}_{\mu\nu}(p)]_{av.} = \hat{t}_{\mu\nu}(p)\,\overline{\Pi}(p^2) \tag{5.10}$$

The condition that the physical mass be zero can now clearly be expressed as:

$$\overline{\Pi}(0) = 0 \tag{5.11}$$

VI. Final Remarks

To conclude, we can list a series of questions which can be raised, and deserve further investigation.

1. Proof of the unitarity of the S-matrix to all orders.
2. Simplification of the method of calculation of the S-matrix, for example by means of a "non-local time-ordering operator", which would enable one to obtain rules analogous to those of Feynman.
3. Problem of the non-renormalizable theories. It is possible that, in this case, the averaging process will lead to new singularities.
4. Physical meaning of the averaging of a field operator over a time-like path.
5. Why does the photon have a vanishing physical mass, if this is not a direct consequence of an invariance principle?
6. Is it possible to account for an eventual "breakdown" of QED at high energy with one single form factor? Is a model based on time-like paths averages experimentally distinguishable from the usual non-local theories involving space-time symmetrical form factors?

References

1. M. Levy, Nucl. Phys. **57**, 152 (1964).
2. J. Schwinger, Phys. Rev. **125**, 397 (1962); **128**, 2425 (1962).
3. G. Feldman and P. T. Matthews, Phys. Rev. **130**, 1633 (1963).
4. G. Wentzel, Phys. Rev. **74**, 1070 (1948).
   G. Källen, Handb. d. Phys. **5**, Part I (Springer, Berlin, 1958).
5. P. Kristensen and C. Møller, Mat. Fys. Medd. Dan. Vid. Selsk. **27**, No. 7 (1952);
   C. Møller, Proc. Int. Conf. of Theor. Phys. Kyoto and Tokyo (1953) p. 13.

6. C. BLOCH, Mat. Fys. Medd. Dan. Vid. Selsk. **27**, No. 8 (1952).
7. M. CHRETIEN and R. E. PEIERLS, Proc. Roy. Soc. **A 223**, 468 (1954).
8. H. BORCHERS, Nuovo Cim. **33**, 1600 (1964).
9. K. JOHNSON, Nucl. Phys. **25**, 431 (1961);
B. S. DE WITT, Phys. Rev. **125**, 2188 (1962);
S. MANDELSTAM, Ann. of Phys. **9**, 1 (1962).
10. E. C. G. STUECKELBERG, Helv. Phys. Acta **11**, 299 (1938).
11. F. DYSON, Phys. Rev. **73**, 929 (1948).
12. G. KÄLLEN, Arch. Phys. **2**, 371 (1950);
C. N. YANG and D. FELDMAN, Phys. Rev. **79**, 972 (1950).
13. W. PAULI, Nuovo Cim. **10**, 648 (1953).
14. C. HAYASHI, Prog. Theor. Phys. **11**, 226 (1954).
15. H. LEHMANN, K. SYMANZIK and W. ZIMMERMANN, Nuovo Cim. **1**, 205 (1955).
16. F. DYSON, Phys. Rev. **82**, 428 (1951); **83**, 608 (1951) and 1207.
17. G. N. WATSON, A treatise on the theory of Bessel functions (Cambridge University Press, Cambridge 1922).
18. J. SCHWINGER, Proc. Nat. Acad. Sci. **37**, 452 (1951).

Theoretical and Experimental Status of Quantumelectrodynamics*

By

J. D. Björken, Stanford, USA.

1. Introduction

What we shall mean in these lectures by "quantum electrodynamics" is the study of electromagnetic interactions of electrons, μ-mesons, and photons with each other and with those external sources which can be understood phenomenologically. It is therefore worthwhile to try to delimit the expected domain of validity of this (narrowly-defined) theory. On the large-distance side, it is probably only gravitation which gets in the way; on the small-distance side, the strong interactions (via vacuum polarization effects) and, in a more serious way, weak interactions. To estimate the breakdown-distance, we may calculate, for example, "weak" corrections to the process of $e^+ e^-$ pair annihilation. In a point Fermi theory, the graph $\delta \mathcal{M}$ goes something like

$$\frac{\delta \mathcal{M}}{\mathcal{M}} \sim \begin{cases} G^2 \Lambda^2 E_{CM}{}^2 \ (?) \\ G^2 E_{CM}{}^4 \quad (?) \end{cases}$$

with the weak interaction cutoff (*a la* Ioffe [1]) $\Lambda \lesssim 300$ BeV: Thus when $G E_{CM}{}^2 \sim 1$, or $E_{CM} \sim 300$ BeV, the weak corrections become important.

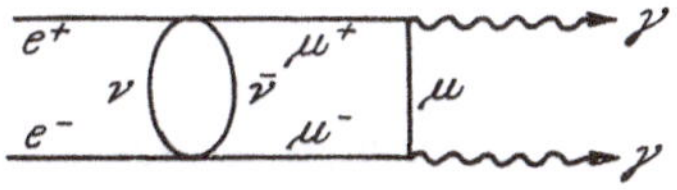

If on the other hand, there are light vector bosons mediating the weak interactions, the high-energy limit is even lower. Take for example [2] the cross-section

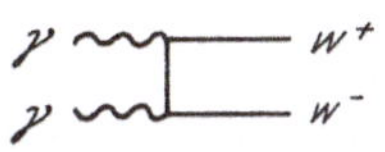

which goes at least as fast as $\sigma \sim \alpha^2 (E_{CM}{}^2/M_W{}^4)$ or

---

* Lecture given at the IV. Internationale Universitätswochen für Kernphysik, Schladming, 25 February—10 March 1965.

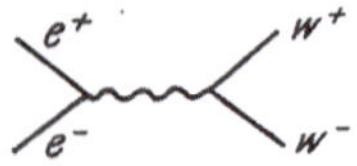

which goes as $\sigma \sim (\alpha^2/M_W^2)$ or faster. These reach strong coupling $(\sigma \sim 1/E_{CM}^2)$ quickly, when

$$E_{CM} \sim \frac{M_W}{\sqrt{\alpha}} \sim 10\, M_W.$$

So for W-mesons of mass $\lesssim$ 30 BeV "standard" electrodynamics break down below 300 BeV. If the W is heavier, we may again reach 300 BeV before weak interactions of leptons become strong (unitarity limited) and quantum electrodynamics thereby bounded.

The important point is that in any case (because of the extreme difficulty in finding a way of keeping weak interactions weak at high energies) quantum electrodynamics *as we know it*, is unlikely to be applicable at center-of-mass energies $\gtrsim$ 300 BeV.

Within this domain of energies, the theoretical status of electrodynamics is excellent. We know how, in principle, to calculate just about anything. There are, of course, classic exceptions: notably m_μ/m_e and the fine-structure constant α, both of which may well depend upon phenomena outside the boundaries which we intend to discuss. But, aside from this, things are in good shape. Convergence of the perturbation expansion (ignoring infrared questions) is in good condition (within [3] the defined domain!!) with the possible exception of the Budini correction [4] to pair annihilation, where an $\alpha \log^2 (E/m)$ appears. This, always troublesome, question of $\alpha \log^2$ corrections still perhaps deserves more attention. If such terms can be found, nonperturbative effects could appear below the weak cutoff $\sim$ 300 BeV. Finally, there is the problem of making a quantum electrodynamics of a vector boson, should such a thing be found. The approaches of LEE [5] and SALAM [6] are both encouraging. They would be even more so if they could somehow be unified.

2. Experimental Status

We turn to the main subject: testing the existing theory against experiment: We divide this into parts (1) Tests of the basic symmetry properties, (2) Tests of the dynamics at the large-distance limit, (3) Tests at the small-distance limit, and (4) Tests for other couplings.

Basic Symmetry Properties:

The most important symmetries are the Lorentz and translation symmetries. It is hard even to define a test in terms of a breakdown model. For example, it would be nice to have a *quantitative* measure of how well the symmetries of translation and Lorentz invariance hold, but I don't know how to do it. Perhaps the best tests, first, are that

(1) high energy accelerators work (2) cosmic ray shower theory accounts for the basic properties of extremely high energy showers ($\lesssim 10^{14}$ MeV) (3) standard relativistic kinematics works, and (4) the "dispersion relation" $\omega = \sqrt{k^2 + m^2}$ is ok. For photons this was recently tested [7] at CERN. They found that the speed of γ rays from π^0 decay between 6 and 20 BeV is c to ~ 1 part in 10^4.

For the case of discrete symmetries, P or T violation is checked by the limits of electric dipole moments of electron and μ-meson. For the μ, the CERN $g - 2$ experiments gives a limit [8]

$$(3 \pm 6) \times 10^{-5} \frac{e\hbar}{m_\mu c}$$

while for the electron, low energy measurements [9] give a limit

$$\sim 10^{-5} \frac{e\hbar}{m_e c}$$

Electron scattering at high momentum transfer from a spin zero target [10] has put a limit on an anomalous moment of $\sim 10^{-5}\,(e\hbar/m_e c)$ at momentum transfers of ~ 200 MeV/c. This is because the cross section without a moment has the form

$$\frac{d\sigma}{d\Omega} \sim \cos^2 \frac{\theta}{2} \times F(q^2) \times \text{(known kinematical factors)}$$

in the high energy limit.

Charge conjugation invariance is more difficult to test directly. Not even the long lifetime of triplet positronium measures C; the 2-photon decay is forbidden by angular momentum conservation and the Bose statistics of the photon alone.

TCP is best tested by the ratios of lifetimes and masses of the $\mu^\pm$. The masses agree [11] to ~ 1 part in 10^5.

3. Large-distance Tests

Turning to more specific dynamical questions, we start with the largest distances and proceed to smaller and smaller. The main large-distance question has to do with the mass of the photon. The best limits here are quite recent: Satellites have measured the magnetic field of the earth out to several earth radii ($\sim 5 \times 10^4$ km.) and found it to go down like r^{-3}. This puts a limit on the mass $m_\gamma \lesssim 10^{-20} m_e$. There has also been detected recently [12] radiation at 8 cps, associated with the resonance cavity (the Q is 5) bounded by the surface of the earth and the ionosphere. If the mass of the photon were greater than 8 cps ($\sim 10^4$ km) such a mode could not exist at that frequency. This limit is comparable with that of the magnetic field measurement.

Going to smaller distances, I cannot resist mentioning a very pretty macroscopic test of quantum electrodynamics invented by CASIMIR [13]. Consider two conducting neutral plates separated by distance d, and calculate the vacuum energy of the electromagnetic field within the plates. One finds, introducing a frequency-dependent cutoff, for the energy per unit area

$$\frac{E}{A} = C_1 d + C_2 + B_4 \frac{\hbar c}{\pi d^3}$$

C_1 and C_2 are cut-off dependent constants and B_4 is a Bernoulli number, $-1/30$. The first term gives a pressure on the plates independent of separation, which is cancelled off by the back-pressure of the vacuum outside the plates. The second term clearly doesn't count. The final term leads to an attractive $1/d^4$ force which has been measured, in good agreement with theory [14].

4. Higher Energy Experiments

We shall first discuss the higher-energy experiments as a test of the theory at small distances, as discussed by DRELL [15] long ago. As a measure of the sensitivity of the experiment to small distance modifications, it has been the fashion to butcher the theory by cutting off the propagators and vertex functions with the simple form $-\Lambda^2/(q^2 - \Lambda^2)$, and using the experiment to limit Λ. This is not to be taken at all seriously, but it is probably better to have some measure of sensitivity than none, if only to provide some impetus to the experimentalists to do the experiments.

We look first at the low energy, high precision experiments. These include the beautiful magnetic moment measurements of μ and e, the Lamb shift, and the hyperfine structure.

The magnetic moment of the electron has been measured by the Michigan group in a long series of increasingly accurate experiments, culminating in a final measurement [16] good to ~ 1 part in 10^9, and verifying the theoretical expression to 2% of the $(\alpha/\pi)^2$ term in the gyromagnetic ratio. This invites speculation on possibilities of measurement and calculation of the α^3 term. The calculation of the 6th order moment is feasible but has not been carried out. DURAND [17] has estimated the contribution of the $2\,\pi$ (ϱ) state in the vacuum polarization to the moment. For the μ its magnitude is around $3\,(\alpha/\pi)^3$. For the electron, it is completely negligible. This 6th order calculation is one of the last challenges of (practical) quantum electrodynamic theory and will probably fall in a few years. Already, DRELL and PAGELS [18] have estimated the coefficient of the 6th order term to be $\sim 0.1(\alpha/\pi)^3$, using a dispersion relation argument.

The sensitivity of the electron moment to breakdown is unfortunately low. A cutoff in propagators leads to a correction to the anomaly $\sim m_e^2/\Lambda^2$. This still gives $\Lambda \gtrsim 200\, m_e \sim 100$ MeV.

Because of the larger μ-mass, the μ-moment is much more sensitive to a breakdown of the party-line sort. Here theory and experiment agree to 3 ± 5 parts per million and just touches the 4th order term in accuracy. The recent fat report of the CERN group [19] summarizes thoroughly the implications of a breakdown in various places in the theory and in various ways.

With 95% confidence limit

	Λ(GeV)
Photon propagator alone	$\gtrsim 0.8$
Vertex alone	$\gtrsim 1.1$
μ Propagator alone	$\gtrsim 2.0$
Sum of all three	$\gtrsim 2.5$

If the dispersion integral for the μ-moment (as function of momentum transfer) is butchered the limits come out to be

$$\text{Sharp cutoff } \Lambda \gtrsim 1.5\,(\text{GeV})$$
$$\text{Smooth cutoff } \Lambda \gtrsim 3.7\,(\text{GeV})$$

This is a good example of why not to take these estimates too seriously. But in any case we get cutoff limits $\sim 1 - 3$ BeV.

The status of bound-state measurements (Lamb shift, hyperfine splitting) was reviewed thoroughly by FEYNMAN [20] two years ago and here we discuss only those features which have changed significantly in the last two years. This means mainly the hyperfine measurements in hydrogen and muonium. The hydrogen (maser) measurements are out to 11 places by now and leave the theory far, far behind. Uncertainty in theory appears at about the fifth place, and comes from two-photon exchange contributions

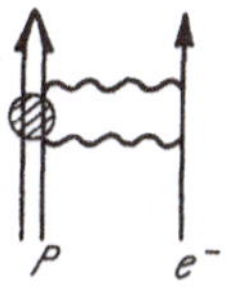

These have been recently restudied theoretically by IDDINGS [21]; they fall into 3 classes.

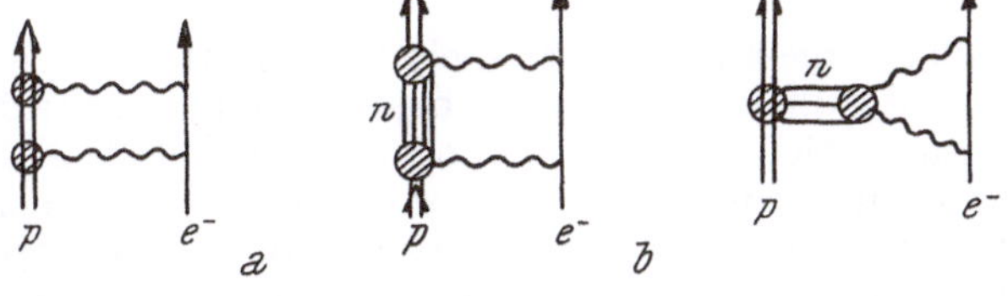

These graphs are to be understood in a dispersion relation sense. Process (a) is perturbation theory, with form factors from electron scattering and gives ~ 50 ppm. The processes (b) (s and u-channel contributions) can be evaluated using an analytic continuation to spacelike photons

only, *a la* Cottingham [22], and probably gives, according to Iddings' only a few parts per million. Using the value of the fine structure constant obtained by Lamb from measurement of fine structure in deuterium, Iddings finds a correction of $\sim -$ (20−30) ppm, with an uncertainty of $\sim$ 20 ppm due to uncertainty in the value of the fine structure constant. This is about 2 standard derivations away (40 ppm) from the measurements, and the blame is now being concentrated on a (t-channel) "subtraction term" behaving roughly like exchange of a 1^+ meson. Good numerical estimates have not, however, been made.

The measurement of hyperfine splitting in muonium ($\mu^+ e^-$) is not as yet a sensitive test of electrodynamics (for photon propagator, μ and e vertex functions $\Lambda \gtrsim 200$ MeV), but is a very useful measurement of α, of the same order at present as the Lamb deuterium measurement. There is agreement between the two determinations here [24], and strengthens the conclusion that something must be done about hydrogen. But in any case this problem is not really quantum electrodynamics, from our point of view here. It is also hoped to improve the fine structure constant determination via the Lamb or hyperfine shift in He, but both more theoretical and more experimental work will be necessary to accomplish this. [25]

5. High Energy Tests

The high energy tests consist of scattering and production processes, ideally with no strongly interacting particles around except perhaps a "spectator" proton. They should be of a variety in order to separately test various conceivable kinds of breakdown.

The cleanest experiment is that of colliding electron and positron beams, designed to test the photon propagator and electron vertex. There are now 2 electron storage rings at Stanford of 500 MeV energy; these are operating [26] with storage times $\gtrsim$ 1 hr. (?). For some time instabilities have plagued the experimentalists, these instabilities, of two classes, are now believed understood.

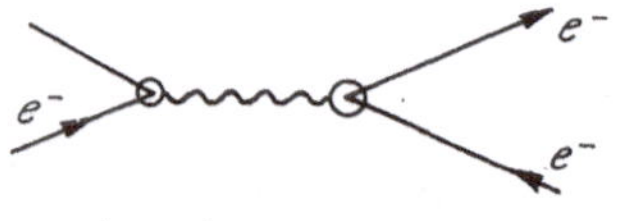

Some actual large-angle electron-electron scatterings have been observed, but it still will take time before numbers appear.

A single ring which stores both positrons and electrons has been built by the Frascati group and installed at Orsay with several interaction regions around the ring. The energy is $\sim$ 1 BeV, and storage times $\sim$ 10 hrs. have been reached. However intensities are low.

An electron-positron ring, with a 1.5 BeV linear accelerator, will also be constructed at Frascati.

The distances attainable in these experiments can be directly estimated. A cutoff of standard form gives, for 90° scattering with 10%

accuracy at 500 MeV, a sensitivity $\Lambda \sim 2$ BeV. Elastic electron-proton scattering gives a test of electrodynamics only down to distances where proton structure enters. This gives the best test [27] of electron vertex and photon propagators, with the cutoff $\Lambda \gtrsim 600$ MeV. In addition at high energies where the mass difference can be neglected, electron and μ elastic scattering and inelastic scattering from protons or any light nucleus should be identical, at least in 1st Born approximation. In higher orders, as long as radiative corrections have no log m terms, this will remain so. No structure-dependent (noninfrared) terms which depend upon m in these higher orders have been found and it is safe to say that at high energies μ and e scattering should approach each other. Experiments on μ scattering from protons have been carried out by the Washington group [28] and also very recently by the Columbia-Brookhaven group [29], while an earlier experiment [30] off carbon was carried out at CERN. These all agree very well with theory. In this case, the measurement tests differences in μ and e vertex functions. If summarized via the famous Λ, the 3 experiments give with 95% confidence level.

Λ	f^{-1}	BeV
CERN	4	~ 0.5
Washington (Bevaton)	6	~ 1
Columbia (Brookhaven)	12	~ 2

Just as $e^+ p$ and $e^- p$ should be equal, needless to say so also should $\mu^+ p$ and $\mu^- p$.

A complementary experiment to this is provided by the process

$$\bar{p} + p \to \begin{cases} \mu^+ + \mu^- \\ e^+ + e^- \end{cases}$$

or in fact the ratio of wide angle Dalitz pairs from anything. This is not easy to measure, but μ's have been recently seen [31] at CERN, with a cross section of a few nano-barns. This, incidentally, is interesting because it says the form factors of the proton are small ($\lesssim 0.1$) at these timelike momentum transfers, and is a significant contribution to the theory of proton structure.

As a final test of TCP (and/or one-photon exchange) one may compare electron-proton and positron-proton scattering. This has been carried out at Stanford [32] by Browman and Pine. Their results are summarized in the figure:

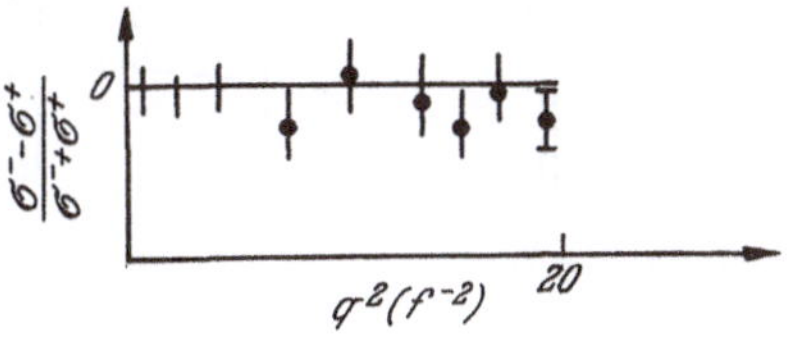

This is consistent with zero and at best would mean a 4% admixture in amplitude of 2nd Born corrections at the largest momentum transfer.

$$\sigma_\pm = |f_2 \pm f_4|^2 \frac{\sigma_- - \sigma_+}{\sigma_- + \sigma_+} \sim \frac{4\,\mathrm{Re}\, f_2 f_4^*}{2 f_2^2} \cong 2 \left|\frac{f_4}{f_2}\right|$$

That is a trifle on the large side. There is now a positron-proton experiment in progress at CEA, which should increase the q^2 somewhat.

A measurement of proton polarization, which is proportional to $\mathrm{Im}\, f_A$, has been done [33] at Orsay; an amplitude of $2 \pm 1\%$ is found at a comparable momentum transfer.

The other ingredient to be tested is the propagator, or more precisely the dependence of processes on how virtual the electron and μ become. This qualification is necessary because if there is an electron propagator in a process there is also a vertex; gauge invariance demands that modifications in the propagator imply modifications in the adjacent vertices. The first pair production test was carried out at Stanford by Richter [34] at low energies ($\sim$ 100 MeV) and low sensitivity because only 20% of the process is sensitive to the propagator. The diagrams are

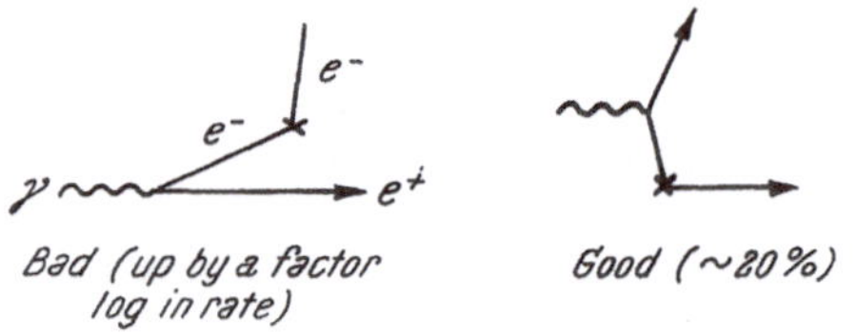

This put a rough limit on electron propagator of $1/\Lambda \gtrsim (0.9\, f)$.

There have been two more experiments on wide angle μ-pairs. The first [35], at Frascati, consisted of detecting with counter telescopes the μ^+ and μ^- at 10° and momenta $\sim$ 300—400 MeV/c from bremsstrahlung γ-rays up to 1 BeV. This was done off carbon, and resolution is not good enough to discriminate inelastic events, i.e., events which break up the carbon. In principle, provided one-photon exchange is ok, it is possible to take this into account if enough is known about inelastic electron-carbon scattering. In both processes all the strong-interaction physics that enters is the total absorption cross-section of a virtual space-like γ-ray on a proton:

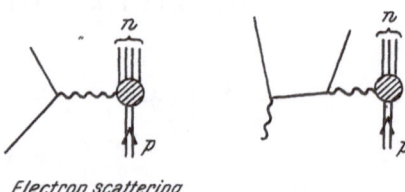

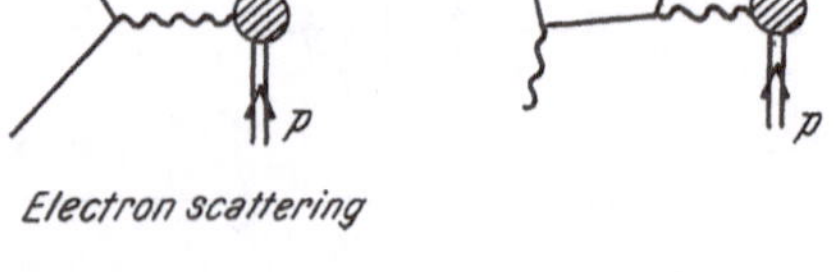

In practice, a sum rule was used with the quasi-elastic contribution $\propto Z(1 - F^2)$. The ratio of observed/expected was $\sim 1.1 \pm 0.1$ which leads to a propagator cutoff $\gtrsim$ 1 BeV. The second μ-pair experiment was recently completed [36] at CEA. This experiment is distinguished by measuring an accurate angular distribution in 9 intervals between 4.5 and 11.5 degrees. The maximum γ-ray energy is 5 BeV and the μ energies restricted between 1.8 and 2.4 BeV.

They quote a ratio of experimental yield to Bethe-Heitler of

$$R = (1.18 \pm 0.15)\,[1 - (0.011 \pm 0.21)\, q_\mu{}^2]$$

and a cutoff limit

$$\left(\frac{1}{\Lambda^2}\right) \lesssim (0.16\ f)^2$$

about the same as that quoted by Frascati. However, the angular distribution measurement provides a great deal more information, and, as we shall see, an important constraint on breakdown models. This experiment also was done in carbon, but the contribution of inelastic channels was at most 8%.

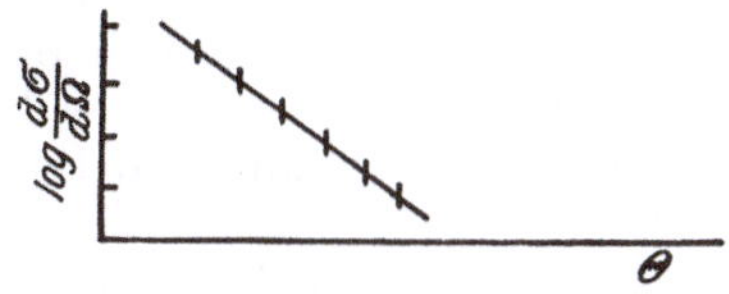

In both experiments a symmetric arrangement in the plane was used.

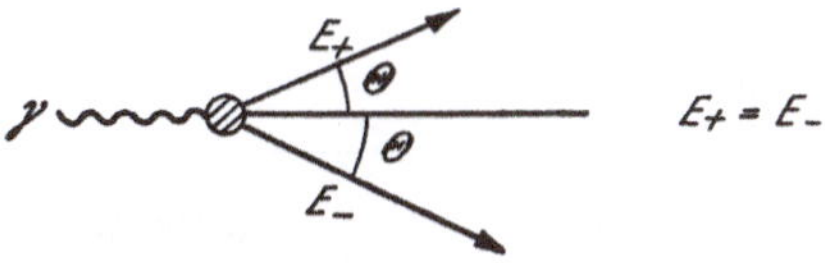

This has several advantages: 1) First, very little momentum ($\lesssim$ 50 MeV) is transferred to the carbon, and nuclear physics complications are minimized. 2) There are diagrams which are essentially Compton scattering followed by Dalitz pair creation, and which are sufficiently unknown so that if they are important, the jig is up.

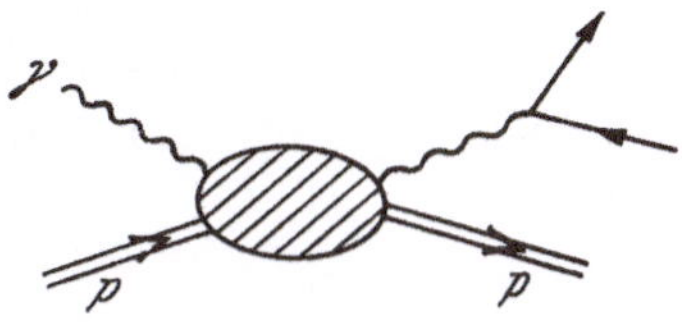

As long as the pairs are detected symmetrically at least the interference between BH and Compton graphs vanishes. For the conditions of the

CEA experiment, the Compton term itself has been estimated to be small. Furthermore, it would lead to an angular distribution $\sim 1/\theta^2$ rather than $\sim 1/\theta^6$ for the BH diagrams, so *experimentally* they are rather well ruled out.

Finally, there has just been completed an experiment by PIPKIN et. al. at CEA on electron pair production, also under symmetric conditions. This experiment does *not* agree with theory. Here the best measurements were made as function of energy of the pair, between 0.2 BeV and 2.5 BeV, with only 3 angle measurements of 4.75, 6.26 and 7.5°.

The observed [37] deviation from the Bethe-Heitler formula depends upon the product $E_+ \cdot \theta_+$, proportional to the virtuality of the intermediate electron or positron. The ratio $R \sim \sigma_{\text{exp}}/\sigma_{\text{BH}}$ varies from ~ 0.8 to 1.5 as the virtuality increases from $E\,\theta \lesssim 100$ MeV to $E\,\theta \sim 400$ MeV. If such an effect were present for μ's, it would almost certainly have been seen by the CEA μ-experiment.

An important constraint on the theorist eager to propose an explanation is the constancy of R with the machine energy (maximum bremsstrahlung energy). Production of an object which decays into electrons would not behave this way; it also probably would give an angular distribution less sharp then $1/\theta^6$. The easiest explanation for the innocent theorist to propose of course is a breakdown of the apparatus, and to wait for new experiments, now being planned. But that's too easy, too passive and too unfair to the experimentalists to be completely satisfying.

So let us summarize the situation at small distances. The "measurable" quantities thus far have been

Quantity	Λ	Experiment
1. Electron propagator (including corrections to adjacent vertex)	$\sim$ 400 MeV	Wide-angle pair production, Lamb Shift [38]
2. μ propagator	$\sim$ 1–2 BeV	Wide-angle pairs $g-2$ experiment
3. Product of e vertex and photon propagator	$\sim$ 600 MeV	Electron-proton scattering
4. Ratio of μ to e vertex	$\sim$ 2 BeV	μ-proton scattering
5. Combination of μ vertex, photon propagator and μ propagator	$\sim$ 1–3 BeV	$g-2$ experiment (cf. previous comments)

We can see that the electron is becoming, from this point of view [39], less wellunderstood than the μ.

6. The Future

What can we expect in the near future? From SLAC and Frascati colliding beams, we should get information on the electron-photon system good to a couple of BeV or so. At SLAC energies ($\lesssim 20$ BeV) we can hope to use $\mu - p$ scattering to improve item 4) a factor of 3 or so, using a secondary μ-beam from SLAC or perhaps (later) a storage ring for μ's.

Colliding electron and positron beams at 3 BeV, as have been proposed, could extend item 3) as well as 1) (from $e^+ + e^- \to 2\gamma$) up to $\sim 10 - 15$ BeV. μ-pair and e-pair production at SLAC is difficult, because the beam pulse is so short that coincidence experiments become heroic efforts due to background problems. Drell [40] has proposed an experiment in which a single extremely high energy μ^- is detected (higher than all π's) in the *forward* direction

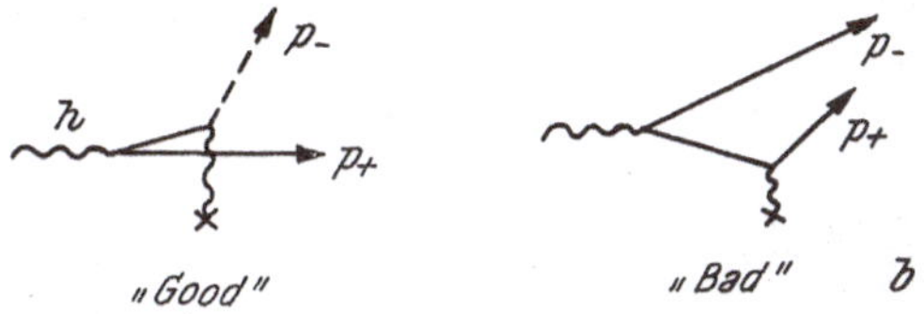

The "good" diagram is of the same order as the "bad" because the Coulomb scattering in a) is at much lower energy and, at a given momentum transfer, smaller. The virtuality of the intermediate μ is of order $(k - p_+)^2 - m_\mu^2 \sim 2\, km_\mu$ for a nonrelativistic μ^+ emerging. For $k \sim 10$ BeV, this gives a sensitivity of $\sim 4 - 5$ BeV in propagator cutoff for a $5 - 10\,\%$ experiment. Extremely good resolution is necessary and it will be a while before the experiment is performed.

7. Tests for New Couplings

Finally, we may also expect in the future explorations along the lines recently suggested [41] by Low, who proposes systematic searches for new couplings of μ, e and γ, i.e. resonant states of one sort or another. The possibilities here are quite varied, but we may classify them in two general categories:

Fermion-fermion resonances: A limit on $\mu^\pm\, e^-$ resonances should be able to be placed from the observed agreement with experiment the stopping power of fast μ mesons traversing matter. For $\mu^-\mu^-$ resonances trident production [42] $(\mu^- + p \to \mu^+ + \mu^- + \mu^- + p)$ should shed light on this process as well as being a test of electrodynamics and the statistics of the μ. Colliding beams, as well as Dalitz-pair mass distributions can test in principle $\mu^+\, \mu^-$ and $e^+\, e^-$ resonant systems [43].

Singly charged resonant systems of integer spin are more difficult. These could for example be $\nu\, \mu$ or $\nu\, e$ resonances (the W? ?) or something more exotic. Again trident production might reveal such a resonance in 3 body $\mu\,\bar{\mu}\,\mu$ or $e\,\mu\,\bar{\mu}$, etc. systems.

Fermion-photon resonances: $\gamma\,e$ or $\gamma\,\mu$ resonances may be explored by looking at the energy spectrum of recoil protons at a given angle from e^- or μ^- scattering: $e + p \to e^* + p$ will then show as a discrete peak. A preliminary search [44] at CEA has not turned up any such peak, for an e^* resonance energy $\lesssim 1$ BeV.

$\gamma\,\nu$ resonances (heavy neutrinos) are difficult but interesting. Perhaps $\nu + p \to \nu^* + p$ or $\mu + p \to \nu^* + N^*$ could shed a little light. The direct effect of these resonances on the electrodynamic tests we have discussed can be made small by making masses large and coupling constants small, *a la* weak interactions. However, unless renormalizable theories of these objects are constructed (a decidedly non-trivial task), sooner or later we can expect their effects to become strong at sufficiently high energy, as discussed for the weak interactions. When this occurs, their presence should become apparent in the party-line experimental test. For example, a heavy e^* could modify the Pipkin pair production process in the way observed due to the graph

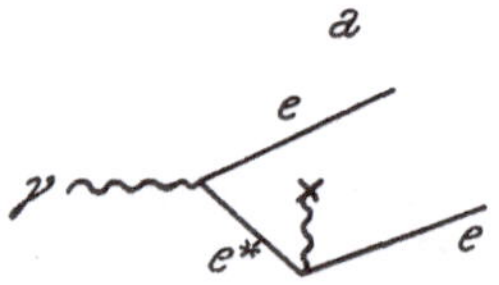

If the couplings are really small, however, the direct observations may be the more feasible. In any case Low's approach suggests a different class of experiments to probe the behavior of leptons and photons, and we may expect in the future a quantitative classification of the kinds of couplings and masses allowed for these kinds of phenomena.

References

1. B. Ioffe, Proc. 1960 Conf. on High Energy Physics, p. 561.
2. We obtain this by using crossing symmetry on the result of Bludman and Young for Compton scattering of a W: Phys. Rev. **126**, 303 (1962).
3. For example, $\alpha/\pi \log \Lambda^2/m_e^2 \lesssim 0.1$ with $\Lambda \lesssim 300$ BeV.
4. Andreassi, Budini, Furlan, Phys. Rev. Letters 8, 184 (1962).
5. T. D. Lee, Phys. Rev. **128**, 899 (1962).
6. R. Delbourgo and A. Salam, Phys. Rev. **135**, 1398 (1964).
7. Alväger, Farley, Kjellman, Wallin, Phys. Lett. **12**, 260 (1964).
8. Charpak, Farley, Garwin, Müller, Sens, Zichichi, Nuovo Cim. **22**, 1043 (1961).
9. Wilkinson, Crane, Phys. Rev. **130**, 852 (1963).
10. Burleson and Kendall, Nucl. Phys. **19**, 68 (1960).
Goldemberg and Torizuka, Phys. Rev. **129**, 2580 (1963).
Rand and Hofstadter, Stanford preprint.

11. CHARPAK et al. Phys. Letters, **1**, **16** (1962).
LATHROP, LUNDY, PENMAN, TELERDI, WINSTON, YOVANOVITCH, BEARDEN, Nuovo Cim, **17**, **114** (1960).
DEVONS, GIDAL, LEDERMAN, SHAPIRO, Phys. Rev. Lett. **5**, **330** (1960).
12. A. PETERSON, private communication via H. W. KENDALL.
13. H. CASIMIR, Proc. kon. Ned. Akad. Wetenshap **51**, **793** (1948).
M. FIERZ, Helv. Phys. Acta **33**, 855 (1960), has calculated temperature-dependent effects.
14. M. SPARNAAY, Physica **24**, **751** (1958).
15. S. DRELL, An. of Phys. (N. Y.) **4**, **75** (1958).
16. WILKINSON and CRANE, op. cit.
17. L. DURAND III, Phys. Rev. **128**, **441** (1962).
18. DRELL and PAGELS, to be published.
19. CHARPAK, FARLEY, GARWIN, MULLER, SENS, ZICHICHI, CERN preprint, 1964.
20. R. P. FEYNMAN, Proc. Solvay Conf. 1962.
21. PLATZMAN and IDDINGS, Phys. Rev. **115**, 919 (1959).
C. K. IDDINGS, Stanford preprint.
22. W. N. COTTINGHAM, Annals of Phys. (N. Y.) **25**, **424** (1963).
23. COHEN, CROWE, DU MOND, The Fundamental Constants of Physics, Interscience, N. Y. (1957).
24. V. HUGHES, report to **1963** Stanford conference:
HOFSTADTER and SCHIFF, Nucleon Structure, Stanford press.
CLELAND, BAILEY, ECKHAUSE, HUGHES, MOBLEY, PREPOST and ROTHENBERG, Phys. Rev. Letters **13**, 202 (1964).
25. To use an electron magnetic moment measurement to improve α requires, in addition to a more accurate $g - 2$ experiment, improvement of the measurement of the proton gyromagnetic ratio, used in mapping the magnetic field in the experiment. See WILKINSON and CRANE, op. cit.
26. B. RICHTER, private communication.
27. S. DRELL, op. cit.
HAND, MILLER and WILSON, Revs. Mod. Phys. **35**, **335** (1963).
28. MASEK, EWART, TOUTONGHI and WILLIAMS, Phys. Rev. Lett. **10**, **35** (1963).
29. ELLSWORTH, MELISSINOS, TINLOT, YAMANOUCHI, LEDERMAN, TANNENBAUM, COOL, MASCHKE, Bull. Am. Phys. Soc. **10**, 80 (1965), and 1964 Dubna conference proceedings.
30. CITRON, DELORME, FRIES, GOLDZAHL, HEINTZE, MICHAELIS, RICHARD, ØVERÄS Phys. Lett. **1**, 175 (1962).
31. Proceed. of 1963 Siena Conference, .
1964 Dubna conference proceedings.
32. BROWMAN and PINE, Nucleon Structure, op. cit.
33. BUON, Siena Converence Proceedings.
34. B. RICHTER, Phys. Rev. Lett. **1**, 114, (1958).
35. ALBERGI-QUARANTA, DE PRETIS, MARINI, ODIAN, STOPPINI, and TAN, Phys. Rev. Lett. **9**, 226 (1962).
36. DE PAGTER, BOYARSKI, GLASS, FRIEDMAN, KENDALL, GETTNER, LARRABEE and WEINSTEIN, Phys. Rev. Lett. **12**, 739 (1964).
37. R. B. BLUMENTHAL, D. C. EHN, W. L. FAISSLER, P. M. JOSEPH, L. J. LANZEROTTI, F. M. PIPKIN and D. G. STAIRS, Phys. Rev. Lett. **14**, 660 (1965).
38. Cf. footnote in BJÖRKEN, DRELL, a. FRAUTSCHI, Phys. Rev. **112**, 1409 (1958).
39. If the cutoff were some multiple of the lepton mass in question, this would not be the case. But that seems a little unlikely.

40. S. Drell, Phys. Rev. Lett. **13**, 257 (1964).
41. F. Low, Phys. Rev. Lett. **14**, 238 (1965).
42. L. Lederman, F. J. M. Farley, private communications.
43. Boson systems coupled to the μ have been discussed rather extensively.
44. H. W. Kendall, private communication.

Lecture given by J. Björken

The Renormalization Group [1]

We first collect together enough of Mitter's and Wess' equations to define a perturbation expansion. We take these to be the following:
I. Photon propagator $D = D_0 + e_0{}^2 D_0 \varrho D$ Indices suppressed.

II. Vacuum polarization

$$\varrho_{\alpha\beta} = (-q^2 g_{\alpha\beta} + q_\alpha q_\beta)\,\varrho(q^2) = \frac{i}{(2\pi)^4} \int d^4p \,\mathrm{Sp}\, \gamma_\alpha S \Gamma_\beta S$$

III. Photon propagator

$$e_0{}^2 D = \frac{e_0{}^2}{q^2(1 + e_0{}^2 \varrho(q^2))} [g_{\mu\nu} + \text{gauge term}]$$

IV. Vertex

$$\Gamma = \gamma + \int \frac{d^4p}{(2\pi)^4} K S \Gamma S$$

V. Ward identity

$$S^{-1}(p') - S^{-1}(p) = q^\mu \Gamma_\mu(p', p) \to S^{-1}(p') = q^\mu \Gamma_\mu(p', p)|_{\gamma p = m}$$

VI. Electron-positron scattering Kernel

$$K = \sum_{\substack{\text{Skeleton}\\ \text{graphs}}} K_{\text{Skel}}(\Gamma, S, D, e_0)$$

VII. S-Matrix elements

$$S = Z_2{}^{-f/2} Z_3{}^{-b/2} \sum_{\text{Skeletons}} S_{\text{Skel}}(\Gamma, S, D, e_0)$$

These may be solved iteratively, by starting with $\Gamma = \gamma$, $S = S_0$, $D = D_0$ then computing VI, IV, V, II, III in this order and iterating. The limiting values of S, D and Γ are:

$$S(p) \to \frac{Z_2}{\gamma p - m} \quad \text{as} \quad \gamma p \to m$$

$$D(q) \to \frac{Z_3}{q^2} [g_{\mu\nu} + \text{gauge term}] \quad q^2 \to 0 \tag{1}$$

$$\Gamma_\mu(p, p)|_{\gamma p = m} = Z_1{}^{-1} \gamma_\mu$$

and the Ward identity is:

$$Z_1 = Z_2$$

Renormalization says we define

$$\begin{aligned} S(p) &= Z_2 \cdot \tilde{S}(p) \\ D(q) &= Z_3 \cdot \tilde{D}(q) \qquad e^2 = Z_3\, e_0^{\,2} \\ \Gamma(p',p) &= Z_1^{-1} \cdot \tilde{\Gamma}(p',p) \end{aligned} \tag{2}$$

so that these equal the bare Green's functions at the poles, etc. Notice that mass renormalization is then handled automatically by use of the Ward identity. Also notice that $e_0^{\,2}\, D$ is *invariant* under renormalization.

The integral equations may be renormalized:

I. $$\tilde{D} = Z_3^{-1}[D_0 + e^2 D_0\, \varrho\, \tilde{D}]$$

II. $$\varrho_{\alpha\beta} = i\, Z_1 \int \frac{d^4 p}{(2\,\pi)^4} \operatorname{Sp} \gamma_\alpha\, \tilde{S}\, \tilde{\Gamma}_\beta\, \tilde{S} + \ldots$$

III. $$e_0^{\,2}\, D = e^2\, \tilde{D} = \frac{1}{q^2 \left\{\dfrac{1}{e_0^{\,2}} + \varrho(0) + [\varrho(q^2) - \varrho(0)]\right\}}$$

With

$$\frac{1}{e^2} = \frac{1}{e_0^{\,2}} + \varrho(0) \tag{3}$$

this gives

$$e^2\, \tilde{D} = \frac{1}{q^2 \left[\dfrac{1}{e^2} + \varrho_c(q^2)\right]} [g_{\mu\nu} + \ldots]$$

or (4)

$$\tilde{D} = \frac{1}{q^2 [1 + e^2\, \varrho_c(q^2)]} [g_{\mu\nu} + \ldots]$$

IV. $$\tilde{\Gamma} = Z_1 \gamma + \int \frac{d^4 p}{(2\,\pi)^4} \tilde{K}\, \tilde{S}\, \tilde{\Gamma}\, \tilde{S}, \qquad \tilde{\Gamma}_\mu(p,p)|_{\gamma p = m} = \gamma_\mu$$

V. $$\tilde{S}^{-1}(p') - \tilde{S}^{-1}(p) = q^\mu\, \tilde{\Gamma}_\mu(p',p)$$

VI. $$K = Z_2^{-2} \sum_{\text{Skel}} K_{\text{Skel}}(\tilde{\Gamma}, \tilde{S}, \tilde{D}, e) = Z_2^{-2}\, \tilde{K}$$

VII. $$S = \sum_{\text{Skel}} S_{\text{Skel}}(\tilde{\Gamma}, \tilde{S}, \tilde{D}, e)$$

This allows then an iteration in powers of e rather than e_0.

This iteration, as is proved with difficulty, leads to an expansion which is finite order by order.

Now in the process of going from unrenormalized to renormalized equations, we did not use strongly the condition of *where* we normalize the Green's functions, for example the condition $q^2 . \tilde{D}(q^2) \to 1$ as $q^2 \to 0$ could just as well have been

$$q^2 \cdot \tilde{D}(q^2, \lambda^2) \to 1 \quad \text{as} \quad q^2 \to \lambda^2 \quad (\lambda^2 \textit{ spacelike}) \tag{5}$$

So upon affixing a *new* dependence of the renormalized functions on *where* they are renormalized, we can see that the structure of the renormalized equations is form invariant with respect to the choice of the normalization point λ. We illustrate this by only rescaling the photon propagator; then for example (III) is given by

$$\text{III'} \qquad e_\lambda^2 D(q^2, \lambda^2) = \frac{1}{q^2\left[\frac{1}{e_\lambda^2} + \{\varrho(q^2) - \varrho(\lambda^2)\}\right]} [g_{\mu\nu} + \ldots]$$

The other equations are the same with the replacement

$$e^2 \to e_\lambda^2, \qquad \tilde{D} \to D(q^2, \lambda^2) \tag{6}$$

Since the equations are form invariant under $\lambda^2 \to \lambda'^2$, we have a symmetry group of transformations here; this is the "renormalization group". Notice, however, that in order to obtain this group we had to generalize a function of one variable to a function of two. So as yet there is little if any physical content in what we have done.

Let us now exploit this group structure. We define, on dimensional grounds,

$$e_\lambda^2 \cdot D_{\mu\nu}(q^2, \lambda^2) = d\left(\frac{q^2}{\lambda^2}, \frac{m^2}{\lambda^2}, e_\lambda^2\right)\left[\frac{g_{\mu\nu} + \ldots}{q^2}\right] \tag{7}$$

with the normalization condition

$$d\left(1, \frac{m^2}{\lambda^2}, e_\lambda^2\right) = e_\lambda^2 \tag{8}$$

According to our discussion above, d is a renormalization invariant; thus for any (spacelike) λ_1^2, λ_2^2

$$d\left(\frac{q^2}{\lambda_1^2}, \frac{m^2}{\lambda_1^2}, e_{\lambda_1}^2\right) = d\left(\frac{q^2}{\lambda_2^2}, \frac{m^2}{\lambda_2^2}, e_{\lambda_2}^2\right) = d\left(\frac{q^2}{\lambda_2^2}, \frac{m^2}{\lambda_2^2}, d\left(\frac{\lambda_2^2}{\lambda_1^2}, \frac{m^2}{\lambda_1^2}, e_{\lambda_1}^2\right)\right) \tag{9}$$

Here is an equation that must hold for all λ_2^2. However there isn't much physics in it yet because we introduced dependence on a new variable in order to get the relation.

To get physics out of this equation, we must make another *assumption*, that for

$$\lambda^2 \gg m^2$$

the m^2 dependence in d can be ignored, i.e.

$$\lim_{m^2 \to 0} d\left(\frac{q^2}{\lambda^2}, \frac{m^2}{\lambda^2}, e^2\right)$$

exists.

Here is the physics in the renormalization group approach. This is a nontrivial assumption, and we may return to this later. Ignoring the mass dependence, we have

$$d\left(\frac{q^2}{\lambda_1^2}, e_{\lambda_1}^2\right) = d\left(\frac{q^2}{\lambda_2^2}, d\left(\frac{\lambda_2^2}{\lambda_1^2}, e_{\lambda_1}^2\right)\right) \tag{10}$$

The solution to this is found as follows:

a) Differentiate with respect to q^2 and set equal to λ_2^2

b) Integrate with respect to λ_2^2 for fixed e^2

$$\frac{\partial}{\partial \lambda_2^2} d\left(\frac{\lambda_2^2}{\lambda_1^2}, e_1^2\right) = \frac{1}{\lambda_2^2} \varphi\left(d\left(\frac{\lambda_2^2}{\lambda_1^2}, e_1^2\right)\right)$$

Therefore:

$$\int_{e_1^2}^{d(q^2/\lambda_1^2, e_1^2)} \frac{d[d]}{\varphi(d)} = \int_{\lambda_1^2}^{q^2} \frac{d\lambda_2^2}{\lambda_2^2} \tag{11}$$

One finds in this way the solution

$$F\left(d\left(\frac{q^2}{\lambda_1^2}, e_1^2\right)\right) - F(e_1^2) = \log \frac{q^2}{\lambda_1^2}.$$

$$d\left(\frac{q^2}{\lambda_1^2}, e_1^2\right) = F^{-1}\left[F(e_1^2) + \log \frac{q^2}{\lambda_1^2}\right] \tag{12}$$

We see that we find a paradox: As $q^2/\lambda^2 \to \infty$

$$d \to F^{-1}[\infty] = e_0^2 \tag{13}$$

That is, independent of the value of the *renormalized* charge, the bare charge is a well-defined number $F^{-1}(\infty)$ (possibly ∞, of course), according to this argument.

We have a dilemma. We used perturbation theory in the bare charge e_0 in setting up the renormalization program. However, now we find either e_0 is ∞ or that e_0 is a definite number. Something seems, therefore, to be wrong with our assumptions, in particular, perturbation theory in e_0^2 or possibly the lack of dependence of d on m.

It is also possible to "improve" perturbation theory using renormalization group arguments. Here the idea is to assume a form for d and demand consistency with the renormalization group equation. For example, if we assume only

$$d(x, e^2) = e^2 + e^4 f_1(x) + O(e^6) \tag{14}$$

we find, inserting into (11) and doing the integral

$$d(x, e^2) = \frac{e^2}{1 + c_1 e^2 \log x + O(e^4 \log x)} \tag{15}$$

Without calculation, we conclude there are no 4th order $\log^2$ terms in the vacuum polarization. Putting in the 4th order calculation of Jost and Luttinger [2] and repeating the procedure, we get the leading 6th order correction completely:

$$d(x, e^2) = \frac{e^2}{1 - \frac{\alpha}{3\pi}\log x - \frac{\alpha^2}{4\pi^2}\log x - \frac{\alpha^3}{24\pi^3}\log^2 x + O(\alpha^3 \log x)} \tag{16}$$

All this depends upon the lack of dependence of D on m^2. Arguments have been given to support this contention by Erikson and Kinoshita.

References

1. The idea is due to M. Gell-Mann and F. Low, Phys. Rev. **95**, 1300 (1954). We follow the method of Bogoliubov and Shirkov, "Introduction to the theory of Quantized Fields", Interscience, New York, 1959.
2. Jost and Luttinger, Helv. Phys. Acta **23**, 201 (1950).

High Energy Quantumelectrodynamics II *

By

K. Johnson**

Copenhagen

I. Introduction

Quantumelectrodynamics is at present the only example of a complete relativistic quantum theory which is known to be relevant to physics. The rules for calculation based upon Feynman diagrams give one an unambiguous method to compare the theory with observation. The agreement so far has been without exception. One likes to believe that there is a basis for the calculational rules in quantum field theory. However, so far the attempt to produce such a field theory either has led to a case in which divergences occured if the perturbation rules were applied universally or to rather complicated "renormalized" versions of the theory which try to circumvent the divergences by formulating the calculational rules in terms of renormalized operators at the expense of an explicit expression for the Hamiltonian in terms of these operators. Usually attempts to modify the theory have only led to situations where the modifications produced more new deseases than the few cured.

What I would like to describe are some recent efforts to try to understand electrodynamics as a normal Hamiltonian theory [1]. The idea behind it is that two domains of approximation to the theory might exist. The first and usual domain, is the ordinary renormalized perturbation theory which yields the empirical successes. This is to be regarded as an approximation scheme valid in the neighbourhood of the physical mass shells of the particles which occur in the theory, photons and electrons. For all quantities where the contribution from virtual particle states far from this mass shell is small, the method should work. However, for those quantities which receive large contributions from such states the method fails. As an example we may consider the case of the photon Green's function which plays the role of the coupling agent. If $D(q^2)$ is the exact unrenormalized photon Green's function and e_0^2 the square of the so-called bare charge, the equations for the

---

* Lecture given at the IV. Internationale Universitätswochen für Kernphysik, Schladming, 25 February—10 March 1965.

** Permanent address: M.I.T., Cambridge, Mass.

Green's functions of the field operators can be written in such a way that only the combination $e_0^2 D = \bar{D}$ appears, with the exception of the equation for $\bar{D}$ itself which can be written:

$$\frac{1}{\bar{D}} = q^2 \frac{1}{e_0^2} + q^2 \varrho(q^2; \bar{D}), \tag{1}$$

where the photon "self-energy part" divided by e_0^2, ϱ, may be regarded as a functional of $\bar{D}$. However, since when $e_0^2 \to 0$, $\bar{D} \to 0$, $\varrho(q^2; 0) \to$ divergent, the above equation cannot be solved by an expansion procedure using the parameter e_0^2. The usual thing to do is to observe as $q^2 \to 0$

$$\frac{1}{\bar{D}} \to q^2 \left(\frac{1}{e_0^2} + \varrho(0; \bar{D}) \right) = q^2 \cdot \frac{1}{e^2}, \tag{2}$$

where

$$\frac{1}{e^2} = \frac{1}{e_0^2} + \varrho(0; \bar{D}) \tag{3}$$

is the square of the so-called physical charge. If we use this equation to remove e_0^2 from the equations then

$$\frac{1}{\bar{D}} = q^2 \frac{1}{e^2} + q^2 (\varrho(q^2; \bar{D}) - \varrho(0; \bar{D})). \tag{4}$$

This equation for $\bar{D}$ can be solved by an expansion in e^2 since the *difference* $\varrho(q^2; \bar{D}) - \varrho(0; \bar{D})$ is finite as $\bar{D} \to 0$. If we call

$$\bar{\varrho}(q^2; \bar{D}) = \varrho(q^2; \bar{D}) - \varrho(0; \bar{D}) \tag{5}$$

then we have

$$\frac{1}{\bar{D}} = q^2 \frac{1}{e^2} + q^2 \bar{\varrho}(q^2; \bar{D}) \tag{6}$$

and e_0^2 is no where to be seen. However, the function $\bar{\varrho}$, when computed by an expansion procedure, remains small for q^2 small, but when $q^2 \to \infty$, each term in the series grows. Consequently, the expansion method converges uniformly only in the neighbourhood of $q^2 = 0$, but fails for large q^2. In particular we get no equation for e_0^2. Since from the exact theory, the canonical commutation rules constrain $D(q^2)$ so that $D(q^2) \to (1/q^2)$ as $q^2 \to \infty$, consequently

$$\varrho(q^2; \bar{D}) \to 0 \qquad \text{as} \qquad q^2 \to \infty.$$

Then equation

$$\frac{1}{e_0^2} = \frac{1}{e^2} - \varrho(0; \bar{D}) \tag{7}$$

can be written

$$\frac{1}{e_0^2} = \frac{1}{e^2} + \lim_{q^2 \to \infty} \bar{\varrho}(q^2; \bar{D}), \tag{8}$$

so giving e_0^2 in terms of e^2. However, since the expansion method fails for q^2 large, so does the utility of this equation. We may in fact remark that the so-called "renormalized" theory stands by itself, and is even quite insensitive to the manner in which it is formulated in terms of field operators. The same equation (6) can be obtained from a theory with no electromagnetic field operators A^μ appearing in it at all as has been remarked by several people in different contexts [2, 3, 4].

Thus, if we take for example a self-coupled Fermi Lagrangian with a contact interaction

$$G j^\mu j_\mu, \tag{9}$$

the equations for Green's functions can be formulated in a way so that they are *identical* with the ones for electrodynamics with the sole exception of the equation for $\bar{D}$ which becomes

$$\frac{1}{\bar{D}} = \frac{1}{G} - \pi(q^2; \bar{D}), \tag{10}$$

where

$$\pi(q^2; \bar{D}) - \pi(0; \bar{D}) \equiv - q^2 \varrho(q^2; \bar{D}).$$

If we then "choose" G so that

$$0 = \frac{1}{G} - \pi(0; \bar{D}), \tag{11}$$

(So the collective mode of virtual pairs of total spin 1 comes with 0 mass) we have

$$\frac{1}{\bar{D}} = q^2 \varrho(q^2; \bar{D}), \tag{12}$$

which aside from the absence of the e_0^2 term is identical to (1). However, it is clear that when expressed in terms of e^2 the equation for $1/\bar{D}$ is identical to (6). Thus, only the equations

$$\frac{1}{G} = \pi(0; \bar{D})$$

and

$$0 = \frac{1}{e^2} + \lim_{q^2 \to \infty} \bar{\varrho}(q^2; \bar{D}) \tag{13}$$

differ from the theory formulated with electromagnetic field operators, and these equations cannot be used, if we make use of the "renormalized" expansion method. The fact that in this theory there is a formal *equation* for e^2, (13), has no content since we have no finite expression for $\lim_{q^2 \to \infty} \bar{\varrho}\,(q^2; \bar{D})$ as a function of e^2. Consequently, we see that the expression of the *renormalized version* of the theory in terms of operators is somewhat arbitrary; the only true content comes when we look at the equations for the renormalized Green's functions or better, when expanded in a power series, the theory of renormalized Feynman graphs. We, however, are more interested in the theory for q^2 large, than small, and as a consequence the renormalized version of the theory is of little use to us unless it is possible to consider all the terms of the perturbation series. Even looking at the dominant terms order by order need not lead to a valid conclusion. We shall try to develope a different technique. Since there will be much in common with the method which we shall suggest with ordinary perturbation theory, we shall also learn something about the later. However, it is important to keep in mind that the method we will outline also has important differences. There are also troublesome points which we shall spend some time on as we proceed.

II. Choice of Gauges

We shall find it convenient to allow ourselves the freedom of working in an arbitrary electromagnetic gauge in the sense that, in terms of Feynman diagrams, we would like to allow the photon lines to designate functions of the form

$$D_{\alpha\beta}(q) = \left(g_{\alpha\beta} + G\frac{q_\alpha q_\beta}{q^2}\right)\frac{1}{q^2}. \tag{14}$$

It is easily shown that all physical quantities which one may compute are insensitive to the choice of G. If one wants an operator basis for the theory it is provided by a Lagrangian of the form [5]

$$\begin{aligned}\mathscr{L} = -\frac{1}{2}F^{\mu\nu}(\partial_\mu A_\nu - \partial_\nu A_\mu) + \frac{1}{4}F^{\mu\nu}F_{\mu\nu} + \\ + \frac{\xi}{2}\mathfrak{J}^2 - \mathfrak{J}\,\partial^\mu A_\mu + J^\mu A_\mu,\end{aligned} \tag{15}$$

which gives rise to the field equations

$$\begin{aligned}F^{\mu\nu} &= \partial^\mu A^\nu - \partial^\nu A^\mu, \\ \xi\,\mathfrak{J} &= \partial^\mu A_\mu, \\ \partial_\mu F^{\mu\nu} &= -J^\nu + \partial^\nu \mathfrak{J}.\end{aligned} \tag{16}$$

We note that

$$-\Box^2\,\mathfrak{J} = -\partial_\nu J^\nu, \tag{17}$$

so that if $\partial_\nu J^\nu = 0$, then $\mathfrak{J}$ is a free field. The equation for A^μ is, as usual,

$$\Box^2 A^\nu - \partial^\nu\,\partial_\mu A^\mu = -J^\nu + \partial^\nu\mathfrak{J} = -J^\nu + \frac{1}{\xi}\,\partial^\nu\partial_\mu A^\mu, \tag{18}$$

so that

$$\left(-\Box^2 g^{\mu\nu} + \left(1 + \frac{1}{\xi}\right)\partial^\mu\,\partial^\nu\right)A_\nu = -J^\mu; \tag{19}$$

as a consequence the "photon Green's" function will obey

$$\left(q^2 g^{\mu\nu} - \left(1 + \frac{1}{\xi}\right) q^\mu q^\nu\right) D_{\nu\lambda} = g_\lambda{}^\mu, \tag{20}$$

or if we put

$$D_{\nu\lambda} = \left(g_{\nu\lambda} + G\frac{q^\nu q^\lambda}{q^2}\right)\frac{1}{q^2}, \tag{21}$$

then

$$\xi = -(1 + G). \tag{22}$$

We may then simply state, without the need of a detailed proof, that:

a) An indefinite metric must be employed.

b) If $\partial_\mu J^\mu = 0$, then, if we start in a subspace with positive norm, we cannot make transitions to states with negative norm.

c) The formalism is canonical for all choices of ξ.

d) Even in the "singular" case $G = -1 (\xi = 0)$, the method works when $\mathfrak{J}$ is regarded as a kind of operator Lagrange multiplier (as was shown by B. Lautrup in Copenhagen).

If we begin in this way we easily obtain the standard Dyson- [6] or Schwinger-equations [7] for Green's functions, written now in an arbitrary covariant gauge. Using the usual method we can show that, as a consequence of the canonical commutation rules, the exact electron Green's function

$$S(p) = \int (dx)\, e^{-ipx}\, i\, \langle 0 |\, T(\psi(x)\, \bar{\psi}(0)) \,| 0 \rangle \tag{23}$$

has the asymptotic property, $p^2 \to \infty$,

$$S(p) \to \frac{1}{\gamma p}, \tag{24}$$

and the photon Green's function

$$D_{\mu\nu}(q) = \int (dx)\, e^{-iqx}\, i \langle T(A_\mu(x)\, A_\nu(0))\rangle_0 \tag{25}$$

is, as $q^2 \to \infty$,

$$D_{\mu\nu}(q) \to \left(g_{\mu\nu} + G\frac{q_\mu q_\nu}{q^2}\right)\frac{1}{q^2}. \tag{26}$$

That is, as a consequence of the equal time relation among the operators, we have a constraint on the Green's functions as $q^2 \to \infty$ $(x \to 0)$. Since the canonical commutation rules are independent of the coupling, it is no surprise that the S and D functions also coincide with the free field Green's functions asymptotically. The basic consistency problem has always been related to this. The divergences which result from a naive application of perturbation theory come as a consequence of integrations over high virtual momenta. In this domain, however, we see that the exact D and S coincide with the free ones and so it would appear as if perturbation theory should work very well for large momenta. Instead it gives divergences. However, the actual problem is more complicated since the commutation rules constrain only the two elementary Green's functions, S and D, while S and D themselves are coupled to a third Green's function, the vertex, and for this we have no such constraint imposed by the commutation rules.

To be explicit, let us write the equations. We have to begin with an equation for

$$\frac{1}{S(p)} = \gamma p + m_0 + i\, e_0^2 \int \frac{(dp')}{(2\pi)^4} D^{\alpha\beta}(p - p')\, \gamma_\alpha\, S(p')\, \Gamma_\beta(p', p). \quad (27)$$

Written graphically, the so-called self-energy part or mass operator is

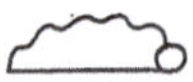

where: —— = exact electron function, ∼∼∼ = exact photon function, 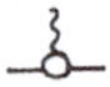 = exact vertex function. For D we have a similar expression which, however, expresses D as a functional of S and Γ:

$$\frac{1}{e_0^2\, D(q^2)} = q^2 \left(\frac{1}{e_0^2} + \varrho(q^2)\right), \quad (28)$$

where

$$\varrho(q^2)\,(g^{\mu\nu} q^2 - q^\mu q^\nu) =$$

$$= -i \int \frac{(dp)}{(2\pi)^4} \operatorname{tr} \left(\gamma^\mu \left(S\left(p + \frac{q}{2}\right) \Gamma^\nu \left(p + \frac{q}{2}, p - \frac{q}{2}\right) S\left(p - \frac{q}{2}\right)\right) + \right.$$

$$\left. + \frac{\partial}{\partial p^\nu} S(p) + \frac{1}{24}\left(q \frac{\partial}{\partial p}\right)^2 \frac{\partial}{\partial p^\nu} S(p)\right). \quad (29)$$

The graph for ϱ is aside from the extra terms needed to maintain current conservation

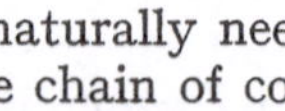

To complete the set, we naturally need an expression for Γ. It is this which leads to an infinite chain of coupled equations for the functions of the theory. However, there is one constraint on Γ_μ which couples it back to S, namely charge conservation, which leads to the identity of Ward [8, 9]

$$q^\mu\, \Gamma_\mu(p + q, p) = S^{-1}(p + q) - S^{-1}(p). \quad (30)$$

This equation also allows us to find the more familiar form of the identity, providing we make the additional, dynamical, assumption that Γ is regular at $q = 0$. If we differentiate (30) with respect to q, we find

$$\Gamma_\mu(p+q, p) + q^\nu \frac{\partial}{\partial q^\mu} \Gamma_\nu(p+q, p) = \frac{\partial}{\partial p_\mu} S^{-1}(p+q).$$

So, if Γ_μ is regular as $q \to 0$,

$$\Gamma_\mu(p, p) = \frac{\partial}{\partial p^\mu} S^{-1}(p). \tag{31}$$

From this fact we see that the assumption of regularity in Γ_μ at $q = 0$ also implies that $\varrho(0)$ is finite, since we see, as $q^2 \to 0$,

$$\varrho(q^2)\,(g^{\mu\nu} q^2 - q^\mu q^\nu) \underset{q^2 \to 0}{\longrightarrow}$$

$$\to -i \int \frac{(dp)}{(2\pi)^4} \operatorname{tr} \left\{ \gamma^\mu \left(S(p)\, \Gamma^\nu(p, p)\, S(p) + \frac{\partial}{\partial p^\nu} S(p) \right) \right\} = \cdot/. .$$

So, using (31),

$$\cdot/. = 0. \tag{32}$$

Thus, we see that $1/D$ vanishes at $q^2 = 0$ as a consequence of current conservation *and* the dynamical assumption that Γ is regular at $q = 0$.

We see also that the subtraction terms present in the *integrand* of (29) are necessary in order to insure that the photon self-energy part has the charge conserving form indicated. Thus, with the use of (30) we find by multiplying the right side of (29) with q^ν

$$-i \int \frac{(dp)}{(2\pi)^4} \operatorname{tr} \left\{ \gamma^\mu \left(S\left(p - \frac{q}{2}\right) - S\left(p + \frac{q}{2}\right) + \right.\right.$$

$$\left.\left. + q \frac{\partial}{\partial p} S(p) + \frac{1}{24} \left(q \frac{\partial}{\partial p}\right)^3 S(p) \right) \right\}. \tag{33}$$

If we expand the integrand in powers of q

$$\simeq \left(q \frac{\partial}{\partial p}\right)^5 S(p) + \ldots$$

and so if we integrate term by term using Gauss' law, we find zero. If the terms

$$\frac{\partial}{\partial p^\nu} S(p) + \frac{1}{24} \frac{\partial}{\partial p^\nu} \left(q \frac{\partial}{\partial p}\right)^2 S(p)$$

were not present in the integrand, we would find divergent surface terms remaining.

As a consequence of Ward's identity, we may obtain one further constraint from the commutation relations, since from (31) we see that as $p \to \infty$

$$\Gamma_\mu(p, p) \to \gamma_\mu$$

or also for any *fixed* q

$$\Gamma_\mu(p+q,p) \to \gamma_\mu \tag{34}$$

as $p^2 \to \infty$.

However, we might note that from Ward's identity we do not obtain any information regarding the asymptotic behaviour of Γ as $q \to \infty$. This is of course to be expected. Even though Ward's identity is a constraint on only the "longitudinal" part of Γ_μ we are able to obtain a condition on Γ_μ itself in the neighbourhood of $q = 0$ (or with p large compared to q), if we make the assumption that Γ is regular at $q = 0$; it is the later condition which ties the non-longitudinal components to the longitudinal part of Γ in the neighbourhood of $q = 0$. No such argument applies when q is large, so from the identity we can learn nothing about Γ_μ. Consequently, we do not obtain sufficient information about the asymptotic behaviour of Γ to determine anything about the form of the integrand of the mass operator. In the case of the integrand of the photon self-energy, we obtain only the information that as $p \to \infty$, keeping the leading terms

$$S\left(p+\frac{q}{2}\right)\Gamma_\nu\left(p+\frac{q}{2},p-\frac{q}{2}\right)S\left(p-\frac{q}{2}\right)+\frac{\partial}{\partial p^\nu}S(p)+$$
$$+\frac{1}{24}\frac{\partial}{\partial p^\nu}\left(q\frac{\partial}{\partial p}\right)^2 S(p) \to \left\{S(p)+\frac{q}{2}\frac{\partial}{\partial p}S(p)+\frac{1}{2}\left(\frac{q}{2}\right)^2\left(\frac{\partial}{\partial p}\right)^2 S(p)\right\}\cdot$$
$$\cdot\{\Gamma_\nu(p,p)+(\Gamma_\nu^q-\Gamma_\nu^0)\}\cdot\left\{S(p)-\frac{q}{2}\frac{\partial S}{\partial p}+\frac{1}{2}\left(\frac{q}{2}\frac{\partial}{\partial p}\right)^2 S(p)\right\}-$$
$$-S(p)\,\Gamma_\nu(p,p)\,S(p)+\frac{1}{24}\frac{\partial}{\partial p_\nu}\left(q\frac{\partial}{\partial p}\right)^2 S(p)\to$$
$$\to\frac{1}{\gamma p}(\Gamma_\nu^q-\gamma_\nu)\frac{1}{\gamma p}+\text{Born Term};\qquad \text{Born Term}\sim\frac{1}{p^4}. \tag{35}$$

Consequently, we see that in general the vacuum polarization integral is expected to diverge like the integral $\int dp/p$, unless the asymptotic form of $(\Gamma_\nu^q-\gamma_\nu)_{p\to\infty}$ is such to cancel the leading term in (35). Naturally no perturbation treatment of the equations for Γ_ν can have this effect since $\Gamma_\nu^q-\gamma_\nu$ is of higher order in e than the term it must cancel.

III. Self-Energy

In the case of the integrand of the mass operator the situation is different. In this instance, it is not precluded that an expansion procedure is allowed which begins with Γ_ν replaced by γ_ν in the first approximation. The reason is that in contrast to the photon self-energy, the electron mass operator depends upon the choice of gauge indicated by the constant G. We find it possible to always choose G so that no divergence occurs in this function, given only that the photon self-energy is finite, that is, given that as $q^2 \to \infty$, $D(q^2) \to (1/q^2)$ (actually, this later restriction can be relaxed somewhat).

To indicate how this can be shown let us first remark that in at most *one* gauge of the class described by the Lagrangian given can the mass operator be finite and S constructed with the asymptotic form $S \to (1/\gamma p)$. We shall then show that in this gauge the mass operator is indeed finite. Since our result will be to show that in only a single gauge can the mass *operator* be finite, it is convenient to use a cutoff, in order that at the *beginning* all gauges will give unambiguous finite results. Since we *assume* that the exact $D(q^2) \to 1/q^2$, we shall cut off this function so that $q^2 \cdot D(q^2) \to 0$ as $q^2 \to \infty$, say

$$q^2 D(q^2) \to \left\{\frac{\lambda^2}{q^2 + \lambda^2}\right\}^{\varepsilon}$$

with $\varepsilon > 0$. It is then very easy to compute the exact dependence of S on G: we find that [10]

$$S_G(x, y) = e^{iGe_0^2(M_\lambda(x-y) - M_\lambda(0))} S_0(x, y), \tag{36}$$

where

$$M_\lambda(x) = \int \frac{dq}{(2\pi)^4} e^{iqx} \left\{\frac{\lambda^2}{q^2+\lambda^2}\right\}^{\varepsilon} \frac{1}{q^2} \frac{1}{q^2+\lambda_0^2}$$

and where λ_0 is an infrared cutoff say of order $< m$. We have also

$$S_{G'}^{\lambda}(x, y) = e^{ie_0^2(G'-G)(M(x-y)-M(0))} S_G^{\lambda}(x, y). \tag{37}$$

Let us suppose that for some value of G, $S_G(x, y)$ is a finite function as $\lambda \to \infty$. Then, it is easy observed that

$$e^{ie_0^2(G'-G) M_\lambda(x-y)} S_G^{\lambda}(x-y) \tag{38}$$

is finite as $\lambda \to \infty$. But, for $\lambda \gg \lambda_0$

$$e^{-ie_0^2(G'-G) M_\lambda(0)} = e^{\frac{\alpha_0}{4\pi}(G'-G)\log\left(\frac{\lambda^2}{\lambda_0^2}\right)}, \qquad \alpha_0 = \frac{e_0^2}{4\pi}. \tag{39}$$

Consequently, $S_{G'}$ will in general require a divergent multiplicative renormalization. Now if the exact $D(q^2) \to 1/q^2$ as $q^2 \to \infty$, then we may also determine, for a fixed value of G, the leading dependence of S upon λ^2, as $\lambda^2 \to \infty$. If we make use of the fact that the multiplicative renormalization will make S a finite function, *and* if we assume that we may learn in the case of S the leading behaviour of S upon λ, by using the perturbation series for first summing up the photon self-energy to the exact function $D(q^2)$ which we assume to be finite, and then allowing λ to be large and sum up the leading terms in this partly resummed perturbation series as $\lambda \to \infty$, we find from the result of Gell-Mann and Low [11] that the dependence of the multiplicative renormalization for S, Z_2, upon λ is given by

$$Z_2 \sim e^{f(\alpha_0, G)\log\frac{\lambda^2}{\lambda_0^2}} \tag{40}$$

for $\lambda \gg \lambda_0$, with f a finite function of α_0.

From this result and the previous one we see that we know the exact dependence of f upon G, namely

$$f(\alpha_0, G) = f(\alpha_0, G') + \frac{\alpha_0}{4\pi}(G' - G). \tag{41}$$

Since f is given as a power series we see that only the coefficient of α_0 depends upon the gauge constant G. Consequently we may always choose a gauge constant G expressed as a power series in α_0 for which $f = 0$ term by term, and hence for which there occurs no Z_2-divergence in the partly resummed perturbation series mentioned. We find that for $G' = -1$, the lowest order term in f vanishes and as a consequence, the expression for G has the form

$$G = -1 + \alpha_0 G_1 + \alpha_0^2 G_2 + \dots . \tag{42}$$

Now even though we have shown that there exists a gauge in which the partly resummed perturbation series for S (summation of photon lines) leads to a S which does not require a multiplicative renormalization (Z_2) we have not yet shown that the mass operator is finite. For, as is well known, there is also in perturbation theory a divergent mass renormalization associated with Σ which is removed formally by letting $m_0 = m - \delta m$, where δm is the so-called self mass, and then choosing δm to cancel the self-energy divergences in the mass operator. This divergence can have nothing to do with a choice of gauge, since δm must be gauge invariant. However, we can try to see if this divergence can be associated with an unjustified expansion. Since we have shown that Z_2 can be made finite with a suitable choice for G it is also true that the *vertex* function will have a finite renormalization since as a consequence of Ward's identity $Z_1 = Z_2$. Consequently, it is reasonable to try to see if we can eliminate Γ in the equation for S by expressing Γ as a power series in the exact S function. We shall see immediately the connection between this assumption and the fact that there is a suitable choice for G which allows for the exact S the asymptotic form $S \to 1/\gamma p$ as $p^2 \to \infty$. If we make this replacement for Γ in the equation for S we find

$$\frac{1}{S(p)} = \gamma p + m_0 + \Sigma(\gamma p; S), \tag{43}$$

where $\Sigma(\gamma p; S)$ is expressed as a functional power series in the exact S. In graphical form Σ is

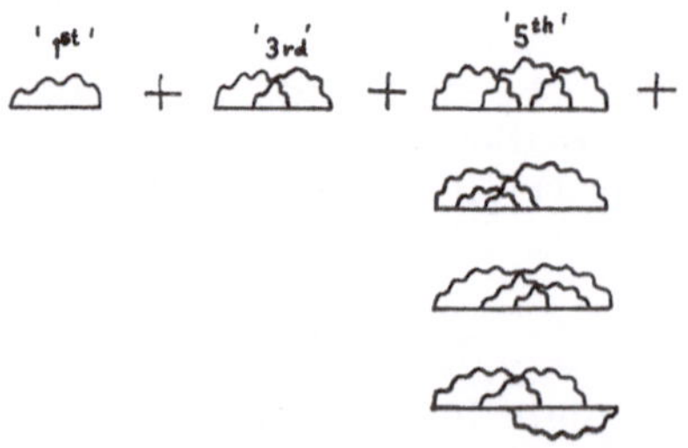

(44)

We can obtain a corresponding equation for Γ itself by imagining that the expansion for Σ is made in the presence of an external electromagnetic potential. It is trivial to show that each term in the above series, when $A^\mu \to A^\mu + \partial^\mu \lambda$, undergoes the transformation

$$T(x, y) \to e^{i e_0(\lambda(x) - \lambda(y))}\, T(x, y). \tag{45}$$

That is, such a gauge transformation does not mix the orders of the various terms in the expansion. Since Γ_μ is defined by

$$\Gamma_\mu = -\frac{\delta}{\delta e_0 A_\mu} S^{-1}, \tag{46}$$

$$\Gamma_\mu = \gamma_\mu - \frac{\delta}{\delta e_0 A_\mu} \Sigma, \tag{47}$$

we obtain from the above expression a linear integral equation for Γ. Written in graphical form with $\Gamma_\mu =$ , we have

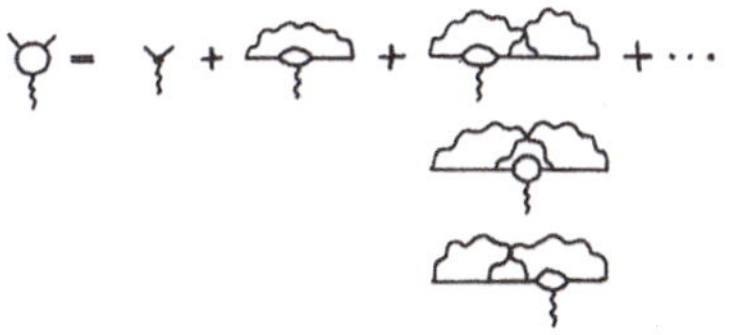

(48)

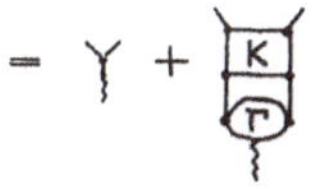

(49)

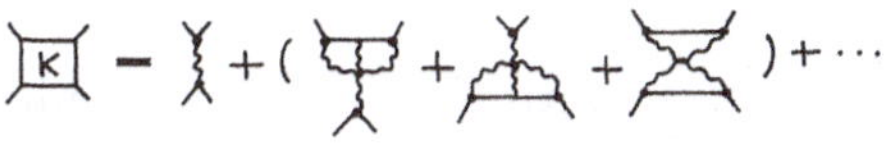

(50)

One should not confuse the expression for K with a power series expansion since the various terms still contain *exact* D and S functions. It is a consequence of the fact that under $A_\mu \to A_\mu + \partial_\mu \lambda$

$$S^{-1}(x, y) \to e^{i e_0(\lambda(x) - \lambda(y))}\, S^{-1}(x, y) \tag{51}$$

and the definition (46) of Γ, that the general form of Ward's identity (written in momentum space with $A^{\text{ext}} = 0$) is

$$q^\mu \Gamma_\mu(q + p, p) = S^{-1}(q + p) - S^{-1}(p). \tag{52}$$

Since the transformation property (51) holds term by term in the series (44) for Σ it also follows that Ward's identity will also hold term by term connecting the expressions for S^{-1} and Γ given by (44) and (48). This result continues to hold even if we reshuffle the terms of the two series by some procedure involving the photon lines. Consequently, if we analyze the behaviour of the Z_1 contributions to Γ coming from the various terms in (49), a similar analysis can be made for the

Z_2 contributions coming from the corresponding terms in the equation for S^{-1}.

Let us first consider the vertex equation with $k = 0$, that is the equation for $\Gamma_\mu(p, p)$. If the gauge G is chosen properly then as $p \to \infty$, $\Gamma_\mu \to \gamma_\mu$. The most slowly vanishing contributions to $\Gamma_\mu(p, p)$ in the integrand come from the asymptotic regions of the integrand, where we know by assumption that $D \to 1/k^2, \Gamma_\mu \to \gamma_\mu$ and $S \to 1/\gamma p$. If the resulting integral is finite, then it must vanish as $p \to \infty$. Consequently, we need only determine whether the integrals are finite. Let us examine the first term. If we put $p = 0$, and look at the asymptotic form of the integrand of the first term we find

$$= -i e_0^2 \int \frac{dk}{(2\pi)^4} D_{\alpha\beta}(k)\, \gamma^\alpha \frac{1}{\gamma k} \gamma_\lambda \frac{1}{\gamma k} \gamma^\beta. \tag{49}$$

To evaluate this we need only realize that the integral is proportional to γ_λ. Hence multiply by γ^λ to get the proportionality constant

$$= + i e_0^2 \int \frac{dk}{(2\pi)^4} D_{\alpha\beta}(k)\, \gamma^\lambda \gamma^\alpha \gamma k \gamma_\lambda \frac{1}{\gamma k} \gamma^\beta \frac{1}{k^2} = \langle (\gamma^\lambda \gamma^\alpha \gamma k \gamma_\lambda) = 2k^\alpha \rangle =$$

$$= i e_0^2 \int \frac{dk}{(2\pi)^4} 2k^\alpha D_{\alpha\beta}(k) \frac{1}{\gamma k} \gamma^\beta \frac{1}{k^2}\,.$$

The integral is finite only if $k^\alpha D_{\alpha\beta}(k) = 0$, that is, if

$$G = -1$$

which defines the so-called Landau gauge [12].

If we think of the equation for Γ written in terms of the kernel

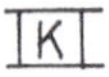 we find that  with $G = -1$ has a finite effect. If we go

to the next term in the kernel

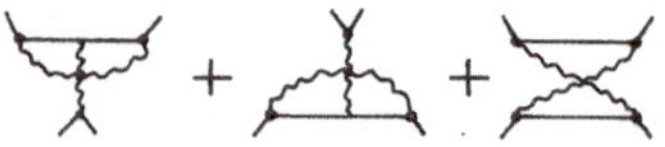 (50)

we see that the first two terms are finite since with $G = -1$ they are just vertex corrections. The last term is also finite. It is easy to verify that the first two terms lead to no divergence in Γ. However, the last term produces a log divergence proportional to α_0^2. If we then let $G = -1 + \alpha_0 G_1$, we find that we also obtain a log divergence proportional to $\alpha_0^2 G_1$ from the first term of the kernel. We may than choose G_1 so that the divergences cancel and we consequently obtain a kernel

 (51)

which also yields a finite Γ. In the above  symbolizes the gauge term

$$\alpha_0 G_1 \frac{k_\alpha k_\beta}{k^2} D(k^2) \tag{52}$$

and  the Green's function in the Landau gauge

$$D_{\alpha\beta} = \left(g_{\alpha\beta} - \frac{k_\alpha k_\beta}{k^2}\right) D. \tag{53}$$

We may continue in this fashion. As we go further in the kernel we each time obtain a finite kernel which produces a single log divergence (this is guaranteed by the Gell-Mann — Low paper) which can then be cancelled with a single log term coming from the first term.

When we also write the corresponding equation for $1/S$,

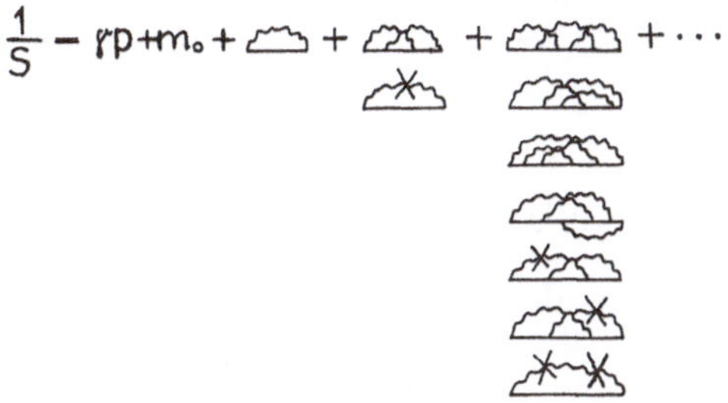

(54)

(where  $\alpha_0{}^2 G_2 (k_\alpha k_\beta/k^2) D$) then term by term the equations for $1/S$ and Γ are connected by Ward's identity.

We are therefore sure that the above equation for S is consistent with the asymptotic behaviour for the exact S, $S \to 1/\gamma p$ as $p^2 \to \infty$. We may now come to the question of the self-energy, since we are not yet certain that the above equation is consistent with a finite Σ. Let us put

$$\frac{1}{S} = \gamma p + M(p) \tag{55}$$

with

$$M(p) = m_0 + \Sigma(p) = \gamma p\, B(p^2) + A(p^2) \tag{56}$$

and assume that as $p^2 \to \infty$ $A(p^2)$ does not grow faster than $(p^2)^\varepsilon$, $\varepsilon < 1$. We know already that $B(p^2) \to 0$ since $1/S \to \gamma p$ as $p^2 \to \infty$. Then

$$S(p) \to \frac{1}{\gamma p} + \frac{A(p^2)}{p^2}.$$

So $(A(p^2)/p^2)$ is the leading term proportional to the Dirac matrix 1 in the function S asymptotically. Now, the most slowly vanishing contributions to $\Sigma(p)$ for $p^2 \to \infty$ come from the region of integration over virtual momenta which are also asymptotic. Consequently, as $p^2 \to \infty$ we may replace in the internal lines the functions $D \to 1/k^2$ and $S \to 1/\gamma p + (A(p^2)/p^2)$. We then obtain an integral equation for $A(p^2)$ valid for $p^2 \gg m^2$, namely

$$A(p^2) = m_0 + \boxed{\bar{K}}\, \frac{A(p'^2)}{p'^2} \tag{57}$$

where

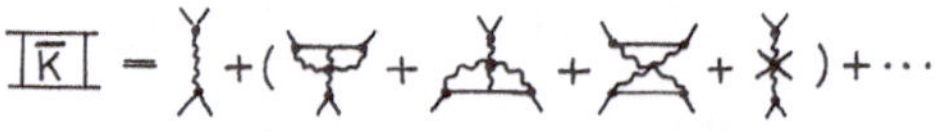

(58)

and where in $\bar{K}$ all lines are replaced by their asymptotes $S \to 1/\gamma p$ and $D \to 1/k^2$. The asymptotic kernel $\bar{K}$ is a power series in $e_0{}^2$ which is *finite* term by term. It is also important to note that there is *no scale* parameter present in the asymptotic kernel.

Now it is not difficult to show that if the above equation has a solution then as $p^2 \to \infty$, $A \to 0$. But if $A \to 0$, then it is also easy to find that

$$A(p^2) - m_0 \to 0 \tag{59}$$

as $p^2 \to \infty$. Consequently the equation has no solution unless $m_0 = 0$. If $m_0 = 0$, of course the equation has the trivial solution $A = 0$. However, because of the scale invariance of the kernel it can also have a nontrivial solution of the form

$$A(p^2) = A_0\, m \left(\frac{m^2}{p^2}\right)^g \tag{60}$$

if $g > 0$ and $A_0 =$ arbitrary constant. If we find $g > 0$, then the above function is a rigorous solution of the integral equation. If we put the above expression into (57) we find an equation for g which has the form

$$1 = \frac{3\,\alpha_0}{4\,\pi}\,\frac{1}{g(1-g)} + \left(\frac{\alpha_0}{\pi}\right)^2 F_2(g) + \left(\frac{\alpha_0}{\pi}\right)^3 F_3(g) + \ldots, \tag{61}$$

where $F_2(g), F_3(g), \ldots$ are respectively transcendental functions of g which are all finite if $g > 0$ and singular as $p \to 0$. It is known by explicit calculation that as $g \to 0$ F_2 and $F_3 \to C/g$ just as the first term. We believe that an analysis of the above equation on a similar basis as that used by Gell-Mann and Low will indicate that this is true generally. In any case, we can simply assume that (61) can be solved with a power series expression for g

$$g(\alpha_0) = \frac{3\,\alpha_0}{4\,\pi} + \left(\frac{\alpha_0}{\pi}\right)^2 \frac{21}{32} + \ldots \tag{62}$$

which we know is consistent at least to order $(\alpha_0/\pi)^3$. We observe that for small $\alpha_0, g > 0$. Consequently, we have given a method of computing the asymptotic corrections to S which yields

$$S(p) \to \frac{1}{\gamma p} + \frac{A_0 m}{p^2}\left(\frac{m^2}{p^2}\right)^{\frac{3\alpha_0}{4\pi} + \frac{21}{32}\left(\frac{\alpha_0}{\pi}\right)^2 + \cdots}, \tag{63}$$

where A_0 is so far an arbitrary constant. It is not unreasonable to assume that the above solution will join to the solution obtained from perturbation theory which is valid only in the region close to the mass shell. Thus, for fixed p we know that as the interaction vanishes we should have a particle with *renormalized* (physical) mass m. Consequently, we see that as $\alpha_0 \to 0$, $A_0 \to 1$.

Finally we may briefly comment on the fact that with $m_0 = 0$ and $A(p^2)$ (or m) $\neq 0$, we have a solution to the equations for the Green's functions of the theory which is not invariant with respect to two groups of transformations which leave the equations invariant. These are scale transformations and γ_5-transformations. The reason that we can find non-invariant solutions to the equations is of course because the equations by themselves do not have unique solutions [**13**]. Thus, the homogeneous linear equation for A has non-trivial solutions for $\alpha_0 > 0$. Consequently, this means that the boundary condition implied by taking matrix elements in the lowest energy state is not sufficient to characterize the solution uniquely. However, with the additional constraint which we have imposed, namely that the physical electron has mass m, and a parity such that $1/S = \gamma p(1 + B) + A$ with no γ_5 term, we obtain a unique solution for the Green's function. Naturally, one might feel that this means that the "same" Hamiltonian operator H has "vacuum states" corresponding to a vacuum filled with virtual pairs of mass m, and also virtual pairs of mass λm for any λ, that is that then would be a continuous infinity of vacuum states. This is of course not really the case. Thus, to express the Hamiltonian operator in terms of the field operators it is necessary to put dots around the field operator products. This must be done in some self-consistent fashion. Consequently, the formal invariance arguments need not apply. One should always realize that in a relativistic theory the formal invariance of the equations need never imply that the solutions carry that invariance. It only means that there may exist solutions which carry the symmetry.

To summarize, we have analyzed the equation for $1/S$ and have found that it yields a finite self-energy operator which asymptotically vanishes as

$$\frac{m}{p^2}\left(\frac{m^2}{p^2}\right)^{\frac{3\alpha_0}{4\pi} + \frac{21}{32}\left(\frac{\alpha_0}{\pi}\right)^2 + \cdots} \cdot [1 + \alpha_0 \mathfrak{A}_1 + \ldots].$$

We see that if we try to express this as a power series in α_0, then

$$``m_0" = m\left[1 + \left[\frac{3\,\alpha_0}{4\,\pi} + \frac{21}{32}\left(\frac{\alpha_0}{\pi}\right)^2 + \ldots\right]\log\left(\frac{m^2}{p^2}\right) + \right.$$

$$\left. + \frac{1}{2}\left(\frac{3\,\alpha_0}{4\,\pi} + \frac{21}{32}\left(\frac{\alpha_0}{\pi}\right)^2 + \ldots\right)^2 \log^2\frac{m^2}{p^2} + \ldots\right](1 + \alpha_0\,\mathfrak{A}_1 + \ldots)$$

which is formally defined as the coefficient of $1/p^2$ as $p^2 \to \infty$. Consequently, as $p^2 \to \infty$, m_0 would diverge term by term; this is just the standard perturbation theory result connecting the bare mass m_0 to the physical mass.

IV. Vacuum Polarization [15]

Let us now consider the question of the vacuum polarization. To make clear the approach that we use we will consider a simple model equation. Let us first make some general remarks. If we compute the function $\varrho(q^2)$ in perturbation theory, it diverges. If ϱ is finite, we believe that it has a spectral representation

$$\varrho(q^2) = \int dm^2\,\frac{\sigma(m^2)}{m^2 + q^2} \tag{64}$$

with $\sigma > 0$ and as a consequence for reasonable σ, we believe that as $q^2 \to \infty$, $\varrho(q^2) \to 0$ monotonically; we have also written the equation relating the bare physical charges, namely

$$\frac{1}{e^2} = \frac{1}{e_0{}^2} + \varrho(0) = \frac{1}{e_0{}^2} + \int dm^2\,\frac{\sigma(m^2)}{m^2}\,. \tag{65}$$

Naturally since $\sigma > 0$ this means $e^2 < e_0{}^2$. The equation we have given for $\varrho(q^2)$ is

$$(g^{\mu\nu} q^2 - q^\mu q^\nu)\,\varrho(q^2) = -i\int \frac{dp}{(2\,\pi)^4}\,\mathrm{tr}\left(\gamma^\mu\left\{S\left(p + \frac{q}{2}\right)\cdot\right.\right. \tag{66}$$

$$\left.\left.\cdot\,\Gamma^\nu\left(p + \frac{q}{2}, p - \frac{q}{2}\right) S\left(p - \frac{q}{2}\right) + \frac{\partial}{\partial p_\nu} S(p) + \frac{1}{24}\frac{\partial}{\partial p_\nu}\left(q\,\frac{\partial}{\partial p}\right)^2 S(p)\right\}\right).$$

We will always carry out our analysis for space-like q^2 where we are allowed to rotate the p^0 contour in (66) and integrate over a Euclidean four dimensional space. If we wish to show that $\varrho(0)$ is finite then we must be able to estimate the integrand (66) as a function of p as $p^2 \to \infty$. If we define

$$C^\nu = S\left(p + \frac{q}{2}\right)\Gamma^\nu\left(p + \frac{q}{2}, p - \frac{q}{2}\right) S\left(p - \frac{q}{2}\right) + \frac{\partial}{\partial p_\nu} S(p) +$$

$$+ \frac{1}{24}\frac{\partial}{\partial p^\nu}\left(q\,\frac{\partial}{\partial p}\right)^2 S(p), \tag{67}$$

then we can write (66) as

$$= \int \frac{(dp)_E}{(2\pi)^4} \operatorname{tr}(\gamma^\mu C^\nu) = \frac{\pi^2}{(2\pi)^4} \int p^2\, dp^2 \operatorname{tr}(\gamma^\mu \langle C^\nu \rangle), \tag{68}$$

where

$$\langle C^\nu \rangle = \frac{1}{2\pi^2} \int d\Omega_p C^\nu \tag{69}$$

is the average of C^ν over all four dimensional angles of the vector p. If we look at perturbation theory with $\Gamma^\nu \to \gamma^\nu$ we find that as $p^2 \to \infty$, (69) becomes

$$\operatorname{tr}(\gamma^\mu \langle C^\nu \rangle) \sim (g^{\mu\nu} q^2 - q^\mu q^\nu) \frac{4}{3} \frac{1}{p^4},$$

so

$$\varrho(0) \sim \frac{1}{12\pi^2} \int^\infty \frac{dp^2}{p^2},$$

and the integral diverges logarithmically. Let us write in general

$$\varrho(0) = \frac{1}{12\pi^2} \int^\infty dp^2\, V(p^2), \tag{70}$$

where, as $p^2 \to \infty$,

$$V(p^2)\,(g^{\mu\nu} q^2 - q^\mu q^\nu) \sim \frac{3}{4} \operatorname{tr}(\gamma^\mu \langle C^\nu \rangle). \tag{71}$$

We have a linear integral equation for Γ^ν and thus for C^ν and hence for V. Let us now consider a simple model. *In our model*, let us suppose that the linear equation has the form valid for $p^2 \gg m^2$

$$V(p^2) = \frac{1}{p^2} - \lambda \int \frac{dq}{\pi^2} D(p-q) \left[V(q^2) \frac{1}{q^2} \right], \tag{72}$$

with $D(p-q) = 1/(p-q)^2$. We shall then show that this equation always has the property that the integral term on the right side always cancels the inhomogeneous term as $p^2 \to \infty$, no matter how small λ is, and V vanishes rapidly enough as $p^2 \to \infty$ so that $\varrho(0)$ is finite. The only thing that is required is that λ be > 0. Nevertheless the solution also has the property that as $\lambda \to 0$ *for fixed* p^2

$$V(p^2) \to \frac{1}{p^2}.$$

The above integral equation can easily be converted to a one dimensional equation by integrating over the angles of q first. Then

$$\int \frac{d\Omega_q}{2\pi^2} D(p-q) = \frac{1}{p_>^2}, \qquad \text{where} \qquad p_>^2 = \begin{bmatrix} p^2 & p > q \\ q^2 & q > p \end{bmatrix}. \tag{73}$$

Thus,

$$V(p^2) = \frac{1}{p^2} - \lambda \int dq^2 \frac{1}{p_>^{\,2}} V(q^2) = \frac{1}{p^2} - \lambda \int\limits_{p^2}^{\infty} \frac{dq^2}{q^2} V(q^2) - \frac{\lambda}{p^2} \int\limits_{m^2}^{p^2} dq^2\, V(q^2), \tag{74}$$

where m^2 is a lower cut-off which is of order of the electron mass. To solve, differentiate with respect to p^2,

$$\frac{dV}{dp^2} = -\left(\frac{1}{p^2}\right)^2 + \frac{\lambda}{(p^2)^2} \int\limits_{m^2}^{p^2} dq^2\, V(q^2)$$

and

$$\frac{d}{dp^2} (p^2)^2 \frac{dV}{dp^2} = \lambda\, V(p^2). \tag{75}$$

We see that the "Born term" *drops out*, i.e., the asymptotic form of V is "self-consistently" determined by V itself. The solution of (75) is easily gotten by putting

$$V(p^2) = \left(\frac{1}{p^2}\right)^{\varepsilon+1};$$

then we get from (75) the equation

$$(1+\varepsilon)\, \varepsilon = \lambda. \tag{76}$$

For *small* λ there is a root near λ and one near -1. It is easy to show that the root near -1 cannot be used in the integral equation. Thus for small λ we have

$$\varepsilon \sim \lambda$$

and

$$V(p^2) = V_0 \left(\frac{1}{p^2}\right)^{1+\lambda} \tag{77}$$

If we insert this expression into the integral equation we get

$$V_0 \left(\frac{1}{p^2}\right)^{1+\lambda} = \frac{1}{p^2} - \frac{\lambda}{p^2} \int\limits_{m^2}^{p^2} dq^2\, V_0 \left(\frac{1}{q^2}\right)^{1+\lambda} - \lambda \int\limits_{p^2}^{\infty} \frac{dq^2}{q^2} V_0 \left(\frac{1}{q^2}\right)^{1+\lambda} =$$

$$= \frac{1}{p^2} - \frac{\lambda}{p^2} \left(\frac{1}{\lambda} \frac{V_0}{(m^2)^\lambda} - \frac{1}{\lambda} V_0 \left(\frac{1}{p^2}\right)^{1+\lambda}\right)$$

(dropping terms of order λ).

We find that the inhomogeneous term is cancelled by the integral term if

$$V_0 = (m^2)^\lambda. \tag{78}$$

Therefore, the solution to the integral equation is

$$V(p^2) = \frac{1}{p^2}\left(\frac{m^2}{p^2}\right)^{\lambda}. \tag{79}$$

We see that for *fixed* p^2 as $\lambda \to 0$, V goes smoothly to the "Born term". We also see that

$$\varrho(0) = \frac{1}{12\,\pi^2}\int_{m^2}^{\infty} dp^2\, V(p^2) = \frac{1}{2\,\pi^2}\int_{m^2}^{\infty} dp^2 \left(\frac{1}{p^2}\right)\left(\frac{m^2}{p^2}\right)^{\lambda}, \tag{80}$$

$$\varrho(0) = \frac{1}{12\,\pi^2\,\lambda}$$

which is finite and *singular* as $\lambda \to 0$. We also see that expressed as a power series

$$V(p^2) = \frac{1}{p^2}\left(e^{\lambda \log \frac{m^2}{p^2}}\right) =$$

$$= \frac{1}{p^2}\left(1 + \lambda \log\left(\frac{m^2}{p^2}\right) + \frac{\lambda^2}{2}\left(\log\frac{m^2}{p^2}\right)^2 + \ldots\right)$$

and, if we tried to use this "perturbation" solution for the calculation of $\varrho(0)$, we would get a series of logarithmically infinite integrals for $\varrho(0)$. Let us now turn away from the model and go back to the actual problem of interest, namely, electrodynamics.

We may again summarize the equations.

$$\varrho(0) = \frac{1}{12\,\pi^2}\int^{\infty} dp^2\, V(p^2),$$

where as $p^2 \to \infty$

$$V(p^2)\,(g^{\mu\nu} q^2 - q^\mu q^\nu) \simeq \frac{3}{4}\,\mathrm{tr}\,(\gamma^\mu \langle C^\nu \rangle)$$

and

$$C^\nu\left(p + \frac{q}{2}, p - \frac{q}{2}\right) = S\left(p + \frac{q}{2}\right)\Gamma^\nu\left(p + \frac{q}{2}, p - \frac{q}{2}\right) S\left(p - \frac{q}{2}\right) + $$
$$+ \frac{\partial}{\partial p_\nu} S(p) + \frac{1}{24}\left(q\,\frac{\partial}{\partial p}\right)^2 \frac{\partial}{\partial p_\nu} S(p). \tag{81}$$

Finally we know that asymptotically $S(p) \to (1/\gamma\, p)$ and also we know that Γ^ν obeys a linear integral equation

(49)

where

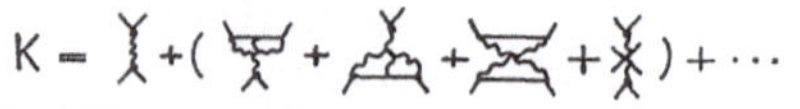

we can then solve (81) for Γ^ν in terms of C^ν and replace Γ^ν by this expression in equation (49). We then obtain an integral equation for C^ν of essentially the same form as that for Γ^ν, except that there is a more complicated inhomogeneous term whose asymptotic behaviour as $p^2 \to \infty$ can however be explicitly calculated. The inhomogeneous term in the equation for C^ν instead of remaining constant as $p^2 \to \infty$ vanishes as $1/p^2$. Thus the equation for C^ν looks very much like the model equation except of course for the complications produced by spin and the considerably more involved kernel function.

Again, since we are interested in the solution of the equation valid in the asymptotic region, and since the most slowly vanishing contributions to the integrals in question come from the region where all virtual lines are asymptotically far from their mass shells, we can replace the kernel K of the integral equation by the "asymptotic" kernel K defined previously. K is *explicitly known* as a power series in e_0^2. Since the *asymptotic* kernel is γ_5-invariant, and because of the property of the inhomogeneous term in the integral equation for C^ν, it follows that asymptotically C^ν has the spinor form

$$C_\nu = -C_{\nu\lambda}\gamma^\lambda + \varepsilon_{\nu\alpha\beta\lambda}q^\alpha p^\beta \gamma^\lambda \gamma_5 C; \tag{82}$$

we see then that

$$V(p^2)\,(g^{\mu\nu}q^2 - q^\mu q^\nu) \simeq 3\,\langle C^{\mu\nu}\rangle. \tag{83}$$

When we investigate the integral equation we find that only two amplitudes are coupled, namely

$$\langle C^{\mu\nu}\rangle \qquad \text{and} \qquad \langle C^{\mu\lambda}\, d_\lambda{}^\nu(p)\rangle, \tag{84}$$

where

$$d_\lambda{}^\nu(p) = 4\,\frac{p_\lambda p^\nu}{p^2} - g_\lambda{}^\nu. \tag{85}$$

We find we can also put

$$\langle C^{\mu\lambda} d_\lambda{}^\nu\rangle = W(p^2)\,(g^{\mu\nu}q^2 - q^\mu q^\nu). \tag{86}$$

Then the integral equations have the form, for $\alpha_0 \ll 1$,

$$\begin{aligned} V(p^2) &= \frac{1}{p^2} + \int dq^2 (k(p^2, q^2)\, V(q^2) + l(p^2, q^2)\, W(q^2)),\\ W(p^2) &= \frac{W_0}{p^2} + \int dq^2 (m(p^2, q^2)\, V(q^2) + n(p^2, q^2)\, W(q^2)), \end{aligned} \tag{87}$$

where k, l, m, n, can be explicitly calculated as power series in α_0.

Now it is easy to show that the kernels l and n vanish more rapidly than $1/p^2$ (like $1/p^2\, q^2/p^2$ for example) for $p^2 \gg q^2$. The reason is that the amplitude W governs the "d" wave and when we integrate over angles $\langle d_{\mu\nu}(p)\rangle = 0$. It is a fact that only the kernels k, m, which come from the "s" wave amplitude V can vanish as slowly as $1/p^2$ for $p^2 \gg q^2$.

Now, it turns out that because of the fact that the asymptotic kernel $\bar{K}$ produces no log divergence in Z_1 (i.e. in the vertex), that it is guaranteed that k also must vanish more rapidly than $1/p^2$ as $p^2 \gg q^2$. No such requirement is imposed on m however. Nevertheless the *singular part of the* W term never couples directly back to V because of a special property of the kernel l which is guaranteed by renormalizability. Thus, we find because of the property of renormalizability, the asymptotic form of $V(p^2)$ is not altered from $1/p^2$ no matter how far out we go in the series for the kernel functions.

We now come to an important qualification. When we look into the "6th order" asymptotic kernel we find a new class of diagrams.

These have the form

In perturbation theory they produce the first "closed loop" corrections to Γ_μ

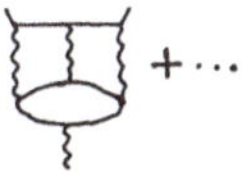

Now it is usually stated in most places that the scattering of light by light diagrams

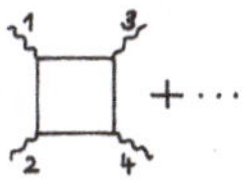

(88)

add up to a finite gauge invariant expression. However, it is the actual case that they add up to a finite but not gauge invariant expression. Thus, the same gauge invariance problem which arises in the case of

 persists when we proceed to fourth order, but in a less dramatic fashion. To see why this happens we can argue as follows. If (88) is gauge invariant it should vanish if we put $k_1 = 0$, for example.

When we put $k_1 = 0$ we can replace

(89)

by $(\partial/\partial q^{\mu_1})\, S(q)$ using Ward's identity. In this case, we easily find that the 6 terms which occur in (88) can be written

$$M_{\mu_1}(0, \ldots) \sim \int \frac{(dq)}{(2\,\pi)^4} \frac{\partial}{\partial q^{\mu_1}} \qquad (90)$$

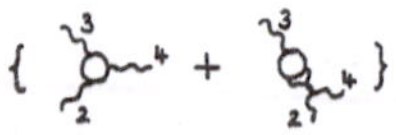

Thus, M is expressed as a "surface term". It is then clear since only $q \to \infty$ contributes to the surface term we can effectively put $k_3 = k_4 = k_2 = 0$ to find

$$\sim \int \frac{(dq)}{(2\,\pi)^4} \frac{\partial}{\partial q^{\mu_1}} \frac{\partial}{\partial q^{\mu_2}} \frac{\partial}{\partial q^{\mu_3}} \operatorname{tr}\,(\gamma^{\mu_4} S(q)) = \qquad (91)$$

$$= (\text{const}) \cdot (g^{\mu_1 \mu_2} g^{\mu_3 \mu_4} + \ldots).$$

Thus, as promised, M is not gauge invariant. Now just as in the case of the loop  the non-gauge invariant part of (88) is a surface term in momentum space, and just as in that case the line integral definition of the current removes such terms. Therefore it should be no surprise that this also occurs in the case of the scattering of light by light diagrams. Consequently, when we use our careful definition of the current, the integral equation for Γ^ν has present in it a subtraction in the "6th order" term in the integral equation so that we actually find

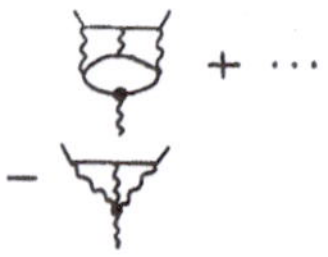

where the subtraction has the form given in (91) with the *exact* $S(q)$ function. However, since the exact $S(q)$ functions is $1/\gamma\, q$ asymptotically and only the asymptotic part of $S(q)$ contributes to the subtraction, the effect is the same as in (91).

It is easy to see that if the subtraction given by gauge invariance were not made then there would be an additional renormalization required on the vertex since the integral

is logarithmically divergent. Thus, we can see that the first time that the asymptotic kernel can produce a logarithmic divergence in Z_1 occurs in the 6th order, but that in reality no such difficulty occurs because of the explicit appearance of a subtraction dictated by gauge invariance which removes the divergence.

Let us now consider the perturbation version [16] of the calculation of the contribution in 8th order of the diagrams

to vacuum polarization. In our present formalism we would find for the relevant 6th order vertex correction

where  indicates the subtracted vacuum loop which is gauge invariant. If we insert this expression into the formula (66) for ϱ we would get

 (92)

It is to be noted that the formalism so far developed does *not* provide a subtraction from the loop on the left. We see that (92) can also be written

 (93)

and this is bad because unitarity with respect to three photon intermediate states would tell us that only the first term in (93) should be present. Consequently, this means that the formula (66) fails when used in perturbation theory in the eighth order.

Now why is this bad? It is the property of the first term in (93) that it is proportional to $\log \lambda^2$ where as the second term behaves like $(\log \lambda^2)^2$. In solving our integral equation for C it is the presence of higher powers of log which would allow us to obtain a finite result for $\varrho(0)$. It is then clear why an objection to the use of the formulation of the theory we have presented is important. In any case it is clear that equation (66) should not be used without careful modification in a renormalized perturbation theory analysis of the theory.

It is still possible however that this objection need not be valid when applied to the non-perturbative treatment.

That is, let us imagine that we compute the solution to (72) first, and then use it in (66). In this case, if the asymptotic behaviour of the solution is improved over the perturbation theory, the formal subtraction which is apparently necessary at the "other end" of the photon self-energy will give no contribution. Hence, as a consequence, we would see a distinction which would appear between a graph by graph analysis of the photon self-energy, and the non-perturbative approach developed here.

Naturally these points are somewhat intricate and require a deeper understanding of the theory than we have at the present.

Note added in proof:

The relation (66) has been shown to be false. If we use the careful definition of the current defined as the limit of the gauge invariant non-local current then an additional factor of the form

\+ 

contributes to (66). It has then been shown that *no* higher powers of log λ^2 appear anywhere in (66). As a consequence the general form of $\varrho(0)$ is $\varrho(0) = f(\alpha_0) \log(\lambda^2)$ where $f(\alpha_0)$ is a power series in α_0. Consequently, the theory can be only finite if $f(\alpha_0) = 0$, that is, for particular values of α_0. This conclusion is the same as that reached by the authors of reference [11].

References

1. K. Johnson, M. Baker, and R. Willey, Phys. Rev. **136**, B 1111 (1964), Phys. Rev. Lett. **11**, 518 (1963).
2. J. D. Björken, Ann. Phys. (N. Y.) **24**, 174 (1963).
3. M. Gell-Mann and F. Zachariasen, Phys. Rev. **124**, 953 (1961).
4. I. Bialynicky-Birula, Phys. Rev. **130**, 465 (1963).
5. This is a trivial generalization of the original Fermi Lagrangian (E. Fermi, Rev. Mod. Phys. **4**, 87 (1932)).
6. F. J. Dyson, Phys. Rev. **75**, 1736 (1949).
7. J. Schwinger, Proc. Nat. Acad. Sci. **37**, 452 and 455 (1951).
8. J. Ward, Phys. Rev. **78**, 182 (1950); Phys. Rev. **84**, 897 (1951).
9. Y. Takahashi, Nuovo Cim. **6**, 371 (1957).
10. B. Zumino, Nuovo Cim. **17**, 547 (1960).
11. M. Gell-Mann, F. Low, Phys. Rev. **95**, 1300 (1954).
12. Landau, Abrikosov, Kalatnikov, Nuovo Cim. Suppl. **3**, 80 (1956).
13. M. Baker, K. Johnson, B. Lee, Phys. Rev. **133**, B 209 (1964).
14. K. Johnson, Phys. Lett. **5**, 253 (1963).
15. The material described in this section is based upon the unpublished work of M. Baker, K. Johnson, H. Mitter and R. Willey.
16. This difficulty was pointed out to us by K. Wilson (private communication).

Review of Consistency Problems in Quantum Electrodynamics*

By

Gunnar Källén

Department of Theoretical Physics, University of Lund, Lund, Sweden

1. Introduction

The question of the consistency of quantized field theories in general and quantum electrodynamics in particular has been the subject of several papers during the last 15 years. In the literature one can find controversial statements ranging from the claim that any renormalized field theory is completely meaningless[1] to the assertion that even unrenormalized quantum electrodynamics has an exact solution with well behaved physical properties.[2] In view of this rather controversial situation, we find it of some interest to review the actual situation in this field. As can perhaps be expected, our conclusion is that the truth lies somewhere in between the two extreme opinions quoted above. One of the corner stones in our discussion is concerned with an argument published more than 10 years ago with the result that at least one of the renormalization constants in ordinary quantum electrodynamics

---

* Lecture given at the IV. Internationale Universitätswochen für Kernphysik, Schladming, 25 February—10 March 1965.

[1] This statement was published in the middle 1950's by L. D. Landau and col aborators. The original papers here are in Russian, but a convenient summary can be found in the article by L. D. Landau "On the quantum theory of fields" in "Niels Bohr and the development of physics". Pergamon Press, London 1955.

[2] K. Johnson, M. Baker, R. S. Willey, Phys. Rev. Lett. **11**, 518 (1963). At the end of this paper one finds the statement: "In conclusion, we find that quantum electrodynamics may be regarded as a perfectly consistent theory. The usual divergences from our point of view arise from an unjustified use of perturbation theory..." In a later paper by the same authors, Phys. Rev. **136** B, 1111 (1964) this somewhat aggressive formulation is considerably modified. To be impartial, we quote also the summary of this last reference. It reads: "A perturbation theory is developed within the usual formalism of quantum electrodynamics which yields a finite unrenormalized electron Green's function and a finite value for the electron electromagnetic self mass in each order. This is subject only to the qualifications in this paper, that the vacuum polarization is also obtained without divergences. Furthermore, the bare mass of the electron vanishes; the electron mass must be totally dynamical in origin."

is infinite.[3] For some time this result was accepted by most research workers in the field. However, later it has been criticized because of its lack of rigour.[4]

We feel very strongly that this criticism is largely based on a misunderstanding both of the mathematical rigour intended by the original work[5] as well as of the actual calculations. In particular, we feel that the specific complaint put forward by GASIOROWICZ et al. is irrelevant, as the particular limit which these authors are concerned with is actually discussed in the original paper.[6]

However, and quite independent of this published criticism, we freely admit that the actual mathematical rigour of the argument is not very high. Therefore, it is not logically excluded that a rather singular solution of the equations with finite renormalization constants could exist where certain formal interchanges of orders of integration etc. would not be allowed. Further, it should be remarked that the formal tools in this field have developed quite extensively since the days when the original paper in ref 3 was written. Therefore, it may be of some interest to review also these formal developments and to return to the old discussion and see, to what extent it can be simplified and made more transparent with the aid of techniques, developed more recently. At the same time, the mathematical rigour of the argument may be slightly improved, but that is not our main concern.

The material presented in these lectures is partly contained in an unpublished CERN manuscript by the author from 1955,[7] partly included in a lecture given at a CERN meeting in 1956[8] and is partly based on material which is supposed to be included in a supplement paper to the old Handbuch article[9].

2. Spectral Representations

For our purposes it is convenient to use a formulation of field theories in general and quantum electrodynamics in particular in terms of spectral representations, especially for the two point functions of

---

[3] G. KÄLLÉN, Dan. Mat. Fys. Medd. **27**, Nr. 12 (1953). A short summary of the argument is included in G. Källén. Quantenelektrodynamik, Handbuch der Physik V_1 (1958).

[4] S. G. GASIOROWICZ, D. R. YENNIE, H. SUURA, Phys. Rev. Lett. **2**, 153 (1959). The remarks of these authors are uncritically repeated in the textbook by S. S. SCHWEBER "An introduction to quantum field theory" (1961) esp. p. 683. Other objections of a slightly different nature have been raised by K. JOHNSON, Phys. Rev. **112**, 1367 (1958). Cf. footnote 26 below.

[5] Cf. esp. the concluding remarks in the first paper quoted in ref. 3.

[6] Cf. esp. pp. 11 and 12 in the first paper in ref. 3.

[7] CERN/T/GK/3 Nov. 1955.

[8] Cf. Proc. of the CERN Symposion on High Energy Accelerators and Pion Physics, June 1956, p. 187.

[9] To be published in Ergebn. d. exakten Naturwissenschaften.

the theory. As part of the material we are going to use is nowadays reasonably well-known, we here limit ourselves to writing down some of the basic formulae and refer the reader to the original literature for a derivation of these expressions.[10] One of the fundamental formulae in our discussion is the following representation for the vacuum expectation value of a product of two electromagnetic current operators

$$\langle 0|\, j_\mu(x)\, j_\nu(x')\, |0\rangle = \frac{1}{(2\pi)^3}\int dp\, e^{ip(x-x')}(p_\mu p_\nu - \delta_{\mu\nu} p^2)\, \Pi(p^2)\, \theta(p)$$

$$= i\int_0^\infty da\, \Pi(-a)\left(\delta_{\mu\nu}\Box - \frac{\partial^2}{\partial x_\mu\, \partial x_\nu}\right)\Delta^{(+)}(x-x', a), \qquad (2.1)$$

$$\Delta^{(+)}(x, a) = \frac{-i}{(2\pi)^3}\int dp\, e^{ipx}\, \delta(p^2+a)\, \theta(p) = \lim_{\varepsilon\to 0}\left[\frac{-a}{8\pi}\frac{H_1^{(1)}(\sqrt{a\,z})}{\sqrt{a\,z}}\right], \qquad (2.1\text{ a})$$

$$z = (x_0 - i\,\varepsilon)^2 - \bar{x}^2; \qquad \varepsilon > 0; \qquad \mathrm{Im}\,(\sqrt{a\,z}) > 0, \qquad (2.1\text{ b})$$

$$\Pi(p^2) = \frac{V}{-3p^2}\sum_{p^{(n)}=p}\langle 0|\, j_\mu\, |n\rangle\langle n|\, j_\mu\, |0\rangle. \qquad (2.1\text{ c})$$

The weight function $\Pi(p^2)$ here is one of our basic quantities. The detailed shape of this function is characteristic for the interaction we are studying, i.e. for quantum electrodynamics. The general shape of the representation (2.1) follows from standard assumptions about Lorentz invariance and the mass spectrum conditions in the theory, i.e. the assumption that the energy momentum vector of every physical state is time like with a positive energy, as well as the existence of a non-degenerate state, the vacuum, which has zero energy and zero momentum. The factor $p_\mu p_\nu - \delta_{\mu\nu} p^2$ appears because of the continuity equation for the current operator. Even if it is not perfectly evident from the formulae given above, the weight function $\Pi(p^2)$ is positive definite and can be written as a sum of positive definite contributions.[11] Consequently, we can obtain a lower bound to the function $\Pi(p^2)$ by

---

[10] Historically, spectral representations of the kind used here and for the particular case of quantum electrodynamics were first published by H. Umezawa and S. Kamefuchi, Progr. of Theor. Phys. **6**, 543 (1951). The particular notation which we are using was introduced in G. Källén, Helv. Phys. Acta **25**, 417 (1952). A few years later (and without any claim of priority) a related discussion was also given by M. Gell-Mann and F. Low in Phys. Rev. **95**, 1300 (1954). About simultaneously with this last paper appeared a paper by H. Lehmann in Nuovo Cim. **11**, 342 (1954). This last paper is widely missquoted as being the first place where spectral representations were introduced.

[11] G. Källén, Helv. Phys. Acta **25**, 417 (1952) and Handbuch der Physik V_1 (1958).

including only a few states in the sum over intermediate states in Eq. (2.1 c). We find it convenient to use the following estimate

$$\Pi(p^2) \geqq \frac{V}{-3p^2} \sum_{q+q'=p} \langle 0| j_\mu |\bar{q}, \bar{q}'\rangle \langle \bar{q}, \bar{q}'| j_\mu |0\rangle. \tag{2.2}$$

Here, the symbol $|\bar{q}, \bar{q}'\rangle$ denotes a state with one positron with three-dimensional momentum $\bar{q}'$ and one electron with three-dimensional momentum $\bar{q}$. In passing, we note that the states considered in Eq. (2.2) are exactly those states, which contribute in a first order perturbation theory calculation of the function $\Pi(p^2)$. Actually, we have the following expression for the first non-vanishing contribution $\Pi^{(0)}(p^2)$ in perturbation theory

$$\Pi^{(0)}(p^2) = \frac{V}{-3p^2} \sum_{q+q'=p} \langle 0| j_\mu^{(0)} |\bar{q}, \bar{q}'\rangle \langle \bar{q}, \bar{q}'| j_\mu^{(0)} |0\rangle =$$

$$= \frac{1}{3p^2}\frac{e^2}{V} \sum_{q+q'=p} \sum_{\text{spin}} \bar{u}^{(-)}(-\bar{q}')\gamma_\mu u^{(+)}(\bar{q})\, \bar{u}^{(+)}(\bar{q})\gamma_\mu u^{(-)}(-\bar{q}') =$$

$$= \frac{\alpha}{3\pi}\left(1 - \frac{2m^2}{p^2}\right)\sqrt{1 + \frac{4m^2}{p^2}}\,\theta(-p^2 - 4m^2); \qquad p^2 = \bar{p}^2 - p_0^2. \tag{2.3}$$

The physical meaning of the function $\Pi(p^2)$ is given by the fact that it appears as the imaginary part of "the dielectric constant of the vacuum" in the phenomenon of vacuum polarization. From standard arguments of causality then follows that the real part of the same dielectric constant can be expressed in terms of a Hilbert transform of the function $\Pi(p^2)$. If we perform the standard renormalization procedure which corresponds to normalizing the dielectric constant of the vacuum to one for zero "frequency", we find the following expression for the vacuum expectation value of the current operator in a system with a very weak external field described by the external current $j_\mu^{\text{ext}}(p)$:

$$\langle 0|\delta j_\mu(x)|0\rangle = \frac{1}{(2\pi)^4}\int dp\, e^{ipx}\,[1 - \Pi_R(p^2) + \Pi_R(0)]\, j_\mu^{\text{ext}}(p), \tag{2.4}$$

$$\Pi_R(p^2) = \int_0^\infty \frac{\Pi(-a)\,da}{(a+p^2)_R} \equiv \int_0^\infty \frac{\Pi(-a)\,da}{(a+p^2)_P} + i\pi\,\Pi(p^2)\,\varepsilon(p) \equiv$$

$$\equiv \overline{\Pi}(p^2) + i\pi\,\Pi(p^2)\,\varepsilon(p). \tag{2.4 a}$$

As is obvious in Eq. (2.4), the dielectric constant inside the square bracket has been normalized in such a way that it is identically equal to one for $p = 0$. This is achieved by the last term $\Pi_R(0) = \overline{\Pi}(0)$. Therefore, this last term corresponds to the charge renormalization constant, while the observable quantity is given by the difference of the two terms in Eq. (2.4) or by

$$\Pi_R(p^2) - \bar{\Pi}(0) = -p^2 \int_0^\infty \frac{da\,\Pi(-a)}{a(a+p^2)_R}, \tag{2.5 a}$$

$$\bar{\Pi}(0) = \int_0^\infty \frac{da}{a}\Pi(-a) = \text{charge renormalization constant.} \tag{2.5 b}$$

These last two formulae exhibit clearly how the "high energy behaviour", i.e., the behaviour of the function $\Pi(p^2)$ for large values of the mass square $-p^2$ is significant for the finiteness of the theory. If the function $\Pi(p^2)$ behaves in such a way that the integral in Eq. (2.5 b) is finite, we conclude that our formalism has a finite charge renormalization. As is well-known, the perturbation theory expression (2.3) as well as all the higher order terms behave in such a way that, in a term by term expansion, the integral in Eq. (2.5 b) is divergent while the integral in Eq. (2.5 a) converges. For the purpose of our discussion here we use that a necessary condition for the renormalized theory to be consistent and finite is that the integral appearing in Eq. (2.5 a) is actually convergent when the exact function $\Pi(p^2)$ is used. Because of the positive definiteness of the function $\Pi(p^2)$ this condition implies that the integral in Eq. (2.5 a) should also be convergent when the lower bound exhibited in Eq. (2.2) is substituted under the integral sign on the right hand side. Corresponding statements hold for the integral in Eq. (2.5 b) if we want to consider the case when not only the renormalized version of the theory is finite and consistent but if we also assume that the charge renormalization described by the formalism corresponds to a finite factor.

For our discussion it is sometimes convenient to write the spectral representations in a slightly different form. If, for the moment, we disregard questions about convergence or, rather, assume that the charge renormalization factor is finite, we can consider the expression

$$H(z) = \int_0^\infty \frac{da\,\Pi(-a)}{a-z}, \tag{2.6}$$

as an analytic function of a complex variable z

$$z = (p_0 + i\,\varepsilon)^2 - \bar{p}^2 \sim -p^2. \tag{2.6 a}$$

Evidently, the function $H(z)$ is regular analytic in the whole complex z-plane except on the positive real axis. Further, every point which is a zero of the function $H(z)$ must lie on the positive real axis. This last statement follows from the positive definitness of the weight function $\Pi(-a)$. For the proof we first of all remark that a real negative value of z in Eq. (2.6) yields an integrand which is positive definite everywhere and, therefore, the function H can never vanish for such values of z. Further, a complex point $z = x + i\,y$ gives the following inequality for the imaginary part of the function $H(z)$:

$$\operatorname{Im} H(z) = y \int_0^\infty \frac{da\, \Pi(-a)}{(a-x)^2 + y^2} \neq 0 \qquad \text{for} \qquad y \neq 0. \tag{2.7}$$

We conclude that only positive real values of z can be zeroes of the function $H(z)$. Next, we remark that it follows from this property that the inverse of the function $H(z)$ is also regular analytic everywhere in the complex z-plane except on the positive real axis. Further, we observe that the function $H(z)$ can never vanish faster than $1/z$ for large values of $|z|$. This is again a consequence of the positive definiteness of the function $\Pi(-a)$. The exact condition for the function $H(z)$ to behave in a certain way at infinity is rather involved if expressed in a rigorous mathematical way. However, intuitively, we can say that if $|z|$ is very large, we can neglect a compared to z in the denominator of Eq. (2.6) and get the intuitive asymptotic behaviour of $H(z)$

$$H(z) \sim -\frac{1}{z} \int_0^\infty da\, \Pi(-a). \tag{2.8}$$

The behaviour indicated in Eq. (2.8) is such that the inverse of the function $H(z)$ is not conveniently bounded for large values of $|z|$ and, therefore, we cannot apply the standard representation technique with a Cauchy integral to this function.

The argument above is essentially correct provided that the integral which appears on the right-hand side of Eq. (2.8) is convergent. However, there are practically important cases where the integral in Eq. (2.8) is divergent while the integral in Eq. (2.6) is convergent. Under those circumstances the function $H(z)$ vanishes slower than $1/z$ for large values of $|z|$ but its inverse may still be unbounded. In any case the requirement that the integral on the right hand side of Eq. (2.8) should be convergent is somewhat foreign to our main philosophy. Therefore, we try to make the intuitive ideas sketched above a little more precise and consistent with our basic assumptions by requiring that only the integral in Eq. (2.6) is convergent. Then, we do not consider the function $H(z)$ itself but rather the expression

$$H_1(z) = 1 - H(z) + H(0) = 1 - z \int_0^\infty \frac{da\, \Pi(-a)}{a(a-z)}. \tag{2.9}$$

We note, in passing, that the right-hand side of Eq. (2.9) is essentially the "dielectric constant of the vacuum" appearing on the right-hand side of Eq. (2.4) but extended to an analytic function of the complex number z. As the imaginary parts of the functions $H_1(z)$ and $H(z)$ are the same, the inequality (2.7) also holds for the imaginary part of $H_1(z)$ and we conclude that the function $H_1(z)$ has no complex zeroes either. Further, its asymptotic behaviour for large values of z is given by

$$H_1(z) = 1 + \int_0^\infty \frac{da\,\Pi(-a)}{a(1-a/z)} \approx 1 + \int_0^\infty \frac{da\,\Pi(-a)}{a} = 1 + H(0) > 1$$

$$\text{for} |z| \to \infty. \tag{2.10}$$

Further, it also follows from the last form of the right-hand side of Eq. (2.9) that $H_1(z)$ is never zero for negative real values of z. Collecting all the statements above we find that the expression

$$H_1{}^*(z) = \frac{1}{H_1(z)} = \frac{1}{1 - H(z) + H(0)}, \tag{2.11}$$

is analytic and bounded everywhere in the complex z-plane except on the positive real axis. Using standard techniques with the Cauchy integral and taking explicit account of the fact that $H_1{}^*(0)$ is equal to 1 we find

$$H_1{}^*(z) = 1 + z \int_0^\infty \frac{da\,\Pi^*(-a)}{a(a-z)} = 1 + H^*(z) - H^*(0), \tag{2.12}$$

$$\Pi^*(-a) = \frac{1}{2\pi i}\left[H_1{}^*(a+i\varepsilon) - H_1{}^*(a-i\varepsilon)\right] =$$

$$= \frac{\Pi(-a)}{[1 - \overline{\Pi}(-a) + \overline{\Pi}(0)]^2 + \pi^2 \Pi^2(-a)}, \tag{2.12 a}$$

$$H^*(z) = \int_0^\infty \frac{da\,\Pi^*(-a)}{a-z}. \tag{2.12 b}$$

The new function $H^*(z)$ defined here has several interesting properties. First of all, according to Eq. (2.12 a) and the assumptions we have made about the convergence of the integral (2.9), the function $\Pi^*(-a)$ behaves in such a way that the integral in Eq. (2.12 b) is convergent. Consequently, taking the limit of very large values for the absolute value of z we find from Eqs. (2.11) and (2.12)

$$H_1{}^*(\infty) = \frac{1}{1 + H(0)} = 1 - H^*(0), \tag{2.13}$$

or

$$\int_0^\infty \frac{da}{a}\Pi^*(-a) = H^*(0) = \frac{H(0)}{1+H(0)} =$$

$$= \int_0^\infty \frac{da}{a}\Pi(-a)\left[1 + \int_0^\infty \frac{da}{a}\Pi(-a)\right]^{-1} \leqq 1. \tag{2.13 a}$$

The result exhibited in this last relation is quite interesting because it shows that also in the limit when the charge renormalization constant in Eq. (2.9) is divergent, the corresponding integral over the function $\Pi^*(-a)$ is convergent and approaches 1. Therefore, both when we have a finite charge renormalization and when we have an infinite charge renormalization in the theory, the integral appearing in Eq. (2.13 a) is convergent.[12]

We close this section by the remark that the construction of something which is essentially the inverse of an analytic function regular analytic in the whole complex plane except the positive real axis can be used with advantage to "divide out" a factor corresponding to the function $H_1(z)$ from any given analytic function $F(z)$ which is regular analytic in a domain which does *not* include the positive real axis. For every given function $F(z)$ of this kind we can define a new analytic function $G(z)$ which is regular analytic in the same domain as $F(z)$ by the following construction

$$G(z) = H_1^*(z)\, F(z) = \frac{F(z)}{1 - H(z) + H(0)}\,. \tag{2.14}$$

For the particular case that $F(z)$ is also regular analytic in the whole complex z-plane except the positive real axis and sufficiently bounded at infinity, it has the representation

$$F(z) = \int\limits_0^\infty \frac{da\, f(a)}{a - z}\,, \tag{2.15}$$

which implies that the function $G(z)$ has a similar representation

$$G(z) = \int\limits_0^\infty \frac{da\, g(a)}{a - z}\,, \tag{2.16}$$

with a weight $g(a)$ which is given by the expression

$$g(a) = \Pi^*(-a) \int\limits_0^\infty \frac{db\, f(b)}{(b-a)_P} + f(a)\,[1 + \overline{\Pi}^*(-a) - \overline{\Pi}^*(0)], \tag{2.16 a}$$

$$\overline{\Pi}^*(x) = \int\limits_0^\infty \frac{da\, \Pi^*(-a)}{(a + x)_P}\,. \tag{2.16 b}$$

This last construction is used in the argument below.

[12] To make this statement into an exact mathematical result the argument above has to be considerably amplified. We do not insist on the details as the reader should be able to work them out for himself along the lines indicated here, assuming the convergence of the integral (2.10) but not of the integral (2.9). Further, it can perhaps be remarked that, in the language of perturbation theory, this introduction of the function Π^* corresponds to the transition from "self energy graphs" to "proper self energy graphs".

3. A Particular Matrix Element of the Current Operator

We have already mentioned above that it is possible to construct a lower bound to the function $\Pi(p^2)$ by using only the particular matrix elements $\langle 0| j_\mu |\bar{q}, \bar{q}'\rangle$ in Eq. (2.2). Therefore, it is of some interest to consider the formal structure of these matrix elements in some more detail. We first remark that standard arguments of Lorentz invariance and charge conservation imply that this matrix element can be expressed in terms of two scalar functions $F_1(p^2)$ and $F_2(p^2)$

$$\langle 0| j_\mu |\bar{q}, \bar{q}'\rangle = \frac{i\,e}{V}\,\bar{u}^{(-)}(-\bar{q}')\left[\gamma_\mu F_1(p^2) + \frac{1}{2m}\,p_\nu\,\sigma_{\mu\nu}\,F_2(p^2)\right]u^{(+)}(\bar{q}), \quad (3.1)$$

$$p = q + q'.^{13} \quad (3.1\text{ a})$$

The values at the origin of the two functions F_1 and F_2 correspond to the static charge and the static anomalous magnetic moment of the electron, respectively. Therefore, we have

$$F_1(0) = 1, \quad (3.2\text{ a})$$

$$F_2(0) = \frac{\alpha}{2\pi} - 0.328\,\frac{\alpha^2}{\pi^2} + \mathcal{O}(\alpha^3). \quad (3.2\text{ b})$$

In an actual calculation of these functions the normalization condition (3.2a) does not come out automatically, but has, essentially, to be achieved by an explicit renormalization procedure. In practice this means that the unrenormalized function which is obtained directly is renormalized by the subtraction of its value at the origin. Therefore, we have

$$F_1^{\text{ren}}(p^2) = F_1^{\text{unren}}(p^2) - F_1^{\text{unren}}(0) + 1. \quad (3.3)$$

We are going to refer to the term $F_1^{\text{unren}}(0)$ as one of the renormalization constants in quantum electrodynamics.[14] No similar renormalization is necessary for the function $F_2(p^2)$ and the finite value given in Eq. (3.2b) comes out from a straight forward application of perturbation theory to the formalism.

It can be proved that the two functions $F_i(p^2)$ are both analytic functions of the variable $z = -p^2$, but the exact shape of the regularity domain of these functions is unknown. Perturbation theory experience here leads us to expect that the analyticity domain of these two func-

---

[13] Cf. the well-known situation in the theory of the electromagnetic form factors of the nucleon. We have here adopted the same notational convention which is used in form factor discussions. The expression F_1 is sometimes referred to as the "Dirac form factor" and F_2 is normally referred to as the "Pauli form factor". In the papers mentioned in ref. 3, linear combinations of these two functions were denoted by $\bar{R}(p^2)$ and $\bar{S}(p^2)$.

[14] This constant is actually a combination of the conventional renormalization constants. Note also the similarity of the algebraic structure of Eq. (3.3) and Eq. (2.4).

tions should be the same as the analyticity domain for the function $H(z)$ in Eq. (2.6). However, one can prove by explicit counterexamples that this large analyticity domain does not follow from the standard assumptions of Lorentz invariance, causality and a reasonable mass spectrum[15].

However, it must be admitted that the three general assumptions just mentioned do not in any way exhaust all the physical properties of the theory. First of all, they make no reference to the very intricate relations between various quantities in a field theory which are sometimes referred to as "generalized unitarity". Further, the general assumptions just mentioned make no allowance for the particular properties which may follow from the specific interaction one is considering in quantum electrodynamics. Also, the fact that the large analyticity domain just mentioned comes out in every order in perturbation theory is rather suggestive. Due to these circumstances it is very often assumed both in practical applications of the theory and in discussions of principle that the functions $F_i(p^2)$ can be extended to analytic functions of the kind just mentioned with the representations

$$F_1(p^2) = 1 - p^2 \int_0^\infty \frac{da\, f_1(-a)}{a(a+p^2)} = 1 + \int_0^\infty \frac{da\, f_1(-a)}{a+p^2} - \int_0^\infty \frac{da}{a} f_1(-a), \tag{3.4 a}$$

$$F_2(p^2) = \int_0^\infty \frac{da\, f_2(-a)}{a+p^2}. \tag{3.4 b}$$

A comparison between Eqs. (**3.4** a) and (**3.3**) shows that the renormalization constant mentioned above can be expressed in terms of the weight functions $f_1(-a)$ by the formula

$$F_1^{\text{unren}}(0) = \int_0^\infty \frac{da}{a} f_1(-a). \tag{3.4 c}$$

As the representations (**3.4**) are extremely convenient and transparent, we shall for the rest of this section and as a preliminary calculation explicitly assume that they are correct. By doing this, we intend to simplify the formal calculations and prepare the ground for a more satisfactory argument to be given in the next section.[16]

The basic technique of the argument in ref. **3** was to use as a starting assumption that all the renormalization constants of quantum electrodynamics were actually finite numbers. By drawing rather heavily on this assumption, we were then led to a contradiction of the form that

---

[15] Cf. R. Jost, Helv. Phys. Acta **31**, 263 (1958).

[16] This simplifying assumption is also made, e.g., in the first paper of ref. 4.

the integral appearing in the charge renormalization constant (2.5 b) could not converge. This contradiction proves that there must be something wrong somewhere in the argument and, assuming that what is wrong is not some obscure point of epsilontics which could make the formal manipulations actually made unreliable, it follows that it is the starting assumption which must be wrong. Therefore, one concludes that at least one of the renormalization constants in electrodynamics must be infinite. It was, however, not possible to make any claim as to whether it was the charge renormalization constant itself or some other integral, the convergence of which was needed for the argument, which actually did diverge. Evidently, it is also possible that all of the renormalization constant integrals are divergent. Intuitively, we should rather expect this latter situation to hold. We now try to repeat this old argument but using the strongly simplifying assumptions mentioned above. Using the lower bound in Eq. (2.2) together with the explicit calculation in (2.3) and the general form (3.1) of the matrix elements we are considering we get

$$\Pi(p^2) \geqq \frac{e^2}{3\,p^2}\frac{1}{V}\sum_{q+q'=p}\;\sum_{\text{spin}} \bar{u}^{(-)}(-\bar{q}')\left[\gamma_\mu F_1(p^2) + \frac{1}{2\,m}\,p_\nu\,\sigma_{\mu\nu}\,F_2(p^2)\right] \times$$

$$\times\, u^{(+)}(\bar{q})\,\bar{u}^{(+)}(\bar{q})\left[\gamma_\mu F_1{}^*(p^2) - \frac{1}{2\,m}\,p_\lambda\,\sigma_{\mu\lambda}\,F_2{}^*(p^2)\right] u^{(-)}(-\bar{q}') =$$

$$= \frac{e^2}{3\,p^2}\frac{1}{(2\,\pi)^3}\int\!\!\int dq\,dq'\,\delta(q^2+m^2)\,\delta(q'^2+m^2)\,\theta(q)\,\theta(q')\,\delta(p-q-q') \times$$

$$\times\,\mathrm{Sp}\left[(i\,\gamma\,q - m)\left[\gamma_\mu F_1{}^*(p^2) - \frac{1}{2\,m}\,p_\lambda\,\sigma_{\mu\lambda}\,F_2{}^*(p^2)\right](i\,\gamma\,q' + m)\,\times\right.$$

$$\left.\times\left[\gamma_\mu F_1(p^2) + \frac{1}{2\,m}\,p_\nu\,\sigma_{\mu\nu}\,F_2(p^2)\right]\right]. \qquad (3.5)$$

The trace appearing here can be computed by straight forward and elementary techniques. One finds

$$\mathrm{Sp}\,[\ldots] = 4\,p^2\left\{\left(1 - \frac{2\,m^2}{p^2}\right)|F_1+F_2|^2 + \left(1 + \frac{4\,m^2}{p^2}\right)\mathrm{Re}\,[F_2{}^*(F_1+F_2)] - \right.$$

$$\left. - \frac{p^2}{8\,m^2}\left(1 + \frac{4\,m^2}{p^2}\right)^2 |F_2|^2\right\}. \qquad (3.6)$$

Insertion of this result in Eq. (3.5) gives

$$\Pi(p^2) \geqq \frac{\alpha}{3\,\pi}\sqrt{1 + \frac{4\,m^2}{p^2}}\,\theta(-p^2 - 4\,m^2)\left\{\left(1 - \frac{2\,m^2}{p^2}\right)|F_1+F_2|^2 + \right.$$

$$\left. + \left(1 + \frac{4\,m^2}{p^2}\right)\mathrm{Re}\,[F_2{}^*(F_1+F_2)] - \frac{p^2}{8\,m^2}\left(1 + \frac{4\,m^2}{p^2}\right)^2 |F_2|^2\right\}. \qquad (3.7)$$

The inequality (3.7) is supposed to be exact. Next, we make use of the representations (3.4) and the same argument as indicated in Eq. (2.10) to give the estimate

$$\lim_{-p^2\to\infty} F_1(p^2) = 1 - \int_0^\infty \frac{f_1(-a)}{a}\, da = 1 - F_1^{\text{unren}}(0), \qquad (3.8\text{ a})$$

$$\lim_{-p^2\to\infty} F_2(p^2) = 0. \qquad (3.8\text{ b})$$

Inserting these results in Eq. (3.7) we find the following lower bound for the asymptotic behaviour of the function $\Pi(p^2)$

$$\lim_{-p^2\to\infty} \Pi(p^2) \geqq \frac{\alpha}{3\pi}\,[1 - F_1^{\text{unren}}(0)]^2. \qquad (3.9)$$ [17]

The results exhibited in Eq. (3.9) can now be substituted in the integral (2.5b) for the charge renormalization constants. Evidently, if follows that this integral cannot be convergent unless the right-hand side of Eq. (3.9) vanishes or we have

$$F_1^{\text{unren}}(0) = 1. \qquad (3.10)$$

In this way we see that the assumption that all renormalization constants are finite in the formalism, leads to the definite relation (3.10) as a necessary self-consistency condition. In passing it may be remarked that the assumptions which really enter into the formalism are first of all the finiteness of the particular combination of renormalization constants which appears in Eq. (3.3) and second the spectral representations for the form factors in Eq. (3.4).

It is of considerable interest to use the estimate just found in Eq. (3.9) not only in the charge renormalization constant integral in Eq. (2.5 b) but also in the integral exhibited in Eq. (2.13 a). We remember that the significance of this quantity is that it has to be convergent not only when all renormalization constants are finite but also in the case when only the observable "dielectric constant of the vacuum" is a finite number. Using Eq. (2.12a) and the formulae (2.14) and (2.16) we obtain

---

[17] In obtaining the estimate (3.9) we have used that not only the function $F_2(p^2)$ goes to zero in the high energy limit but that also the term $-p^2|F_2|^2$ can be neglected compared to the right-hand-side of Eq. (3.8 a). A more careful investigation which we do not give here shows that the high energy behaviour of the function $-p^2 F_2(p^2)$ is not worse than some power of $\log(-p^2)$. Consequently, the estimate (3.9) is correct. To obtain this result one has to assume certain well-behaved properties of the weight functions $f_2(p^2)$ but these smoothness properties appear very reasonable from the point of view of physics. Some of the details of this argument are given in the appendix of the first paper in ref. 3, but we do not want to give them here. However, it can be mentioned that the explicit form of the weight functions $f_i(p^2)$ computed in perturbation theory fulfill these smoothness conditions.

$$\lim_{-p^2 \to \infty} \Pi^*(p^2) = \lim_{-p^2 \to \infty} \frac{\Pi(p^2)}{|H_1(-p^2)|^2} \geqq \frac{\alpha}{3\pi} \lim_{-p^2 \to \infty} |G(p^2)|^2, \quad (3.11)$$

$$G(p^2) = F_1(p^2)\, H_1^*(-p^2) = \int_0^\infty \frac{da\, g(-a)}{a + p^2}, \quad (3.11\text{ a})$$

$$g(p^2) = f_1(p^2)\,[1 + \overline{\Pi}^*(p^2) - \overline{\Pi}^*(0)] +$$

$$+ \Pi^*(p^2)\left[1 - p^2 \int_0^\infty \frac{da\, f_1(-a)}{a(a+p^2)_P}\right]. \quad (3.11\text{ b})$$

We note in particular that the estimate in Eq. (3.11) makes no use of any finiteness of renormalization constants. The only thing which is assumed is that the observable difference $\overline{\Pi}(p^2) - \overline{\Pi}(0)$ exists and that the function $F_2(p^2)$ can be disregarded in high energy limit. This last argument is based on the condition of the finiteness of the anomalous magnetic moment (3.2 b) and is in no way related to the magnitude of the renormalization constants. The requirement that the integral (2.13 a) should be convergent now yields the condition that the function $G(p^2)$ has to vanish in the high energy limit. Further, the number $G(0)$ is a finite number (and equal to 1 according to Eq. (3.11 a))

$$G(0) = \int_0^\infty \frac{da}{a} g(-a) = \int_0^\infty \frac{da}{a}\Big[f_1(-a)\,\{1 + \overline{\Pi}^*(-a) - \overline{\Pi}^*(0)\} +$$

$$+ \Pi^*(-a)\left\{1 + a \int_0^\infty \frac{db\, f_1(-b)}{b(b-a)_P}\right\}\Big] < \infty. \quad (3.12)$$

Using the obvious relation[18]

$$\Phi(z) = \int_0^\infty \frac{da\, f(a)}{a - z} \int_0^\infty \frac{db\, g(b)}{b - z}.$$

Eq. (3.13) is the explicit version of the formula

$$\mathrm{Re}\, \Phi(x) = \frac{1}{\pi} \int_0^\infty \frac{\mathrm{Im}\, \Phi(a)\, da}{(a - x)_P}, \qquad x \text{ real.}$$

---

[18] Eq. (3.13) is most easily obtained by considering the conventional Hilbert transformation relation between the real and imaginary parts of the analytic function

$$\int_0^\infty \frac{da}{(a-x)_P}\left\{f(a)\int_0^\infty \frac{db\, g(b)}{(b-a)_P} + g(a)\int_0^\infty \frac{db\, f(b)}{(b-a)_P}\right\} =$$

$$= \int_0^\infty \frac{da\, f(a)}{(a-x)_P}\int_0^\infty \frac{db\, g(b)}{(b-x)_P} - \pi^2 g(a)\, f(a), \tag{3.13}$$

we find

$$G(0) = \int_0^\infty \frac{da\, f_1(-a)}{a}\,[1-\overline{\Pi}^*(0)] + \overline{\Pi}^*(0) =$$

$$= F_1^{\text{unren}}(0)\,[1-\overline{\Pi}^*(0)] + \overline{\Pi}^*(0) = F_1(0)\, H_1^*(0) = 1. \tag{3.14}$$

We emphasize once more that Eq. (3.14) has been derived without any assumption about the finiteness of the renormalization constants but only the finiteness of the integral (2.13 a). We now have two cases to consider. First, if the charge renormalization constant is indeed finite, then $\overline{\Pi}^*(0)$ is a finite number between zero and one. Consequently, it follows from (3.14) that also the particular combination of renormalization constants entering in Eq. (3.10) is equal to one. This is the previous result which we recover here. Next, we remark that if the charge renormalization constant is infinite, then the number $\overline{\Pi}^*(0)$ is exactly equal to one according to the discussion in section 2. Consequently, we have the relation

$$F_1^{\text{unren}}(0)\,[1-\overline{\Pi}^*(0)] = 0. \tag{3.15}$$

Therefore, we find that in this case the number $F_1^{\text{unren}}(0)$ may indeed be infinite but so weakly that the quantity exhibited in Eq. (3.15) can be put equal to zero. In terms of the original charge renormalization constant in Eq. (2.5 b) we have

$$F_1^{\text{unren}}(0)\,[1+\overline{\Pi}(0)]^{-1} = 0. \tag{3.15 a}$$

Summarizing the discussions so far we have found that in the case when the charge renormalization constant is finite, we must have Eq. (3.10) while an infinite value of this constant implies Eq. (3.15 a). It remains to relate these statements about the particular combination of renormalization constants appearing in $F_1^{\text{unren}}(0)$ to the more conventional renormalization constants of quantum electrodynamics. The next section is devoted to a discussion of this problem.[19]

[19] A discussion similar to the argument presented here but related to the behaviour of the vertex part for a scalar meson theory has been published by H. Lehmann, K. Symanzik, B. Zimmermann, Nuovo Cim. **2**, 425 (1955). This paper appeared at rather exactly the time when the manuscript mentioned in ref. 7 was issued. However, the authors just mentioned did not discuss the relation between their result and the conventional renormalization constants. Cf. in this connection also the paper by K. W. Ford, Phys. Rev. **105**, 320 (1957).

4. Relation between the Results in Section 3 and Conventional Renormalization Constants

Standard quantum electrodynamics involves several renormalizations which are necessary to extract physically significant result from the formalism. First of all, we have a mass renormalization for the electron. This operation is not going to concern us here. Second, we have a charge renormalization as well as field operator renormalizations for the Dirac field and for the electromagnetic field. We use the same notation as in refs. 3 and 11

$$\psi(x) = \frac{1}{N}\,\psi^{\text{unren}}(x), \tag{4.1 a}$$

$$A_\mu(x) = \frac{1}{\sqrt{1-L}}\,A_\mu{}^{\text{unren}}(x), \tag{4.1 b}$$

$$e = \sqrt{1-L}\;e^{\text{unren}}. \tag{4.1 c}$$

The constant N is a normalization constant for the Dirac field and the constant L is the charge renormalization constant. It is related to the previous integral in Eq. (2.5 b) through[20]

$$\frac{1}{1-L} = 1 + \overline{\Pi}(0) = 1 + \int_0^\infty \frac{da}{a}\,\Pi(-a). \tag{4.2}$$

As is well-known, there is actually one more renormalization necessary in quantum electrodynamics, viz. the normalization of the vertex function (2.1) for all external momenta on the mass shells. In terms of the formulae in the previous section, this is just the renormalization which is handled by the constant $F_1{}^{\text{unren}}(0)$. However, it is also well-known that, because of the Ward identity, no extra renormalization constant is necessary to achieve this. A little more in detail we remark that the unrenormalized current is given by

$$j_\mu{}^{\text{unren}}(x) = \frac{i}{2}\,e^{\text{unren}}\,[\bar{\psi}^{\text{unren}}(x), \gamma_\mu\,\psi^{\text{unren}}(x)]. \tag{4.3}$$

Performing the three renormalizations indicated in Eqs. (4.1) as well as the mass renormalization which we do not exhibit explicitly, we find that the renormalized current becomes

$$j_\mu(x) = \frac{N^2}{1-L}\,\frac{i\,e}{2}\,[\bar{\psi}(x), \gamma_\mu\,\psi(x)]. \tag{4.4}$$

---

[20] Actually, Eq. (4.1 b) is oversimplified as the relation between the renormalized and the unrenormalized electromagnetic potentials is more involved and different for different values of the index μ. For our present discussion, these complications are irrelevant. The notation L comes from the German expression for charge renormalization, viz. "Ladungsrenormierung".

It is the matrix element of this renormalized current operator which enters into Eq. (3.1).[21]

Our experience from perturbation theory tells us that the renormalizations indicated above are enough to yield finite matrix elements for the renormalized current operator in Eq. (4.4) and also that the function F_1 turns out to be automatically normalized to one at the origin. This means that the renormalization constant in Eq. (3.3) is a special combination of the constants N and L. Perturbation theory experience also tells us here that this relation reads

$$F_1^{\text{unren}}(0) = 1 - \frac{N^2}{1-L}, \tag{4.5}$$

i.e., the multiplicative renormalization factor in Eq. (4.4) appears also as counterterm in the particular matrix elements of the current operator which we are considering here. The relation between the renormalization constants indicated in Eq. (4.5) is equivalent to the well-known Ward identity.[22] Actually, this identity can be proved by formal arguments independent of perturbation theory. Even if we do not have time to give all these details here, we should like to mention a few points in the derivation. Using what is nowadays standard arguments with reduction formulae, one first expresses the matrix element of the current operator in terms of a vacuum expectation value of a retarded commutator and anticommutator plus a few "surface terms". As the details of this argument are rather involved and technical, we here just give the result.[23]

$$\begin{aligned}\langle 0|\, j_\mu(x)\,|\, q, q'\rangle = \text{const} \times \int\!\!\int dx'\, dx''\, \langle 0|\, \bar{\psi}^{(\text{in})}(x')\,|\,\bar{q}'\rangle \times \\ \times \Lambda_\mu(x'-x, x-x'')\,\langle 0|\,\psi^{(\text{in})}(x'')\,|\bar{q}\rangle + \langle 0|\, j_\mu^{(\text{in})}(x)\,|\bar{q},\bar{q}'\rangle \left[\frac{N^2}{1-L} - \Pi_R(p^2)\right],\end{aligned} \tag{4.6}$$

$$\langle 0|\, j_\mu^{(\text{in})}(x)\,|\bar{q},\bar{q}'\rangle = \frac{ie}{V}\,\bar{u}^{(-)}(-\bar{q}')\,\gamma_\mu\, u^{(+)}(\bar{q}), \tag{4.6 a}$$

$$\begin{aligned}\Lambda_\mu(x'-x, x-x'') = \theta(x-x')\,\theta(x'-x'')\,\langle 0|\,\{[f(x'), j_\mu(x)], \bar{f}(x'')\}\,|0\rangle + \\ + \theta(x-x'')\,\theta(x''-x')\,\langle 0|\{f(x'), [j_\mu(x), \bar{f}(x'')]\}|0\rangle.\end{aligned} \tag{4.6 b}$$

---

[21] Note that we get a factor $1 - L$ in the denominator in Eq. (4.4) and not the square root of this number. This happens because we both have to express the unrenormalized charge in terms of the renormalized charge e and divide the whole current operator by the renormalization factor in Eq. (4.1 b). For $\mu = 4$ complications similar to those indicated in footnote 20 also occur, but we do not consider them here. For details we refer to the papers in ref. 11.

[22] J. C. WARD, Phys. Rev. **78**, 182 (1950). The first order perturbation theory expression for this identity was used already by J. SCHWINGER, Phys. Rev. **76**, 790 (1949). A proof of the Ward identity without reference to perturbation theory was given by G. KÄLLÉN, Helv. Phys. Acta **26**, 755 (1953) and, later, by T. TAKAHASHI, Nuovo Cim. **6**, 371 (1957).

[23] The actual calculation is given, e.g., in ref. 3.

Because of the canonical formalism the expression appearing in Eq. (4.6 b) is covariant, i.e., it transforms like the matrix γ_μ under a Lorentz transformation.[24]

The operator $f(x)$ in Eq. (4.6 b) denotes the right-hand side of the renormalized Dirac equation. To obtain the result just indicated, it is essential that the calculations are performed in a gauge which is both covariant and such that the vacuum state is a true no-particle state. Further, the canonical commutation rules for the potentials have to hold in this gauge. The class of gauges where the formula above holds is, therefore, quite restricted and the only practically convenient gauge of this kind known today is the Gupta-Bleuler gauge.[25] Therefore, the renormalization constant N for the Dirac field which appears in the formula above is this renormalization constant in the Gupta-Bleuler gauge. If we for the moment accept the point of view that the term $F_1^{\text{unren}}(p^2)$ in Eq. (3) includes the term $\Pi_R(p^2)$ in Eq. (4.6) as well as the whole contribution from the term (4.6 b), we are immediately led to the Ward identity in Eq. (4.5). Further, it can be remarked that the constant $F_1^{\text{unren}}(0)$ which appears in Eq. (4.5) because of its definition with the aid of the current operator is a gauge invariant quantity. Therefore, Eq. (4.5) says that a gauge independent constant is expressed in terms of the gauge dependent quantity N but evaluated in a particular gauge, viz. the Gupta-Bleuler gauge. Evidently, a relation of this kind does not mean that the formalism violates gauge invariance.[26]

If the statement made above is accepted, one finds by a simple substitution of the relation (4.5) in Eq. (3.15 a) the following consistency condition for quantum electrodynamics

$$N^2 = 0. \tag{4.7}$$

Referring back to Eq. (4.1 a) we see that this result implies that the field operator renormalization of the Dirac field in the Gupta-Bleuler gauge is necessarily infinite. This result holds independent of whether or not the charge renormalization in Eqs. (4.1 b) and (4.1 c) is given by a finite factor. The possibility that both the charge renormalization and the field operator renormalization for the Dirac field in the Gupta-Bleuler gauge are infinite, is, of course, not in any way excluded by our argument. We repeat once more that the result given here is basically a result for the gauge independent quantity $F_1^{\text{unren}}(0)$ but that it so happens that it is also conveniently expressable in terms of the Dirac

---

[24] As before, we disregard some formal complications appearing for the special case $\mu = 4$. These complications make the algebra somewhat more involved than what is indicated here, but do not in any way influence the final result. They are described in ref. 3, esp. the last paper, Section 46.

[25] S. Gupta, Proc. Phys. Soc. London A **63**, 681 (1950); **64**, 850 (1951). K. Bleuler, Helv. Phys. Acta **23**, 567 (1950).

[26] We mention this elementary point explicitly because it appears to have caused some confusion in the literature. K. Johnson, Phys. Rev. **112**, 1367 (1958).

field operator renormalization in a particular gauge, viz. in the Gupta-Bleuler gauge.

5. Completion of the Argument

The discussion above is unsatisfactory in two respects. First of all, it makes use of the spectral representations (3.4) and it is known that even if these representations are true to each order in perturbation theory, it is not possible to prove them by more general arguments[15]. Further, at the end of section 4 we just made the intuitive suggestion that the unrenormalized function $F_1^{\text{unren}}(p^2)$ originates from the vacuum expectation value in Eq. (4.6 b) while the subtraction term comes from the explicit counter-terms in Eq. (4.6). To make the argument reasonably closed, it is evidently necessary to amplify the discussion at these points. The general goal to be achieved in this discussion is to establish, within reasonable limits of rigour, the result that the Fourier transform of the function (4.6 b) approaches zero in the high energy limit in the following way

$$\Lambda_\mu(q, q') = \int\int d\xi\, d\xi'\, e^{-iq\xi - iq'\xi'}\, \Lambda_\mu(\xi, \xi'), \tag{5.1 a}$$

$$\lim_{\substack{-(q+q')^2 \to \infty \\ q^2 = q'^2 = -m^2}} \Lambda_\mu(q, q') = 0. \tag{5.1 b}$$

In the original paper of ref. 3, the result (5.1 b) was derived with the aid of the explicit θ-functions appearing in the definition of the function $\Lambda_\mu(x' - x, x - x'')$. As is well-known, these step functions which correspond to the retarded character in the x-space formalism appear as energy denominators in p-space. Consequently, if the Fourier transform of the function without the step function is reasonably well behaved at infinity, the explicit energy denominators in the function Λ_μ is going to yield the property shown in Eq. (5.1 b). Further, the condition for the Fourier transform of the vacuum expectation value without step functions to be reasonably behaved at high energies is essentially the condition that the renormalization constants in the theory should be finite quantities. Intuitively, all these properties of the function Λ_μ can be understood from a comparison with the retarded current commutator appearing in Eq. (2.4 a).

The explicit step function in the definition of the induced current in Eq. (2.4) corresponds to the energy denominator $a + p^2$ in Eq. (2.4 a). Further, the condition that the retarded function $\Pi_R\,(p^2)$ should vanish in the high energy limit is, as shown by the argument in Eq. (2.10) intimately related to the high energy behaviour of the function $\Pi(p^2)$, i.e., to the magnitude of the charge renormalization constant in Eq. (2.5 b). However, it is this part of the argument which is critized by Gasiorowicz et al. More explicitly, these authors complain that the limit considered in ref. 3 does not correspond to the limit indicated in

Eq. (15.1 b) but rather to a limit where the space parts of the vectors q and q' are kept constant while only the energies are allowed to approach infinity. We do not agree with this criticism and should like to point out that the limits as performed in the original papers of ref. 3 do correspond to the correct limit in Eq. (5.1 b).[27] Therefore, we feel that the paper in ref. 4 is actually based on a misunderstanding.[28] Nevertheless we certainly agree that the original argument in ref. 3 is rather formal in its nature and contains a large number of interchanges of orders of integration, summations over an infinite number of states etc. Such arguments can be notoriously unreliable. However, arguments of this kind are usually felt to be acceptable in theoretical physics unless there are strong physical arguments to believe that they lead to incorrect results.[5]

During the more than ten years which have elapsed since the publication of the original paper the formal tools in this field have been considerably developed. What is particularly relevant for our discussion here is that one has gained a much deeper insight into the structure of the vacuum expectation value of a product of three operators[29]. With the aid of these techniques it is possible to give an alternative discussion leading to the result in Eq. (5.1 b). Compared to the original version of the argument, this new calculation has the advantage that it is more explicitly invariant and does not rely on a non-covariant handling of the θ-functions. Unfortunately, the details also of this simplified argument are too involved to be given explicitly here. They will be published elsewhere.[9]

The first step in the modernized version of the derivation of Eq. (5.1 b) is to consider the vacuum expectation value of products of the fields $\psi(x')$, $\bar{\psi}(x'')$ and $A_\mu(x)$ and write it in the following form

$$\langle 0|\, \psi_\alpha(x')\, A_\mu(x)\, \bar{\psi}_\beta(x'')\, |0\rangle =$$

$$= \sum_{\varrho,\varrho'=0,1} \left(\gamma \frac{\partial}{\partial x'} + m\right)^{\varrho}_{\alpha\varepsilon} \left(\gamma \frac{\partial}{\partial x''} + m\right)^{\varrho'}_{\delta\beta} \Big\{\gamma_\mu F_1{}^{\varrho\varrho'}(x'-x, x-x'') +$$

$$+ \frac{\partial}{\partial x_\mu{}'} F_2{}^{\varrho\varrho'}(x'-x, x-x'') + \frac{\partial}{\partial x''} F_3{}^{\varrho\varrho'}(x'-x, x-x'')\Big\}_{\varepsilon\delta}, \quad (5.2)$$

$$F_i{}^{\varrho\varrho'}(\xi, \xi') = \int\limits_{(m+\mu)^2}^{\infty} da \int\limits_{(m+\mu)^2}^{\infty} db \int\limits_{\sqrt{ab}}^{\infty} dc\, G_i{}^{\varrho\varrho'}(a, b, c)\, \Delta_3{}^{(+)}(\xi, \xi'; a, b, c), \quad (5.2\,\text{a})$$

[27] Cf. esp. Eqs. (42.a), (42.b) and (42.c) and the remarks immediately after these equations in the original paper.

[28] A short note to this effect was communicated to the Phys. Rev. Lett. in the summer of 1959. However, the note has never appeared in print.

[29] Some of the relevant papers in this connection are G. Källén, A. Wightman, Dan. Mat. Fys. Skr. **1**, Nr. 6 (1958); R. Oehme, Phys. Rev. **111**, 1430 (1951); **117**, 1151 (1960); G. Källén, J. Toll, Helv. Phys. Acta **33**, 753 (1960).

$$\Delta_3^{(+)}(\xi, \xi'; a, b, c) = \frac{1}{(2\pi)^6} \int\int dp\, dp'\, e^{ip\xi + ip'\xi'} \times$$
$$\times\, \delta(p^2 + a)\, \delta(p'^2 + b)\, \delta(p p' + c)\, \theta(p)\, \theta(p') =$$
$$= \frac{2i}{(4\pi)^3} \frac{\sqrt{c^2 - ab}}{R} \theta(c - \sqrt{ab})\, [H_0^{(1)}(\sqrt{Q + \sqrt{R}}) - H_0^{(1)}(\sqrt{Q - \sqrt{R}})], \tag{5.2 b}$$

$$Q = -a\xi^2 - b\xi'^2 - 2c\xi\xi', \tag{5.2 c}$$
$$R = 4\,[(\xi\xi')^2 - \xi^2\xi'^2]\,[c^2 - ab]. \tag{5.2 d}$$

The representation given in Eq. (5.2) corresponds closely to the representation for the two point function in Eq. (2.1). The three variables a, b, and c in Eq. (5.2 a) correspond to the mass square a in Eq. (2.1). The function $\Delta_3^{(+)}(\xi, \xi'; a, b, c)$ in Eq. (5.2 b) is the three point generalization of the function $\Delta^{(+)}(x, a)$ in Eq. (2.1 a).

The rather large number of amplitudes appearing in Eq. (5.2) are caused by the somewhat involved transformation properties of the left-hand-side.

It follows from the representation (5.2) that the left-hand-side can be understood as the boundary value of an analytic function of three complex variables in the same way as the function appearing in, e.g., Eq. (2.1) can be understood as the boundary value of an analytic function of the complex variable $z = -(x - x')^2$. Three convenient variables for Eq. (5.2) are given by $z_1 = -(x - x')^2$, $z_2 = -(x - x'')^2$ and $z_3 = -(x' - x'')^2$. The exact analyticity domain of this function of three complex variables is rather complicated and we do not want to discuss it in detail here.[29] However, we remark that if the three operators $\psi(x')$, $\bar{\psi}(x'')$ and $A_\mu(x)$ are multiplied together in some order different from the order in Eq. (5.2), we obtain, a priori, other analytic functions of the three complex variables z_1, z_2, and z_3. Because of the fact that the operators $\psi(x')$ and $\bar{\psi}(x'')$ anticommute for spacelike separations and the operator $A_\mu(x)$ commutes with the other two operators for space like separation, we can, by standard arguments, conclude that all the six analytic functions which are obtained in this way are different representations of the *same* analytic function. What is of particular interest for our discussion here is the behaviour of this analytic function when, e.g., the point x' approaches the point x''. To see the main idea in this connection it is convenient to return to the simpler case of the two-fold vacuum expectation value shown in Eq. (2.1). To simplify the discussion even further and bring out the essential idea we even permit ourselves to consider the vacuum expectation value of the product of two scalar fields.

$$\langle 0|\, \varphi(x)\, \varphi(x')\, |0\rangle = i \int_0^\infty da\, G(-a)\, \Delta^{(+)}(x - x', a) = F(x - x'). \tag{5.3}$$

As before, the function $F(x - x')$ is a boundary value of an analytic function of the complex variable $z = -(x - x')^2$. From Eq. (2.1 a) follows that the function $\Delta^{(+)}(x, a)$ has a singularity like $1/z$ when z approaches zero independently of the value of a. It follows that the function $F(x - x')$ has a singularity like $(x - x')^{-2}$ when the distance between x and x' is light like. The residue of this singularity in Eq. (5.3) is proportional to the integral

$$I = \int_0^\infty da\, G(-a). \tag{5.4}$$

More precisely, we conclude that if the integral in Eq. (5.4) is convergent, the singularity of the function F close to the light cone corresponds to a simple pole at the origin for the variable z. However, if the integral in Eq. (5.4) is divergent, the singularity of the functions F is stronger than z^{-1}. On the other hand, if the integral in Eq. (5.4) happens to vanish, the singularity at the origin of the function F is weaker than the simple pole $1/z$.

From the representation (5.3) we can calculate a similar representation for the commutator between the fields $\varphi(x)$ and $\varphi(x')$. One finds

$$\langle 0|\, [\varphi(x), \varphi(x')]\, |0\rangle = -i \int_0^\infty da\, G(-a)\, \Delta(x' - x, a). \tag{5.5}$$

By taking the time derivative of Eq. (5.5) and putting the two times x_0 and x_0' equal after the differentiation has been performed, we find from Eq. (5.5)

$$\langle 0|\, [\dot\varphi(x), \varphi(x')]\, |0\rangle|_{x_0 = x_0'} = i\, \delta(\bar{x} - \bar{x}') \int_0^\infty da\, G(-a) = i\, I\, \delta(\bar{x} - \bar{x}'). \tag{5.6}$$

Taken literally, the calculation indicated in Eq. (5.6) assumes that the integral I in Eq. (5.4) is convergent. In this way we see that the residue of the pole at the origin for the function F in Eq. (5.3) appears as a coefficient in the canonical commutation relation in Eq. (5.6). If this integral is divergent, it means that the renormalization of the field $\varphi(x)$ in Eqs. (5.3) and (5.6) is infinite, and we find that there is a very intimate connection between the magnitude of the renormalization constant of the field $\varphi(x)$ and the nature of the singularity at the origin for the function $F(x - x')$. Evidently, the discussion given here is closely related to the representation (2.5 b) for the charge renormalization constant.

The argument which we have given in some detail here for the two point function can now be repeated for the three point function. By calculating various commutators and anticommutators of the operator $\psi(x')$, $\bar\psi(x'')$ and $A_\mu(x)$ in Eq. (5.2) and requiring that the canonical commutation relations are not too singular, we obtain various integrals over the weights $G_i^{\varrho\varrho'}(a, b, c)$ which have to be convergent. In this way the assumption that all the renormalization constants in the formalism are finite yields rather strong (and invariant) conditions on the weight functions G. On the other hand, the Fourier transform of the retarded commutator and anticommutator in Eq. (4.6 b) is obtained from the functions G by calculating a convolusion integral with the Fourier transforms of the step functions $\theta(x - x')$ etc. These Fourier transforms correspond to the energy denominators mentioned earlier and we recover in this way in a more invariant and elegant way the old result of ref. 3 that the Fourier transform in Eq. (5.1 a) has to fulfill the condition (5.1 b). From the point of view of rigorous mathematics, this argument may not be very much better than the original discussion in ref. 3 because, just as before, we are using formal convolution integrals and performing various interchanges of orders of limits. However, the organisation of the calculation is more explicitly invariant than what was the case before. In any case, the result is the same as in ref. 3.

6. Comparison with Perturbation Theory

It is of some interest to make a comparison between the result expressed in Eq. (5.1 b) or in Eqs. (3.8) with the result obtained from a first order perturbation theory calculation. When such a comparison is made, one must keep in mind that one of the basic assumptions that goes into the derivation of the result (5.1 b) is the assumed finiteness of the renormalization constants of the theory. This basic assumption is not fulfilled in perturbation theory. Therefore, the limit expressed in Eq. (5.1 b) may not exist. If one wants to make a comparison of this result with perturbation theory, it is necessary to rely on cut-off arguments. As is well-known, it is rather dangerous to introduce arbitrary cut-offs in a theory like quantum electrodynamics as soon as one encounters a divergent integral in the formalism. In this way, one very easily violates some basic property of the theory like Lorentz invariance or gauge invariance.[30] The first non-trivial order of perturbation theory yields the following expressions for the two functions $f_1(p^2)$ and $f_2(p^2)$ in Eqs. (3.4)

$$f_1(p^2) = \left\{ -\frac{\alpha}{3\pi}\left(1 - \frac{2m^2}{p^2}\right)\sqrt{1 + \frac{4m^2}{p^2}} + \frac{\alpha}{2\pi} \times \right.$$

[30] For an illustration of these remarks it is of interest to follow the historical development of the theory as reflected, e.g., in the following papers: N. M. Kroll, W. E. Lamb, Phys. Rev. **75**, 388 (1949); J. B. French, V. F. Weisskopf, Phys. Rev. **75**, 1240 (1949) and G. Wentzel, Phys. Rev. **74**, 1070 (1948).

$$\times \left[\frac{1 + \frac{2m^2}{p^2}}{1 + \frac{4m^2}{p^2}} \log \left(1 - \frac{p^2 + 4m^2}{\mu^2} \right) - \frac{3}{2} \right] \times$$

$$\times \sqrt{1 + \frac{4m^2}{p^2}} + \frac{\alpha}{\pi} \frac{m^2}{p^2} \frac{1}{\sqrt{1 + \frac{4m^2}{p^2}}} \Bigg\} \theta(-p^2 - 4m^2), \qquad (6.1\text{ a})$$

$$f_2(p^2) = -\frac{\alpha}{\pi} \frac{m^2}{p^2} \frac{\theta(-p^2 - 4m^2)}{\sqrt{1 + \frac{4m^2}{p^2}}}. \qquad (6.1\text{ b})$$

If we use the expression given in Eq. (6.1 a) in the Ward identity in Eq. (4.5) we find that the integral on the left hand side is divergent. This only reflects the well-known fact that all the renormalization constants are divergent in a first order perturbation theory approach. If we try to handle these divergences by introducing a cut-off in the integral exhibited, e.g., in Eq. (3.4 c) we find that the left-hand-side of Eq. (4.5) behaves as the square of the logarithm of this cut-off quantity. On the other hand, it is also well-known that the two constants N and L behave only as a single power of the logarithm of the cut-off in a similar approach. However, it is not possible to conclude from this fact that the relation (4.5) is incorrect. Actually, Eq. (4.5) corresponds to the well-known Ward identity and the argument indicated here only proves that it is impossible to introduce a cut-off energy as soon as one encounters a divergent integral and to put this cut-off equal at all places in the formalism. By doing this one runs the risk of violating some fundamental relation in the theory.[30] For the particular case under discussion here one should rather use the basic relation (4.5) to define a connection between the various cut-off energies. Evidently, the same problem exists also in higher orders.[31]

Another point of logic should perhaps also be mentioned in this connection. Even if it is elementary, it seems to have caused some confusion in the literature. We remember that the argument presented here goes essentially the following way. First, we assume that all the renormalization constants are actually finite. After a rather long calculation this argument is seen to lead to a contradiction and we conclude that the starting assumption must have been wrong. Consequently, the intermediate steps in the argument which are derived using the starting assumption, later on proved to be incorrect, are not necessarily reliable. Therefore, a possible contradiction between some of the intermediary results and explicit low order perturbation theory

---

[31] B. Zumino, Nuovo Cim. **27**, 547 (1960).

calculations is not relevant for the internal consistency of the reasoning.[32]

It is perhaps appropriate to devote a few words to a discussion of the work by Johnson et al. referred to in footnote 2. What these authors really do is to give a new scheme of successive approximations which is different from ordinary perturbation theory. In the original paper, the authors essentially calculated only the first approximation in this iteration scheme and, when no explicit contradiction was arrived at in this way, they claimed to have found a self consistent solution of quantum electrodynamics. In the second paper mentioned in footnote 2, the authors have abandoned this position and only assert that their approximation scheme gives finite values for all quantities, including renormalization constants, in each order. Before the existence or non-existence of solutions can be seriously discussed in this way, the convergence of the approximation scheme must be further investigated. As of today, this convergence is an open question and before that problem is settled, the authors can make no statement about the existence of exact solutions in ordinary quantum electrodynamics. However, one more point should be made in this connection. To get finite values for all quantities in each approximation, the authors find it necessary to introduce the assumption that the bare mass of the electron is zero and that the total electron mass is of electrodynamic origin. At the first moment, this idea might seem very appealing, but second thought shows that it is really quite drastic and that it violates the so called γ_5-invariance of the theory. As is fairly well-known, the Dirac equation with zero mass is invariant if the Dirac field $\psi(x)$ is replaced by the quantity $\gamma_5\,\psi(x)$. If one assumes that the physical states in the theory have the same multiplicity when the charge is different from zero as when the charge vanishes, it is then easily shown that also the physical mass of the electron must be zero. Even if the argument is rather well-known, we want to indicate it here for completeness.

The equations of motion for electrodynamics with zero mass of the electron are

$$\gamma \frac{\partial}{\partial x} \psi(x) = i\, e\, \gamma\, A(x)\, \psi(x), \tag{6.2 a}$$

$$\Box\, A_\mu(x) = -\frac{i\,e}{2}\,[\bar{\psi}(x), \gamma_\mu\, \psi(x)]. \tag{6.2 b}$$

Introducing new operators $\psi'(x)$ and $\bar{\psi}'(x)$ from the relations

$$\psi'(x) = \gamma_5\, \psi(x), \tag{6.3 a}$$

$$\bar{\psi}'(x) = [\psi'(x)]^*\, \gamma_4 = -\,\bar{\psi}(x)\, \gamma_5, \tag{6.3 b}$$

---

[32] This logical situation appears to have been overlooked in the paper by B. Zumino, referred to in the previous footnote as well as in the paper by O. Fleischman, Nuovo Cim. **29**, 1098 (1963).

we find by direct algebraic substitutions that these new operators fullfil the same equations of motion as the original quantities

$$\gamma \frac{\partial}{\partial x} \psi'(x) = i e \gamma A(x) \psi'(x), \tag{6.4 a}$$

$$\square A_\mu(x) = -\frac{i e}{2} [\bar{\psi}'(x), \gamma_\mu \psi'(x)]. \tag{6.4 b}$$

Next, consider the vacuum state $|0\rangle$ and the one-particle state $|\bar{q}\rangle$ in the theory. If the one-particle state corresponds to a particle with mass m, it follows from standard arguments of invariance that the matrix element between the vacuum and the one-particle state of the operator $\psi(x)$ must fullfil

$$\left(\gamma \frac{\partial}{\partial x} + m\right) \langle 0| \psi(x) |\bar{q}\rangle = 0. \tag{6.5}$$

As Eqs. (6.4) look exactly the same as Eqs. (6.2), it follows that the operator $\psi'(x)$ must fullfil a similar relation

$$\left(\gamma \frac{\partial}{\partial x} + m\right) \langle 0| \psi'(x) |\bar{q}\rangle = 0, \tag{6.5 a}$$

with the same mass appearing in Eqs. (6.5) and (6.5 a). As we have already assumed that the vacuum state in the theory is one unique state, it follows that the vacuum appearing in Eqs. (6.5) and (6.5 a) must be the same. Further, our basic assumptions about the mass spectrum of the theory imply that the one-particle states are also essentially unique, i.e., that they have no higher degeneracy than what is required by the spin and charge coordinates. Consequently, the physical states appearing in Eqs. (6.5) and (6.5 a) must be the same. Using the definition (6.3 a) we can reformulate Eq. (6.5 a) to read

$$\left(\gamma \frac{\partial}{\partial x} - m\right) \langle 0| \psi(x) |\bar{q}\rangle = 0. \tag{6.5 b}$$

Comparing Eqs. (6.5) and (6.5 b) we find that the physical mass of the theory has to vanish which is a requirement which is obviously not fullfilled in nature. The conclusion is that it is not possible to obtain a finite physical mass of the electron starting from Eqs. (6.2) and using conventional field theory.

The only way out of this situation is to introduce the ad hoc hypothesis that the degeneracy of at least one physical state is different in a theory with interaction and in a theory without interaction.[33] Such a situation would violate the basic assumptions we have started from

[33] Although no explicit mentioning of these facts can be found in the published papers referred to in footnote 2, Dr. Johnson has informed the author in a private letter that he is aware of these circumstances. Dr. Johnson also suggests that the physical vacuum state should be degenerate.

in this paper where we assume, e.g., that the vacuum is one unique invariant state with minimum energy and that the one particle states have the same multiplicity for the interacting fields as for the free fields. Even if we agree that the multiplicity of a state could possibly change with the interaction for the very complicated theories we are dealing with here, we still feel it is of interest to discuss consistency problems in electrodynamics on a somewhat more conservative basis and without introducing new ad hoc assumptions.

7. Further Remarks

The main discussion presented so far was only concerned with the question whether or not the unrenormalized theory can be considered as a satisfactory mathematical scheme. We arrived at the conclusion that this is not the case. There remains the more interesting problem to decide whether or not the theory is consistent after renormalization. What is known so far is only that each term in a perturbation theory expansion always gives a finite number and that the first few terms in such an expansion agree very well with observations. However, as was the case with one of the attempts mentioned earlier, this does not imply that a consistent solution of the theory exists. First if it can be proved that the perturbation theory is convergent can one claim that a satisfactory solution of the problem has been found. So far it has not been possible to make any serious discussion of the convergence of the perturbation theory of fulfledged quantum electrodynamics. Of course, it is always possible to take the attitude that we here have a theory which evidently contains physical truth because of its excellent agreement with observations. From this point of view, the discussion of further consistency problems is at most a mathematical game. However, we feel that this position is really justified only if it turns out that electrodynamics, after renormalization, is a completely consistent theory. In such a case nothing more can be learnt from its study. However, if it should happen that the formalism, taken very seriously, contains inconsistencies at high energies, there is always the hope that we can learn something from a study of these problems. The most optimistic attitude to be taken here is that the presumed basic failure of electrodynamics or any quantized field theory at high energies should bear some relation to the beautiful theory of the future, which we know so little about today.

From this point of view it is tempting to try to speculate if the argument presented above cannot be generalized and turned around in such a way that it tells us something also about the properties of renormalized quantum electrodynamics. The first thing that comes to ones mind is then that the estimate used for the function $\Pi(p^2)$ in Eq. (2.2) is only a lower bound to the exact expression. However, there is no reason to expect that this lower bound should give a good description of the high energy behaviour of the function in question. Rather, it must be expected that in the limit of very high energies a very large number of states should contribute in a significant way. This raises the question of how to estimate the high energy behaviour of more complicated matrix elements than those discussed in section 3. Using a reduction formula technique of the same kind as the expression in Eq. (4.6) it is possible to express arbitrarily complicated matrix elements of the current operator between the vacuum and a state with very many particles in terms of the vacuum expectation value of a generalized retarded commutator. After such a reduction we get the same number of operators in the vacuum expectation value as we have particles in the state under consideration. The mathematical properties of these general vacuum expectation values are very complicated and very little is known about the details as of today. However, it is possible to try to interpret the result derived in Eq. (3.8) in an intuitive way and then to try to generalize this guess also to more complicated states. To this purpose we remark that the actual high energy behaviour of the particular matrix element of the current operator investigated in sections 3, 4 and 5 can be summarized in the simple formula

$$\lim_{-(q+q')^2\to\infty} \langle 0|\, j_\mu x)\, |\bar{q},\bar{q}'\rangle = \frac{N^2}{1-L} \langle 0|\, j_\mu^{(\mathrm{in})}(x)\, |\bar{q},\bar{q}'\rangle. \tag{7.1}$$

Remembering the relations (4.1) between the renormalized and unrenormalized fields we can state that Eq. (7.1) shows that the high energy behaviour of the matrix element between the vacuum and a state with one electron positron pair of the current operator is essentially given by the Born approximation for the same quantity expressed in terms of unrenormalized fields. In physical language, this means that in the limit of high energies the actual particles are able to penetrate the virtual clouds of pairs and photons which surround them according to conventional field theory. Therefore, at very high energies, the bare particles acquire a physical significance.[34] In view of this, it is tempting to speculate that also more complicated matrix elements of the current operator in the high energy limit should be given in terms of Born approximations for the bare fields. Whether or not such an idea is correct is an open question as of today. Nevertheless, if we for the moment accept this idea as an intuitive guide, it is possible to give more stringent lower bounds on the function $\Pi(p^2)$. We have no time to enter into the details here and they have been published elsewhere several years ago.[8] Let it be enough to mention that

[34] Apart from the particular matrix element considered here, similar properties can be shown to hold exactly for the operators of the Lee model and for the operators in the Wentzel pair theory. These are the only four-dimensional models of field theories where the exact solution is known, at least to some extent. For the two-dimensional Thirring model, where the exact solution is also known, the situation is slightly ambiguous but the explicit formulae derived do not violate the intuitive idea suggested here.

estimates of this kind can be carried through and give as a result that the lower bound of the function $\Pi(p^2)$ increases faster than any power of the quantity $-p^2$. This would imply that not only the integral describing the charge renormalization constant but also the integral describing the observable dielectric constant of the vacuum would be badly divergent. This divergence comes about essentially because when energy conservations allow more and more real particles to appear in the sum over intermediate states, more and more terms contribute and they all add with the same sign. If these intuitive ideas are correct, they would imply that there is a definite mathematical inconsistency in quantum electrodynamics which would only show up at high energies. For completeness, it should also be remarked that this effect would never show up in any finite order of perturbation theory as the possible number of particles there is essentially limited by the order at which the power series expansion in perturbation theory is broken off. However, it should not be forgotten that these arguments are to a very large extent based on guess work and it has so far not been possible to decide whether or not the intuitive idea about the Born approximation for bare fields at high energies is physically correct.

In this connection it may be worth while to discuss also another effect, viz. the infrared divergence problems which always occur in quantum electrodynamics. We should first of all like to mention that there seems to be a general belief among research workers in this field that although the infrared problems in themselves are very interesting and not without practical significance, they still have very little to do with the basic consistency problems in this field. The easiest way to avoid any explicit discussion of infrared problems is to introduce a small artificial photon mass in the theory. In general, it is then to be expected that this photon mass should remain in the final result and it has to be interpreted differently for each separate case. In our particular discussion here, such a small photon mass should also remain, e.g., in the high energy limit indicated in Eq. (7.1). More explicitly, it is to be expected that the constant N should depend on the photon mass and in such a way that N probably goes to zero when the photon mass is allowed to vanish. However, it is also well-known that the same result which is obtained for a finite photon mass can be reproduced if one considers states with not only one electron-positron pair, but also with a large number of very weak photons. To achieve this, one has to sum over all states where the total energy of all the weak photons is the same as the artificial photon mass μ. This kind of problem has never caused any serious difficulty in electrodynamics. Therefore, this is an effect where a certain class of states with a very large number of particles can be considered as relatively harmless.

If the intuitive argument which is presented in this section, happens to be correct, we have arrived at a final result which is rather similar to the conclusions of the Russian authors mentioned in footnote 1. However, the way in which the conclusion is reached is entirely different and the exact shape of the break down of the theory is not the same. Actually, the Russian result, expressed in the language used here, leads to the conclusion that the integral (**2.13** a) over the functions $\Pi^*(p^2)$ is actually divergent. As this integral from general arguments must always be smaller than one in a consistent theory, this would indicate a serious inconsistency in the formalism. However, this conclusion is arrived at through a series of approximations which we feel to be very unreliable.[35] Apart from the general remark that any inconsistent formalism by suitable manipulations can be made to yield arbitrary contradictions, we therefore cannot claim that we are in agreement with the authors of ref. 1.

[35] Cf. further the paper by S. Kamefuchi, Dan. Mat. Fys. Medd., **31**, Nr. 6 (1957).

S-Matrix Theory of Electromagnetic Interactions*
(with Topics in Weak and Gravitational Interactions)

By

A. O. Barut[1]

International Atomic Energy Agency
International Centre for Theoretical Physics, Trieste, Italy

With 11 Figures

Contents

---

* Lecture given at the IV. Internationale Universitätswochen für Kernphysik, Schladming, 25 February—10 March 1965.

[1] Permanent address: University of Colorado, Boulder, Colorado, USA.

I. Introduction

The S-matrix theory, as we see it, is a relativistic formulation of interactions of fundamental particles based on their *particle properties* (not fields), that is, the formulation of the laws of physics in terms of the c-number scattering matrix elements. The scattering matrix itself is defined in terms of the (free physical) particle properties such as momenta, spin, and other quantum numbers which are numbers labelling the representations and states of symmetry groups. We shall give a precise mathematical definition of "particles" and of "scattering". There is no need to make a fundamental distinction between the S-matrix theory and the relativistic quantum field theory as they both essentially lead to the same results. In one case the basic analyticity properties of the S-matrix are (partly) derived from field axioms and perturbation theory, and in the other case, partly from unitarity condition and are partly postulated. The difference at this stage is perhaps a practical and didactic one; the S-matrix approach is a more direct, computationally and conceptually simple one, using only quantities very close to observed ones.

The important thing is the use of analyticity and unitarity in making calculations, and methods have been developed on this basis which go over and beyond the practical methods of Lagrangian field theory.

If the S-matrix theory is to be applied to electromagnetic interactions a conceptual question arises as to how an asymptotic charged particle is defined, with or without its soft photons around it and its long-range interaction. This problem is also present of course in field theory in one form or another, and has to do with the way a charged *physical* particle is defined (see **I.3**), and measured. (If an electron is measured with an energy between E and $E + \Delta E$, then one has to account for the soft photons present in this range in the particular experiment.) Can one then formulate an S-matrix theory in the case of long range interactions? [1] However, we shall consider the scattering matrix in the case of charged particles, between the *reducible* representations of the Lorentz group, i.e. representations containing charged particles and photons. And we shall show that due to this fact and due to the smallness of the coupling constants of electromagnetic and gravitational interactions the S-matrix theory can be applied, at least for all practical purposes, in much the same manner as in the case of short-range strong interactions. In these lectures we are interested mainly in showing the equivalence with the perturbation theory and we reproduce by S-matrix methods the results of renormalized perturbation theory including radiative corrections.

It is interesting to see, in a theory where there are no fields or Lagrangian, what principles determine uniquely a definite interaction term of the form

$$j_\mu(x)\, A^\mu(x),$$

for electromagnetic interactions, or a local four-fermion term for weak interactions[2].

It is also interesting to find the particle analogues of such principles as gauge invariance (a concept which one always associates with fields), $V - A$ theory, conserved vector current hypothesis, universality and local action principles, etc. It will be shown that all these can be done in a natural way in terms of the S-matrix. An important point is the use of the Lorentz group (Poincaré group) in its irreducible form, in particular in the case of zero-mass particles. The condition of masslessness turns out to be the most important property characterizing electromagnetic, weak and gravitational interactions. This point is not directly and economically expressed if fields are used so that one has to use, in general, complicated subsidiary conditions. Aside from economy of the formulation (for didactic purposes, the absence of background knowledge on field theory, field quantization, etc.) one has in S-matrix theory the advantage that no renormalization problem arises.

Strictly speaking, all particles interact with each other and all interactions are simultaneously present. It is not always meaningful to separate strong, electromagnetic and weak interactions as, for example, in the case of weak decays of strongly interacting particles, or in the limitations of quantumelectrodynamics due to the interactions of other particles. In many cases, however, it is a good approximation to consider these interactions separately. In terms of the S-matrix, this means that one confines oneself to a certain given set of particles and certain conservation laws. For example, in electrodynamics we shall consider charged massive particles and photons, and parity and charge conjugation invariance, but not isotopic spin invariance. These are the inputs. As output one is interested in calculating the cross-sections, bound states and resonances, lifetimes and magnetic moments. We shall consider some typical problems of this category. In the process of calculations one other input data is introduced, namely the value of the mass-shell amplitude for a vertex, the (renormalized) *coupling constant*. It is however, quite reasonable that the coupling constants are, just like any other amplitude, determined by the masses and quantum numbers, and can be — in some cases — obtained from consistency of the existence of particles in nature. This is not yet the case, however, for e^2.

In Chapter II we review the formulation of general principles in terms of the S-matrix [2]. This is done to discuss as clearly as possible all the assumptions made in the development.

[2] In general there are a large number of possibilities in writing down the coupling of fields. For example, one can use the Dirac field, or the two-component Feynman-Gell-Mann field for fermions to write down the coupling. These give different results, at least in low orders. The S-matrix approach implies of course a unique coupling up to possible undetermined subtraction constants.

In Chapter III we consider the electromagnetic interactions and, in some detail, some typical processes. Then we consider, in Chapters IV and V some related problems with spin 1/2 and spin 2 massless particles.

II. Review of some Properties of the S-Matrix (with emphasis on massless particles)

1. Amplitudes and Quantum Principle

We consider an arbitrary process involving n particles and make no distinction between elementary or composite particles. The S-matrix of the process is inferred from the relative frequencies (probabilities) of transitions from the initial state i to the final state f characterized by a complete set of measurements. The *quantum transition process* is defined by the *conditional probability amplitudes* $S(f, i)$, independent of the magnitude (intensity) of the initial and final state amplitudes ("linear quantum process"). The sum of probabilities being equal to unity, we can deduce immediately the unitarity of the S-matrix:

$$SS^+ = S^+S = I. \tag{2.1}$$

The S-matrix must be defined as a function of those parameters of the particles which are asymptotically constants of the motion (quantum numbers). These constants of the motion arise from the symmetry of the system. More precisely, we *define* a "particle" to be an irreducible (see exceptions below) unitary representation of the symmetry group. Such a representation is characterized by the values of the invariant Casimir operators (which give the "name" of the particle). In each representation we have a number of states characterized by the values of the diagonalizable operators, and these are the "states of the particles". The S-matrix is defined by a quantum transition process between such states. Thus, the S-matrix cannot be defined explicitly unless some invariance principles are known.

The full infinite dimensional S-matrix splits by the selection or superselection rules into blocks so that only transitions within each block are allowed. The *superselection rule* [3] means that one cannot prepare states with coherent superposition of certain properties, such as charge or baryon number. For example, there is no state which is a coherent superposition of states of charge zero and charge 1 particles, with definite phase relations between them. This means that in an experiment with such a beam we have to use a direct product of representations reduced in the form

$$\begin{pmatrix} \mathcal{D}_1 & & & 0 \\ & \mathcal{D}_2 & & \\ & & \ddots & \\ 0 & & & \mathcal{D}_n \end{pmatrix}$$

and that this matrix cannot be mixed by a similarity transformation. In contrast to this, one can prepare states which are coherent superposition of energy or angular momentum, for example, in a given irreducible representation (that is, also with the same charge and baryon number, etc.).

If a sequence of states $i \to j \to f$ actually occurs as a sequence, in time, of processes, then the S-matrix elements S_{ij}, S_{jf} are defined only if the state j is sufficiently long lived. In such a case of succession of localized states (in space and time) one can a posteriori introduce space-time coordinates into the S-matrix theory [4].

If, on the other hand, the state j is of short duration, then only the overall S-matrix element S_{if} is defined. In this connection, the concept of "final state interaction" means that such a separation in a state j is approximately meaningful; clearly, once an S-matrix is defined, the final particles are, by definition, non-interacting. We shall extend the definition of S-matrix to involve unstable particles by introducing suitable *reducible* representation (see II, 2 A).

One place though where space-time is introduced in S-matrix theory is in the calculation of observed cross-sections and lifetimes because the actual experiments are done in definite space and definite time intervals.

The diagonal elements of the S-matrix ($f = i$) refer to the elastic scattering in which case $S - I = R$ is the elastic scattering amplitude. The non-diagonal elements ($f \neq i$) are the various inelastic amplitudes.

2. Invariance Principles in General and Relativistic Invariance

Let the S-matrix be defined as a function of some parameters α transforming under some symmetry group as

$$\alpha' = \lambda \alpha \tag{2.2}$$

and let the S-matrix have the transformation property

$$S'(\alpha') = T\, S(\alpha), \tag{2.3}$$

under (2.2). Then the invariance of the observed probabilities under the symmetry group means that the transition amplitudes can differ only by a phase ω

$$S(\alpha) = \omega\, S'(\alpha). \tag{2.4}$$

Combining now (2.4) with (2.3) and (2.2) we see that the S-matrix must satisfy the requirement

$$S(\alpha) = \omega\, T\, S(\lambda^{-1} \alpha). \tag{2.5}$$

Because of the unitarity condition (2.1), T must be unitary. We shall *define* each asymptotic "particle" as the set of states belonging to an *irreducible unitary representation* of the symmetry groups. Because the particles in the initial and final states are non-interacting, T is a direct product of unitary irreducible representations, and the S-matrix is a "covariant" matrix under T. We note that if all particles in a scattering experiment are drawn as "ingoing", Eq. (2.5) includes the so-called anti-unitary transformations as well.

In the case of invariance under a compact group the explicit form of (2.5) is simple. For example, under $S\,U_2$, each "particle" (= irreducible unitary representation) is characterized by one quantum number I and is a collection of $d(I) = 2\,I + 1$ states, transforming by the matrix $\mathscr{D}^I$. The S-matrix is then an invariant (or isotropic) tensor, satisfying (in matrix notation)

$$S(I_1,\ldots,I_n) = \omega \prod_{j=1}^{n} \mathscr{D}^{I_j} \qquad S(I_1,\ldots,I_n)\,; \tag{2.6}$$

the indices α, β of $\mathscr{D}_{\alpha\beta}{}^I$ are the eigenvalues of I_3 ranging from $+I$ to $-I$ in integral steps.

In the case of an exactly satisfied $S\,U_3$, each particle is characterized by two quantum numbers λ_1, λ_2 and is the collection of $d(\lambda_1, \lambda_2) = \frac{1}{2}\,(\lambda_1 + 1)\,(\lambda_2 + 1)\,(\lambda_1 + \lambda_2 + 2)$ states. Eq. (2.6) becomes

$$S(\lambda_1{}^i, \lambda_2{}^i\,|i = 1, 2 \ldots n) = \omega \prod_{j=1}^{n} \mathscr{D}^{(\lambda_1{}^j,\,\lambda_2{}^j)}\, S(\lambda_1{}^i, \lambda_2{}^i\,|i = 1 \ldots n). \tag{2.7}$$

Here the indices of $\mathscr{D}^{(\lambda_1\,\lambda_2)}$ are the eigenvalues of I, I_3 and Y, for example. In these cases the S-matrix is finite dimensional, and is clearly unitary[3].

For invariance under a non-compact group, such as the inhomogeneous Lorentz group, a particle (= irreducible unitary representation) is a collection of infinitely many states and a transition is possible from any of these states to any other. And the S-matrix is labelled by discrete and continuous parameters of the irreducible representation, one for each particle. Thus, mathematically, the scattering process is defined by the direct product of unitary representations and the S-matrix is the invariant "tensor" defined by Eq. (2.5).

There are many ways of expressing the unitary irreducible representations of the Poincaré group. It is convenient for our purpose to use the form given by WIGNER [5]. The particles are characterized by mass square m^2 and spin S. For a given m^2 and S, the infinitely many states may be labelled by momenta k^μ ($k^2 = m^2$) and the spin component S_3.

Instead of writing a continuously infinite matrix, analogous to (2.6) and (2.7), we can use a functional equation for the continuous parameters k^μ as follows: The irreducible (unitary) representations of the Poincaré group are obtained by the irreducible (unitary) representations of translations, i.e. a factor

$$e^{i k^\mu a_\mu}, \tag{2.8}$$

multiplied, *not* by the representations of the homogeneous Lorentz group, but by the representations of the so-called *little group of momen-*

[3] An invariant function transforming according to unitary representations is unitary.

tum k^μ (defined by (2.8)). The little group of a four-vector k^μ is the set of homogeneous Lorentz transformations which leave k^μ invariant. Clearly it depends whether $k^2 > 0$ or $k^2 = 0$. (We do not consider the cases $k^2 < 0$ and $k^\mu = 0$ here.) In the former case the little group is isomorphic to the three-dimensional rotation group R_3; in the latter case it is isomorphic to the two-dimensional euclidian group E_2 (rotations and translations of the plane). The representations of the little groups express the only degree of freedom allowed (in the rest frame of the particle) under relativistic invariance, namely the spin projection. This shows clearly how the spin orientations of the particle arise from relativistic invariance. The covariance of the S-matrix take then the following form:

$$S(k_1, \ldots k_n) = \omega\, e^{i \sum_j k_j a} \prod_{j=1}^{n} Q^{(S_j)}[A'(k)]\, S(\Lambda^{-1} k_1, \ldots \Lambda^{-1} k_n). \tag{2.9}$$

In this equation:

(1⁰) Λ is a proper orthochronous Lorentz transformation;

(2⁰) $Q^{(S)}$ is the (finite dimensional) unitary representation of the little group operating on the spin indices. For massive particles $Q^{(S)} = \mathcal{D}^S$, the $(2S+1)$-dimensional unitary irreducible representation of the rotation group. For massless particles $Q^{(S)} = \exp[i\, S\, \theta(k, \Lambda)]$, a one-dimensional representation [6]. The last result means that a massless particle of any spin S has only *one* direction of polarization (two if parity is conserved). We can keep, however, $\mathcal{D}^S$ even for massless particle by suitably defining the argument A' for this case;

(3⁰) $A'(k)$ is the so-called "Wigner rotation", a unitary matrix belonging to the little group of the momentum vector in its rest frame, p^μ, for example:

$$A'(k) = B_{k \leftarrow p}{}^{-1} A\, B_{q \leftarrow p}$$
$$k = \Lambda(A)\, q. \tag{2.10}$$

Here we use the correspondence (2-to-1) between the unimodular group $SL(2, C)$ and the real proper homogeneous Lorentz group $L_+^\uparrow$. The unimodular group is the covering group of $L_+^\uparrow$. The representations up to a phase ω can first be reduced to representations up to a sign ± 1. Further, by using the covering group, the sign can be reduced to $+1$. Thus, in (2.10), $\pm A$ generate a Lorentz transformation, written $\Lambda(A) = \Lambda(-A)$, and $\pm B_{k \leftarrow p}$ generate a Lorentz transformation taking p_μ into k_μ. In general $B_{k \leftarrow p}$ is defined by

$$B_{k \leftarrow p}\, p^\mu \sigma_\mu\, B_{k \leftarrow p}{}^\dagger = k^\mu \sigma_\mu \equiv k \cdot \sigma, \tag{2.11}$$

where we use covariant Pauli matrices.

$$\sigma_\mu = (I, \boldsymbol{\sigma}), \qquad \sigma^\mu = (I, -\boldsymbol{\sigma}), \qquad \tilde{\sigma}_\mu = \sigma^\mu, \qquad \tilde{\sigma}^\mu = \sigma_\mu. \tag{2.12}$$

The most general solution of (2.11) for $k^2 = p^2 = m^2 \neq 0$ is given by

$$B_{k \leftarrow p} = \sqrt{\frac{k \cdot \sigma}{m}}\, U \sqrt{\frac{p \cdot \tilde{\sigma}}{m}}\,, \tag{2.13}$$

where $(k \cdot \sigma/m)^{1/2}$ is defined as a hermitian square root, and U is an arbitrary 2-by-2 unitary matrix. For massive particles the last factor in (2.13) can be taken to be one, because one can choose $p = (m, 0, 0, 0)$.

For $k^2 = p^2 = 0$ we can put $p = \lambda(1001)$, $\lambda =$ arbitrary, and the most general solution of (2.11) can be written as

$$B_{k \leftarrow p} = F(k)\, U(k)\, V, \tag{2.14}$$

where $U(k)$ is a rotation taking the 3-axis to the direction of the unit vector $\hat{\boldsymbol{k}}$,

$$U(k)\, \frac{1 + \sigma_3}{2}\, U(k)^\dagger = \frac{1 + \boldsymbol{\sigma} \cdot \hat{\boldsymbol{k}}}{2}\,; \tag{2.15}$$

$F(k)$ is a dilatation,

$$F \frac{1 + \boldsymbol{\sigma} \cdot \hat{\boldsymbol{k}}}{2} F^\dagger = \frac{k^0}{\lambda} \frac{1 + \boldsymbol{\sigma} \cdot \hat{\boldsymbol{k}}}{2} = \frac{k \cdot \sigma}{2\,\lambda}$$

$$F(k) = \frac{1}{2\sqrt{k^0/\lambda}} \left[\frac{k^0}{\lambda} + 1 + \left(\frac{k^0}{\lambda} - 1 \right) \boldsymbol{\sigma} \cdot \hat{\boldsymbol{k}} \right] \tag{2.16}$$

and becomes identity for $\lambda = k^0$:

$$F(k) = I, \qquad \lambda = k^0. \tag{2.16'}$$

Finally, V is an arbitrary element of the little group of the nullvector p.

In Eq. (2.9), $Q_{(A')}{}^{(S)}$ for a massless particle is given by

$$Q^{(S)}(A') = e^{i s \theta (k, A)} = Q^{(S)}(B^{-1} A\, B), \tag{2.17}$$

where the parameter $\theta(k, A)$ is to be calculated from

$$B^{-1} A\, B = (F(k)\, U(k)\, V)^{-1} A\, F(k)\, U(k)\, V = \begin{pmatrix} e^{i\frac{\theta}{2}} & (x + i\, y)\, e^{-i\frac{\theta}{2}} \\ 0 & e^{-i\frac{\theta}{2}} \end{pmatrix}. \tag{2.18}$$

We can obtain the amplitude for a massless particle in a similar way as that of massive particle by writing [7], for each massless particle,

$$\mathscr{D}_{\mu\nu}{}^{(S)} \left(\frac{1 + \sigma_3}{2} \right) S_\nu(k), \tag{2.19}$$

i.e. the projection operator $\mathscr{D}^{(S)}(1 + \sigma_3/2)$ acting on the spinor index of the massless particle picks up just the first element of $(2\, S + 1)$-

column symbol S_ν so that the massless particle has one direction of polarization. If we write now for the amplitude (2.19) the equation (2.9)[4]:

$$\begin{aligned}\mathscr{D}^{(S)} \frac{(1+\sigma_3)}{2} S(k) &= \mathscr{D}^{(S)}(B^{-1} A\, B)\, \mathscr{D}^{(S)} \frac{(1+\sigma_3)}{2} S(\Lambda^{-1} k) \\ &= \mathscr{D}^{(S)} \left(B^{-1} A\, B \frac{1+\sigma_3}{2}\right) S(\Lambda^{-1} k) = \\ &= \mathscr{D}^{(S)} \left(e^{i\frac{1}{2}\theta(k, A)} \frac{1+\sigma_3}{2}\right) S(\Lambda^{-1} k) = \\ &= e^{iS\,\theta(k, A)}\, \mathscr{D}^{(S)} \frac{(1+\sigma_3)}{2} S(\Lambda^{-1} k). \end{aligned} \tag{2.20}$$

which is indeed the required transformation property according to (2.17). In the last calculation we made use of Eq. (2.18) and the fact that $\mathscr{D}^{(S)}(c\, A) = c^{2S}\, \mathscr{D}^{S}(A)$, if c is a complex number.

2. A. Remark on the Realization of States of Irreducible Representations

There is no problem in defining and constructing the S-matrix, even for massless particles, between the states of irreducible representations of symmetry groups. The problem is whether one can prepare the particles experimentally in these states. First of all, continuous variables such as momenta, are never measured exactly so that what are measured are not individual S-matrix elements but their superpositions. Secondly, and more seriously, there are some soft photons (or soft gravitons) always present in the initial and final states and one has a combination of different irreducible representations, or a reducible representation. Then new parameters must be introduced into the S-matrix, on the one hand for the coefficients of the superposition of states (wave packets), and on the other hand, parameters (or quantum numbers again), which tell you how many times the irreducible representations occur in the reducible representation. This is the only way to characterize a general reducible representation. This latter point deserves more attention. We have defined a particle as an irreducible representation of the symmetry group. There is no reason for this, except that it is fitting what an "elementary" structure should be. But if we are not sure what is really elementary, the new quantum numbers of the reducible representations should be considered more seriously. We shall restrict ourselves, except in the electromagnetic scattering of charged particles, to irreducible representations; it is then clear that all S-matrix elements which enter into the theory are the so-called mass-shell amplitudes, i.e. $k_i^2 = m_i^2$.

---

[4] ($\mathscr{D}^{(S)}$ can be defined even for non-unitary arguments; it is equal to $\mathscr{D}^{(OS)}$.)

A particularly important example of reducible representation is an *unstable particle*. Here a one-parameter gaussian distribution of irreducible representations in mass is introduced. The new quantum number is related to the lifetime [8].

3. Consequences of Relativistic Invariance

A. Translational Invariance

From translational invariance alone we obtain

$$\sum_{j=1}^{n} k_j = 0. \tag{2.21}$$

(Note that all particles have been drawn as ingoing.) Consequently the simplest scattering amplitude must involve three particles: $n = 3$. Such a process is physical only if $m_1 \geqslant m_2 + m_3$ (or permutations of this relation). There are only two linearly independent four vectors and none other can be constructed out of these vectors. For $n = 4$, there are three linearly independent four-vectors, but a fourth can always be constructed out of these three by the skew product

$$\varepsilon^{\mu\nu\lambda\varrho} k_{1_\nu} k_{2_\lambda} k_{3_\varrho} \equiv [k_1 k_2 k_3]^\mu. \tag{2.22}$$

For $n > 4$, one can choose four momenta, all others are linear combinations of these four with scalar coefficients.

B. Number of Fermions

If we use in our fundamental equation (2.9) the identity transformation $\Lambda(-I)$ we obtain, because $\mathscr{D}^S(-A') = (-1)^{2S} \mathscr{D}^S(A')$ and $\Lambda(-A) = \Lambda(A)$,

$$S(K) = (-1)^{2\sum_j S_j} \prod_j \mathscr{D}^{(S_j)}(I)\, S[\Lambda(-I)\, K]$$

$$= (-1)^{2\sum_j S_j} S(K),$$

hence

$$\sum_j S_j = \text{even}, \tag{2.23}$$

i.e. the total number of fermions in the initial plus the final states must be even [9]. This is a consequence of relativistic invariance, not a superselection rule.

C. Scalar Invariants

For scalar particles, Eq. (2.9) reduces to

$$S(k_1, \ldots, k_n) = S(\Lambda^{-1} k_1, \ldots, \Lambda^{-1} k_n), \tag{2.24}$$

i.e. S is an invariant function under proper *homogeneous* Lorentz transformations. If S is *analytic* (in the real) then it must be a function of the scalar products formed out of the momenta. In general, however, one could have an extra dependence on the *sign of* k^0 which is also an invariant of the proper Poincaré group. Such a dependence would destroy however the $C\,P\,T$-invariance as we shall see.

For $n = 3$ all scalar products of momenta $(k_i \cdot k_j)$ are determined by the three masses. Hence the value of the S-matrix is a constant determined by the quantum numbers (giving the lifetime of the unstable particle). If one of the particles is not on the mass shell then the S-matrix is a function of this mass, or of one scalar product. We note that if the condition $m_3 \geqslant m_1 + m_2$ is not satisfied, the condition $\Sigma_j k_j = 0$ with $k_j^2 = m_j^2$ can be satisfied for complex values of k_μ.

For $n \geqslant 4$, the configuration of n four-vectors is determined by $4\,n$ real parameters, 10 parameters can be fixed by going to a special frame and we have n mass shell conditions, $k_i^2 = m_i^2$. Thus, the total number of scalar products is

$$3\,n - 10. \tag{2.25}$$

The *physical region* of a process in the momentum space is the product of $(n - 1)$ mass shell conditions $k_i^2 = m_i^2$, plus the energy momentum conservation law.

4. Conservation Laws

The conservation of additive quantum numbers can be obtained exactly in the same way as that of momenta, because the representations of abelian groups are one-dimensional phase factors $\exp(i\,q\,\varphi)$ (q = additive quantum number).

In the case of compact groups, Eq. (2.6) and (2.7), we can reduce out the products of representations $\mathcal{D}$ into its irreducible parts; if we do this for the ingoing and outgoing particles separately we can define amplitudes corresponding to total isotopic spin, for example,

$$S^{II'} = \mathcal{D}^I\, S^{II'}\, \mathcal{D}^{I'\dagger}. \tag{2.26}$$

It follows then by an application of Schur's lemma, because $\mathcal{D}^I$'s are unitary and irreducible, that

$$S_{mm'}{}^{II'} = S^I\, \delta^{II'}\, \delta_{m\,m'}; \tag{2.27}$$

that is, the total isospin is conserved and the amplitude is independent of the third component of total isospin. Under $S\,U_3$-invariance, the total $S\,U_3$ quantum numbers λ_1, λ_2 after reduction, must be conserved; and so on.

In the case of (2.9) we see immediately that the total spin is not strictly conserved, because of the appearance of $\Lambda^{-1}\,k$ in the arguments of S on the right-hand side, as well as the k-dependence of the arguments of $\mathcal{D}^{(S)}(\Lambda'(k))$. These, of course, distinguish spin and isospin; and express the so-called *spin-orbit coupling*. In order to express the conservation of total angular momentum J we must use a different basis for the representations of the Poincaré group, in which we have

$$S(k_i \cdot k_j) = \mathscr{D}^J\, S(k_i \cdot k_j)\, \mathscr{D}^{J\dagger}, \tag{2.28}$$

and the same argument as in (2.26) can be applied.

5. Spinorial Amplitudes

We shall define now amplitudes which transform according to representations of the homogeneous Lorentz group (see Ref. 9). We start from the observation that in (2.9) we can write

$$\begin{aligned} \mathscr{D}^S(B_{k\leftarrow p}{}^{-1}\, A\, B_{q\leftarrow p}) &\equiv \mathscr{D}^{OS}(B_{k\leftarrow p}{}^{-1}\, A\, B_{q\leftarrow p}) \\ &= \mathscr{D}^{OS}(B_{k\leftarrow p}{}^{-1})\, \mathscr{D}^{OS}(A)\, \mathscr{D}^{OS}(B_{q\leftarrow p}). \end{aligned} \tag{2.29}$$

Here $\mathscr{D}^{OS}(A)$ are the representations of the *homogeneous* Lorentz group. [The general finite-dimensional non-unitary representations of the homogeneous Lorentz group of dimension $(2\,S+1)\,(2\,S'+1)$ are the matrices $\mathscr{D}^{S\,S'}(A)$ where A is not necessarily unitary.] If we insert the form (2.29) into (2.9) for all the $\mathscr{D}^S$-factors we can see that the new amplitudes defined by[5]

$$M(k) = \prod_j \mathscr{D}^{OS_j}(B_{k\leftarrow p})\; S(k) \tag{2.30}$$

satisfy the spinorial transformation property

$$M(k) = \prod_j \mathscr{D}^{OS_j}(A)\; M(\Lambda^{-1}\, k). \tag{2.31}$$

These equations hold also for mass zero case. In this case, because of (2.19), i.e. $S^{(\text{physical})} = \mathscr{D}^{OS}[(1+\sigma_3)/2]\; S(k)$, we have

$$\mathscr{D}^{(OS)}(B^{-1})\, M(k) \equiv \mathscr{D}^{OS}\, \frac{(1+\sigma_3)}{2}\, \mathscr{D}^{OS}(B^{-1}),$$

or

$$M(k) = \mathscr{D}^{OS}\left(B\,\frac{1+\sigma_3}{2}\,B^{-1}\right) M(k) = \mathscr{D}^{OS}\left(\frac{k\cdot\sigma}{2\,\lambda}\right) M(k). \tag{2.32}$$

Thus, every spinor index belonging to a mass zero particle has to satisfy Eq. (2.32). A spinor $M_\alpha(k)$ satisfying Eq. (2.32) will be called a *null-spinor*.

We can re-express the meaning of the M amplitudes as follows. An arbitrary representation of the Poincaré group can be expressed as a product of a spin zero representation of the Poincaré group mul-

---

[5] Notice that the connection between M and S involves by (2.13) $\sqrt{(k\cdot\sigma)/m}$ factors; thus the analytic properties of M and S are quite different.

tiplied with a $\mathcal{D}^{0S}$ representation of the homogeneous Lorentz group (non-unitary), provided in the latter case the states are transformed by $B_{k\leftarrow p}{}^{-1}$. (These states correspond to Foldy-Wonthuysen states.)

It is natural that any analytic properties of the amplitude hold for the covariant spinorial amplitudes, because the singularity structure of such amplitudes are independent of the Lorentz frame.

6. Helicity Amplitudes

Eq. (2.9) defines in a unified manner other classes of amplitudes depending on the choice of the rotation U in (2.13). We shall denote the amplitudes with $U = I$ for all particles by R. This choice corresponds to a measurement of spin components of all particles with respect to one and the same fixed axis $\mathfrak{z}$. The *helicity amplitudes* are so defined that the spin component of each particle is measured with respect to the direction of the spatial momentum vector, k, which is just the helicity of the particle. We shall denote these amplitudes by $H_{(\lambda)}$. If the momentum direction, $\hat{\mathbf{k}}$, is specified by angles θ and φ with respect to the fixed axis $\mathfrak{z}$, then the rotation U is given by

$$U = e^{-i\frac{\sigma_3}{2}\varphi}\, e^{i\frac{\sigma_2}{2}\theta}\, e^{i\frac{\sigma_3}{2}\varphi}. \tag{2.33}$$

For massless particles spin is already directed along $\hat{k}$, so that the helicity corresponds to the choice $V = I$ in (2.14) and (2.18).

7. Analytic Decomposition of the Spin Amplitudes

We have seen that for spinless particles the amplitude is a single invariant function of scalar products, Eq. (2.24). The general spin case can be reduced to the scalar case if we expand the amplitude in terms of a set of linearly independent basis amplitudes:

$$M(K) = \sum_i A_i(k_j \cdot k_l)\, Y_i(K), \tag{2.34}$$

where $K \equiv (k_1 \ldots k_n)$ and $Y_i(K)$ are definite, explicitly known spinorial functions carrying the transformation property of M and A_i are scalar amplitudes satisfying (2.24). Physically, the sum in (2.34) represent all the independent possible transitions. The number of such transitions for massive particles is

$$\prod_{j=1}^{n} (2\,S_j + 1), \qquad n_j \geqslant 4. \tag{2.35}$$

For massless particles and under discrete symmetries (P, C, T) this number is considerably reduced as we shall see.

The basis functions $Y_i(k)$ are simple polynomials in momenta and, at the same time, spinors. The fundamental basis functions, out of which all others are constructed, are associated with two spin 1/2 particles. For one massive spin 1/2 particle in the final state and one in the initial state — all other particles being spinless — Eq. (2.31) reduces to

$$M(K) = A\, M(\Lambda^{-1} K)\, A^{\dagger}. \tag{2.36}$$

The most general solution of (2.36) is given by

$$M(K) = \sum_{i=1}^{N} A_i\, V_{\mu}^{i}\, \sigma^{\mu}, \tag{2.37}$$

where $N = 4$ for $n \geqslant 4$, and $N = 2$ for $n = 3$, and V_{μ}^{i} are linearly independent four-vectors constructed out of momenta k_{μ} [6].

We shall formulate the analyticity properties in terms of the covariant M amplitudes. Thus, we eliminate some of the square root branch points of S, associated with factors $\mathscr{D}^{OS}(B^{-1})$. There is still the possibility that the expansion (2.34) might introduce additional *kinematical singularities* in A_i which are not present in M. These can happen at the points where some of the Y_i vanish. This in turn can happen at the boundary of the physical region (e.g. forward or backward scattering, or threshold points) where some of the momenta are linearly dependent.

In practice we shall choose the amplitudes according to the unitarity condition, in particular through the one-particle intermediate states. We conjecture that the amplitudes so chosen (which carry the pole terms) are also free of kinematical singularities [10].

8. Discrete Symmetry Transformations

Any discrete symmetry group, such as C, P, T or their combinations, has two elements. Consequently all irreducible representations are one-dimensional phase factors, one for each particle. The corresponding quantum numbers are thus multiplicative.

A. Parity

Parity invariance is expressed by the condition

$$S(k) = \eta_p\, S(\tilde{k})\,;\; \tilde{k} : (k^0, -\boldsymbol{k}), \tag{2.38}$$

where η_p is the *parity of the process* and is equal to ± 1; η_p is the product of individual parities of the particles; but these individual parities are clearly not uniquely defined if there are additive quantum numbers. For together with η_j; also $\eta_j \exp(i\,\lambda\, q_j)$ (for ingoing particles $\eta_j^{-1}\, e^{-i\lambda q_j}$) give the same parity for the process. For spinorial amplitudes the requirement of parity invariance is obtained from (2.38) and (2.30):

$$M(K) = \eta_P \prod_f \mathscr{D}^{OS_f}\!\left(\frac{k\cdot\sigma}{m_f}\right) M(\tilde{K}) \prod_i \mathscr{D}^{OS_i}\!\left(\frac{k_i\cdot\sigma}{m_i}\right)^{\dagger}, \qquad \tilde{K} : (k^0, -\boldsymbol{k}). \tag{2.39}$$

---

[6] The spinorial amplitudes for arbitrary processes can be obtained from the combinations of factors of the type $V\cdot\sigma$. Only two-component spinors need to be used. For the four-component forms see Appendix I.

B. *Time Reversal and Antiparticles*

We have drawn all particles as ingoing (or outgoing) and have the relation $\Sigma k_j = 0$. There are in general $(2^n - 2)/2$ distinct *physical reactions* (with distinct physical ranges) associated with this equation, if we change the sign of a group of particles such that $\Sigma_f k_j = \Sigma_i k_j$. If we draw all particles as ingoing, the actual outgoing particles have negative momenta and negative additive quantum numbers. Quite generally, if the outgoing particles are the irreducible representations $\mathcal{D}$ the ingoing ones must be the irreducible representations $\mathcal{D}^*$ (at least in the case of the rotation group). [This follows also from the definition of S-matrix elements in the form $\langle f | S | i \rangle$.] Thus, if we start with the "all-in" amplitude $S(k_1, \ldots, k_n)$ then the amplitudes with various $(-)$ signs, $S(k_1, -k_2, \ldots, -k, \ldots k_n)$ etc., represent physical processes in which the particles with negative momenta are outgoing particles with opposite additive quantum numbers, and also with opposite signs of the third components of spin and isospin, i.e. *antiparticles.* The last fact follows from the transformation equation (equivalence) of $\mathcal{D}^S$ into $\mathcal{D}^{S*}$:

$$\mathcal{D}^S = \mathcal{D}^S(C)\, \mathcal{D}^{S*}\, \mathcal{D}^S(C^{-1}) \tag{2.40}$$

$$\left[\text{for } S = \frac{1}{2} : A = C A^* C^{-1}, \qquad C = \begin{pmatrix} 0 & -1 \\ 1 & 0 \end{pmatrix}\right],$$

where

$$\mathcal{D}^S(C)_{mn} = (-1)^{S+m}\, \delta_{m,-m}; \qquad \mathcal{D}^S(C^{-1})^{mn} = (-1)^{S-m}\, \delta^{-m,-n} \tag{2.41}$$

are the lowering and raising spinors of rank S.

Because particles have been defined with $k^0 > 0$, *time reversal* must be understood as changing an outgoing particle into an ingoing particle. Above we have actually the process $C\,P\,T$ on each particle. T and C invariance separately are:

$$C:\ M_{m_f \dot{m}_i}(k_f, k_i) = \eta^C M_{\dot{m}_f m_i}(k_f, k_i)$$

$$T:\ M_{m_f \dot{m}_i}(k_f, k_i) = \eta^T M_{\dot{m}_f m_i}(-k^0\, \boldsymbol{k}_f;\ -k^0{}_i, \boldsymbol{k}_i). \tag{2.42}$$

Notice the change of dotted to undotted indices, which will be affected by the mixed metric spinor $G_{\dot{a}a}(k)$, given later in Eq. (3.5).

C. *C P T Theorem and Analyticity*

If we have the amplitudes for a process in the form

$$M = \sum_i A_i\, Y_i(k_1, \ldots, k_n),$$

then the basis functions $Y_i(k_1, \ldots \pm k, \ldots, k_n)$ — with all possible sign changes, represent the bases for the "crossed antiparticle processes" defined previously. Among all these "crossed processes" there is one corresponding to $Y_i(-k_1, \ldots, -k_n)$, all ingoing particles are changed into outgoing antiparticles and *vice versa*. For a reaction and its $C\,P\,T$ equivalent one we note that (1^0) the physical regions are the same; (2^0) the arguments of the scalar amplitudes A_i are the same; (3^0) the basis functions satisfy either

$$Y_i(-K) = Y_i(K), \quad \text{or} \quad Y_i(-K) = -Y_i(K), \quad \text{for all } i. \tag{2.43}$$

This follows from the fact that Y_i are polynomials in k's and for a given process all Y_i have either an even or an odd power of k. We shall assume that scalar amplitudes A_i defined over exactly the same physical region (defined by masses) are unique and the same. This implies immediately the $C\,P\,T$-theorem that a reaction and its $C\,P\,T$ equivalent have the same amplitudes up to a phase. A process and another crossed process have always disjoint physical regions. More strongly we shall assume that A_i are analytic functions; Y_i are clearly analytic polynomials in k^μ. Thus we can continue the M-functions analytically, even to complex values of k. In this way, we can even use complex Lorentz transformations and the $C\,P\,T$ invariance follows then from the invariance under the transformation $\Lambda = -I$ which in the complex case is continuously connected to identity [11, 12].

D. Connection between Spin and Statistics

Another discrete symmetry of importance is the *permutation symmetry* in the case of identical particles. If we have two identical particles (i.e. direct product of two equal representations), then the states are degenerate with respect to the total quantum numbers. We shall now postulate that the two degenerate states corresponding to the permutation of two *identical* particles, both in the initial or both in the final states, are in fact to be counted as one state. This means, according to our general procedure, that we require the S-matrix to be invariant under the representations of the permutation group.

In the case of permutation of *two* identical particles, the unitary irreducible representations of the group are one-dimensional phase factors and we obtain

$$S(\ldots k_1, k_2 \ldots) = \pm S(\ldots k_2, k_1, \ldots), \tag{2.44}$$

if k_1 and k_2 are the momenta of the identical particles. If there are more identical particles present, we could have in principle higher dimensional representations of the permutation group (which would correspond to so-called parastatistics). It is possible that analyticity and the usual crossing properties would eliminate parastatistics.

We shall use in Chapter III the normal connection between spin and statistics.

9. Unitarity and Extended Unitarity

The unitarity condition (2.1) written explicitly in terms of the R-amplitudes $(S - I)$ is

$$R_{fi} + R_{fi}{}^{\dagger} = -\sum_{N_j} \int \varrho_{N_j}\, R_{fj}\, R_{ji}{}^{\dagger}, \tag{2.45}$$

or, in terms of the spinorial amplitudes

$$M(k_f, k_i) + M^\dagger(k_i, k_f) = -\sum_{N_j} \int \varrho_{N_j} M(k_f, k_i)\, \mathscr{D}^{0 s_j}\left(\frac{k_j \cdot \tilde{\sigma}}{m_j}\right) M^\dagger(k_i, k_j), \tag{2.46}$$

where the invariant phase space element is given by

$$\varrho_N = \prod_{j=1}^{N} \frac{dk_j}{(2\pi)^4} = \prod_{j=1}^{N} \frac{d^3 k_j}{(2\pi)^3} \frac{1}{E_j}. \tag{2.47}$$

The unitarity condition is a nonlinear equation relating the real part of an amplitude to the product of other amplitudes (all on the mass shell). In conjunction with the analytic continuation we must use the unitarity as an analytic equation. In particular, it must hold below threshold and for complex values of k. If there are unphysical processes allowed by intermediate states for which the conservation and mass shell relations can be satisfied for complex k_μ they must be included in the unitarity condition. This is the meaning of *extended unitarity* [13].

It seems likely that both analytic continuation to crossed channels and extended unitarity can be justified by analyticity requirements in larger processes [14].

However, we shall take here the extended unitarity as a further assumption. It can be understood by either saying that if the momenta are complex the particular intermediate state below threshold is reached; or, one can continue analytically in the mass of one of the particles, bring the process to the physical region — where it exists by unitarity — and then continue back to the actual mass so that the analytic equations

$$\sum_f k_f = \sum_i k_i, \qquad k_j^2 = m_j^2,$$

hold. [One could write more generally $k_j^2 = m_j^2 + f(k)$; but $f(k) = 0$ for real k's, hence, if $f(k)$ is analytic it must vanish identically.]

10. Summary of Assumptions

1. *Definition:* A *particle* is an irreducible unitary representation of the symmetry groups. In the presence of zero mass particles we have to use reducible representations; for example, representations containing charged particles and photons.

2. *Definition:* A scattering process is a linear quantum process characterized by the transition amplitudes between the direct product states of initial particles and those of final particles.

Theorem: The S-matrix is unitary.

3. *Theorem:* As a result of relativistic invariance the spinorial amplitudes have the analytic decomposition

$$M(K) = \sum A_i(s, t, \ldots)\, Y_i(K);$$

where the basis functions $Y_i(K)$ describe all crossed antiparticle channels in their respective channels.

4. *Assumption:* The scalar amplitudes $A_i(s, t, \ldots)$ describe in the region of the crossed channels the actual antiparticle processes (Crossing; $C\,P\,T$-theorem).

5. *Assumption:* The unitarity holds below the physical threshold and all unphysical mass-shell amplitudes realizable for complex momenta must be included in the unitarity. (Extended unitarity).

6. *Assumption:* To every order of unitarity the scalar amplitudes are analytic functions of momenta (or, of invariants $s, t, \ldots$) except for singularities due to unitarity *to that order.* (Analyticity.)

III. Electromagnetic Interactions

1. Survey

We shall consider in this chapter the most general form of the interaction of charged particles with photons. The framework is, however, from the beginning, general enough to account in principle for the effect of all other particles, by considering sufficient terms in the unitarity condition. The scope of the electromagnetic effects can be roughly classified as follows:

(1^0) Electromagnetic interactions of leptons: Scattering processes, $e\,e \to e\,e$, $e\,\gamma \to e\,\gamma$, $\gamma\,\gamma \to \gamma\,\gamma$ with all the crossed channels, inelastic reactions $e\,e \to e\,e\,\gamma, \ldots$ etc., anomalous magnetic moments of e and μ.

(2^0) Electromagnetic interactions in the presence of or via the weak interactions: $\mu \to e\,\gamma$, $\mu \to e\,\bar{\nu}\,\nu$, $\nu\,\gamma \to \nu\,\gamma, \ldots$.

(3^0) Electromagnetic interactions of strongly interacting particles: Scattering and decay processes $e\,N \to e\,N$, $\pi^0 \to 2\,\gamma$, $\Sigma^0 \to \Lambda + \gamma, \ldots$ electromagnetic form factors; electromagnetic mass differences.

We shall discuss some typical examples, but begin with the important point: the description of the spin one particle and the photon (in its irreducible form).

It should be remarked that dispersion relations have been in use even in quantumelectrodynamics for some time [15—17]: the imaginary part of the amplitude has been evaluated from Feynman graphs and the real part via the Cauchy relations. In our treatment the imaginary parts are calculated from the extended unitarity.

2. Description of Spin 1 Particles and Photon

Consider a process with one outgoing spin 1 particle, the other $(n - 1)$ particles being of spin zero, for the time being. Eq. (2.31) for the spinorial amplitude takes the form

$$M(K) = \mathcal{D}^{01}(A)\, M(\Lambda^{-1} K). \tag{3.1}$$

If we define an irreducible spinorial object[7] transforming as

$$\mathcal{D}^{01}(A)\,\varepsilon_\mu(\Lambda^{-1}k) = \Lambda_\mu{}^\nu\,\varepsilon_\nu(k), \tag{3.2}$$

where k is the momentum of the spin 1 particle, we can write the most general solution of (3.1) in the form

$$M(K) = \sum_{i=1}^{4} A_i\,v_i{}^\mu\,\varepsilon_\mu(k), \qquad n \geqslant 4, \tag{3.3}$$

where, as before, v_i are the four-linearly independent four-vectors and A_i scalar functions. [The case $n = 3$ will be treated in Section **3.**] The four amplitudes in (3.3) reduce however automatically to three, as they must because spin 1 particles have three directions of polarization. To see this we derive the explicit form of $\varepsilon_\mu(k)$ from (3.2). Let us decompose $\mathcal{D}^{01}$ into the product of two $\mathcal{D}^{0\,(1/2)}(A) = A$:

$$\mathcal{D}^{01}(A)_m{}^n = \begin{bmatrix}\frac{1}{2} & \frac{1}{2} & 1\end{bmatrix}_m^{\alpha\beta} \begin{Bmatrix}\frac{1}{2} & \frac{1}{2} & 1\end{Bmatrix}_{\gamma\delta}^n A_\alpha{}^\gamma\,A_\beta{}^\delta, \tag{3.4}$$

where the brackets are the Clebsch-Gordan coefficients with the upper and lower indices as shown. Inserting (3.4) into (3.2) and using the orthogonality of the Clebsch-Gordan coefficients we obtain

$$A_\alpha{}^\gamma \begin{Bmatrix}\frac{1}{2} & \frac{1}{2} & 1\end{Bmatrix}_{\gamma\delta}^n \varepsilon_n{}^\mu(\Lambda^{-1}k)\,A_\beta{}^\delta = \Lambda_\nu{}^\mu \begin{Bmatrix}\frac{1}{2} & \frac{1}{2} & 1\end{Bmatrix}_{\alpha\beta}^m \varepsilon_m{}^\nu(k).$$

This equation shows that the product

$$\begin{Bmatrix}\frac{1}{2} & \frac{1}{2} & 1\end{Bmatrix}_{\alpha\beta}^n \varepsilon_n{}^\mu(k)$$

transforms exactly like $\sigma_{\alpha\beta}{}^\mu(k)$. These Pauli spinors with two undotted indices can be obtained from the usual Pauli matrices $\sigma_{\alpha\dot\beta}{}^\mu$ (see Eq. (2.36)) by the use of the metric spinors $G_\beta{}^{\dot\beta}(k)$ changing a dotted index into an undotted one:

$$\sigma_{\alpha\beta}{}^\mu(k) = G_\beta{}^{\dot\beta}(k)\,\sigma_{\alpha\dot\beta}{}^\mu = \sigma_{\alpha\dot\beta}{}^\mu\,G_\beta{}^{T\dot\beta}(k) = \sigma^\mu\frac{k\cdot\tilde\sigma}{m}C, \tag{3.5}$$

where C is the metric spinor, $C = \begin{pmatrix}0 & -1\\ 1 & 0\end{pmatrix}$, and, in the second term, we have inserted the explicit form of G. Thus, apart from a scalar function which can be incorporated into the scalar amplitudes, ε^μ is given by, with a normalization factor $1/\sqrt{2}$,

$$\varepsilon_n{}^\mu(k) = \frac{1}{\sqrt{2}}\begin{bmatrix}\frac{1}{2} & \frac{1}{2} & 1\end{bmatrix}_n^{\alpha\beta}\left(\sigma^\mu\frac{k\cdot\tilde\sigma}{m}C\right)_{\alpha\beta}. \tag{3.6}$$

---

[7] There are other types of polarization vectors used in the literature; we shall use the precise definition (3.2) via the irreducible representation $\mathcal{D}^{01}(A)$ of the homogeneous Lorentz group.

From this form it follows immediately that

$$\varepsilon^{\mu}(k)\, k_{\mu} = 0, \tag{3.7}$$

because $k \cdot \sigma\, k \cdot \tilde{\sigma} = m^2$ and

$$\text{trace}\left\{\left[\frac{1}{2}\,\frac{1}{2}\,1\right] C\right\} = 0,$$

so that the four amplitudes in (3.3) are indeed reduced to one.

A more convenient formalism which has Eq. (3.7) built in is to introduce the three k-independent irreducible matrices

$$\tau_1^{\mu\nu}, \qquad \tau_0^{\mu\nu}, \qquad \tau_{-1}^{\mu\nu}, \tag{3.8}$$

which are antisymmetric

$$g_{\mu\nu}\, \tau_n^{\mu\nu} = 0 \tag{3.9}$$

and satisfy

$$\mathcal{D}^{01}\, \tau^{\mu\nu} = \Lambda_{\mu'}{}^{\mu}\, \Lambda_{\nu'}{}^{\nu}\, \tau^{\mu'\nu'}. \tag{3.10}$$

The amplitudes M, (3.3), have then the form

$$M = \sum A_i\, t_{\mu\nu}{}^{i}(K)\, \tau^{\mu\nu}, \tag{3.11}$$

where the tensors $t_{\mu\nu}{}^{i}$ constructed out of momenta are also antisymmetric.

If $k^2 > 0$ there are indeed three independent space-like vectors $\varepsilon_n{}^{\mu}(k)$, $n = 1, 0, -1$, uniquely determined up to 3-dimensional rotations, which are orthogonal to k_{μ}. If now, $k^2 = 0$, there are only two linearly independent space-like vectors orthogonal to k_{μ}. Furthermore, ε_{μ} are now not uniquely determined. For ε_{μ} can be replaced by $\varepsilon_{\mu} + \lambda\, k_{\mu}$, λ arbitrary, without violating the condition $\varepsilon \cdot k = 0$. Because all such polarization vectors $(\varepsilon + \lambda\, k)$ are physically equivalent, they should not be distinguishable and we require the amplitudes to be invariant under the substitution

$$\varepsilon \to \varepsilon + \lambda\, k. \tag{3.12}$$

This is one way of formulating *gauge invariance* in S-matrix theory; it shows the limiting process for a spin 1 particle when mass is made zero. Of course, one can also use directly the representation $e^{i\,\varphi(k,A)}$, $[(m = 0, S = 1)]$, instead of $\mathcal{D}^{01}$ [18]. By "gauge invariance" we shall mean in the following always that we are using the zero mass representation of the Lorentz group with one direction of polarization.

3. Vertex Amplitudes Involving Photons

In this section we consider amplitudes of the type shown in Fig. 1 with in general complex momenta on the mass shell and denote them by their spins by

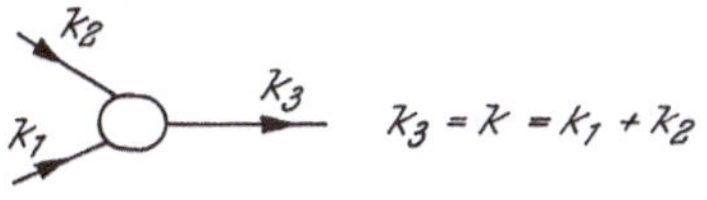

Fig. 1

$\langle S_1 S_2 S_3 \rangle$. There are only two linearly independent momenta in the problem which we choose conveniently as $(k_1 + k_2)$ and $(k_1 - k_2)$.

A. $\langle 001 \rangle$: There is a single amplitude

$$M = A(k_1 - k_2) \cdot \varepsilon \tag{3.13}$$

valid for all values of masses of the three particles. (Thus parity is necessarily conserved.) If now the spin 1 particle has zero mass, we get from (3.12),

$$A(k_1 - k_2) \cdot (k_1 + k_2) = 0,$$

hence[8]

$$A = 0, \qquad \text{if} \qquad k_1^2 \neq k_2^2. \tag{3.14}$$

A may or may not be zero if $k_1^2 = k_2^2$. We observe here the remarkable fact that the above vertex amplitude with one photon is a non-analytic function of the masses of the other two.

The non-analyticity (or the discrete nature of the amplitude) being related to the charge degree of freedom. The energetically possible reactions like $K \to \pi + \gamma$, $\eta \to \pi + \gamma, \ldots$ are forbidden.

B. $\left\langle \frac{1}{2} \frac{1}{2} 1 \right\rangle$: From particles 1 and 2 we can form the following five tensorial forms with transformation property and parity indicated

$$\begin{array}{lll}
S & \left(\dfrac{k_1}{m_1} + \dfrac{k_2}{m_2}\right) \cdot \sigma & \eta_p = + \\[2ex]
P & \left(\dfrac{k_1}{m_1} - \dfrac{k_2}{m_2}\right) \cdot \sigma & - \\[2ex]
V & \left(\sigma^\mu + \dfrac{k_2 \cdot \sigma}{m_2} \tilde{\sigma}^\mu \dfrac{k_1 \cdot \sigma}{m_1}\right) & + \\[2ex]
A & \left(\sigma^\mu - \dfrac{k_2 \cdot \sigma}{m_2} \tilde{\sigma}^\mu \dfrac{k_1 \cdot \sigma}{m_1}\right) & - \\[2ex]
T & \dfrac{k_2 \cdot \sigma}{m_2} (\tilde{\sigma}^\mu \sigma^\nu - \tilde{\sigma}^\nu \sigma^\mu) + (\sigma^\mu \tilde{\sigma}^\nu - \sigma^\nu \tilde{\sigma}^\mu) \dfrac{k_1 \cdot \sigma}{m_1}. & +
\end{array} \tag{3.15}$$

Thus we have for the vertex amplitudes, under parity conservation,

$$\begin{aligned}
M = {} & A_1 \left(\frac{k_1}{m_1} + \frac{k_2}{m_2}\right) \cdot \sigma (k_1 - k_2) \cdot \varepsilon + A_2 \left(\sigma^\mu + \frac{k_2}{m_2} \cdot \sigma \tilde{\sigma}^\mu \frac{k_1}{m_1} \cdot \sigma\right) \varepsilon_\mu + \\
& + A_3 \left[\frac{k_2 \cdot \sigma}{m_2} (\tilde{\sigma}^\mu \sigma^\nu - \tilde{\sigma}^\nu \sigma^\mu) + (\sigma^\mu \tilde{\sigma}^\nu - \sigma^\nu \tilde{\sigma}^\mu) \frac{k_1 \cdot \sigma}{m_1}\right] (k_1 - k_2)_\mu \varepsilon_\nu = \\
= {} & A_1 Y_1 + A_2 Y_2 + A_3 Y_3.
\end{aligned} \tag{3.16}$$

[8] Eq. (3.14) is valid for an arbitrary choice of (3.13): $M = A\, v^\mu(K)\, \varepsilon_\mu$.

The last term can be written as $Y_3 = a\, Y_1 + b\, Y_2 + \Delta m\, Y_3'$, where $\Delta m = m_1 - m_2$, so that the term Y_3' is not analytic in the masses.

Let now the spin 1 particle be a photon. The amplitude $A_2\, Y_2$ is already gauge invariant, the amplitude A_1 must vanish, as before, unless $k_1{}^2 = k_2{}^2$; i.e. it is also non-analytic in the masses. We rewrite (3.16), using the identity[9],

$$\left[(n\cdot\sigma\,\varepsilon\cdot\tilde{\sigma} - \varepsilon\cdot\sigma\, n\cdot\tilde{\sigma})\frac{k_1\cdot\sigma}{m_1} + \frac{k_2\cdot\sigma}{m_2}(n\cdot\tilde{\sigma}\,\varepsilon\cdot\sigma - \varepsilon\cdot\tilde{\sigma}\, n\cdot\sigma)\right] =$$
$$= 4\left(\varepsilon\cdot\sigma + \frac{k_2\cdot\sigma}{m_2}\varepsilon\cdot\tilde{\sigma}\frac{k_1\cdot\sigma}{m_1}\right) - 2\left(\frac{k_1}{m_1} + \frac{k_2}{m_2}\right)\cdot\sigma\left(\frac{k_1}{m_1} + \frac{k_2}{m_2}\right)\cdot\varepsilon$$
$$\left(n = \frac{k_2}{m_2} - \frac{k_1}{m_1}\right) \tag{3.17}$$

in the form corresponding to the usual electric and magnetic couplings

$$M = e\left(\varepsilon\cdot\sigma + \frac{k_2\cdot\sigma}{m_2}\varepsilon\cdot\tilde{\sigma}\frac{k_1\cdot\sigma}{m_1}\right) + \mu\left[(n\cdot\sigma\,\varepsilon\cdot\tilde{\sigma} - \varepsilon\cdot\sigma\, n\cdot\tilde{\sigma})\frac{k_1\cdot\sigma}{m_1} + \right.$$
$$\left. + \frac{k_2\cdot\sigma}{m_2}(n\cdot\tilde{\sigma}\,\varepsilon\cdot\sigma - \varepsilon\cdot\tilde{\sigma}\, n\cdot\sigma)\right] + A_3\, \Delta m\, Y_3', \tag{3.18}$$

with

$$e = 2\, A_1 + A_2, \qquad \mu = -\frac{1}{2}A_1.$$

The μ-term is non-analytic in the masses, so that the only term analytic in the masses is the electric coupling

$$\mathcal{Y}_e = e\left(\varepsilon\cdot\sigma + \frac{k_2\cdot\sigma}{m_2}\varepsilon\cdot\tilde{\sigma}\frac{k_1\cdot\sigma}{m_1}\right). \tag{3.21}$$

This may be the formulation of "*minimal electromagnetic coupling*" in the S-matrix theory.

---

[9] Note that the plus combinations gives

$$\left[(n\cdot\sigma\,\varepsilon\cdot\tilde{\sigma} + \varepsilon\cdot\sigma\, n\cdot\tilde{\sigma})\frac{k_1\cdot\sigma}{m_1} + \frac{k_2\cdot\sigma}{m_2}(n\cdot\tilde{\sigma}\,\varepsilon\cdot\sigma + \varepsilon\cdot\tilde{\sigma}\, n\cdot\sigma)\right] =$$
$$= 2\left(\frac{k_1}{m_1} + \frac{k_2}{m_2}\right)\cdot\sigma\, n\cdot\varepsilon.$$

In these and other calculations we make use of the relations

$$\sigma^\mu\,\tilde{\sigma}^\nu + \sigma^\nu\,\tilde{\sigma}^\mu = 2\, g^{\mu\nu}$$
$$\sigma^\mu\,\tilde{\sigma}^\lambda\,\sigma^\nu = i\,\varepsilon^{\varrho\mu\lambda\nu}\,\sigma_\varrho + \sigma^\mu\, g^{\lambda\nu} - \sigma^\lambda\, g^{\mu\nu} + \sigma^\gamma\, g^{\lambda\mu}, \tag{3.19}$$

hence

$$a\cdot\sigma\, b\cdot\tilde{\sigma}\, c\cdot\sigma = i\,[a\, b\, c] + (b\cdot c)\, a\cdot\sigma - (a\cdot c)\, b\cdot\sigma + (a\cdot b)\, c\cdot\sigma$$
$$b\cdot\sigma\, a\cdot\tilde{\sigma}\, c\cdot\sigma = 2(a\cdot b)\, c\cdot\sigma - a\cdot\sigma\, b\cdot\tilde{\sigma}\, c\cdot\sigma \tag{3.20}$$

If masses are unequal we have anyway a single amplitude. Thus reactions like $\mu \to e + \gamma$, $\Lambda \to n + \gamma$, $\Sigma^0 \to \Lambda + \gamma$, $\Sigma^+ \to p + \gamma$ are described by the single amplitude (3.21) and conserve parity.

C. $\langle 110 \rangle$. The independent tensors formed out of the momenta of the spin 1 particles, k_1 and k_2, are

$$k_i^{\mu} k_j^{\nu} (i, j = 1, 2), \qquad \varepsilon^{\mu\nu\lambda\varrho} k_{1\lambda} k_{2\varrho}. \tag{3.22}$$

Hence the spinorial amplitude is given by

$$M = A_1 k_1 \cdot \varepsilon_2 k_2 \cdot \varepsilon_1 + A_2 \varepsilon_1 \cdot \varepsilon_2 + A_3 \varepsilon_{\mu\nu\lambda\varrho} k_1^{\lambda} k_2^{\varrho} \varepsilon_1^{\mu} \varepsilon_2^{\nu}. \tag{3.24}$$

The last term is gauge invariant and has opposite parity relative to the first two terms. The masslessness condition (3.12) gives then, if one or both of the spin 1 particles is a photon (e.g. $\varrho \to \pi\gamma$, $\omega \to \pi\gamma$; $\pi^0 \to 2\gamma$, $\eta \to 2\gamma$)

$$A_2 = -(k_1 \cdot k_2) A_1.$$

Thus,

$$M = A_1 [k_1 \cdot \varepsilon_2 k_2 \cdot \varepsilon_1 - (k_1 \cdot k_2) \varepsilon_1 \cdot \varepsilon_2]. \tag{3.25}$$

If parity is not conserved we have to add the third term in (3.24). Note that under parity $\varepsilon_{\mu}(k)$ given by (3.6) transforms into

$$\varepsilon_n^{\mu(p)}(k) = \frac{1}{\sqrt{2}} \begin{bmatrix} \frac{1}{2} \frac{1}{2} 1 \end{bmatrix}_n^{\alpha\beta} (\tilde{\sigma}^{\mu} k \cdot \sigma C)_{\alpha\beta}.$$

D. $\langle 111 \rangle$: We have now

$$M = \sum A_i t_{\mu\nu\lambda}^{i} \varepsilon_1^{\mu} \varepsilon_2^{\nu} \varepsilon_3^{\lambda}. \tag{3.26}$$

The independent tensors $t_{\mu\nu\lambda}^{i}$ formed out of k_1 and k_2 are

$$k_i^{\mu} k_j^{\nu} k_l^{\lambda} (i, j, l = 1, 2); \qquad g^{\mu\nu} k_i^{\lambda}, g^{\mu\nu} k_i^{\nu}, g^{\nu\lambda} k_i^{\mu}; \qquad \varepsilon_{\mu\nu\lambda\varrho} k_i^{\varrho}. \tag{3.27}$$

Under parity conservation we obtain then the four amplitudes (note: $k_1 \cdot \varepsilon_3 = -k_2 \cdot \varepsilon_3$)

$$\begin{aligned} M = {} & A_1 k_1 \cdot \varepsilon_2 k_2 \cdot \varepsilon_1 k_1 \cdot \varepsilon_3 + A_2 \varepsilon_1 \cdot \varepsilon_2 k_1 \cdot \varepsilon_3 + \\ & + A_3 \varepsilon_1 \cdot \varepsilon_3 k_1 \cdot \varepsilon_2 + A_4 \varepsilon_2 \cdot \varepsilon_3 k_2 \cdot \varepsilon_1. \end{aligned} \tag{3.28}$$

If two of the particles are massless, the condition (3.12) on particles 1 and 2 gives

$$A_1 (k_1 \cdot k_2) + A_2 + A_3 = A_4. \tag{3.29}$$

The remaining three amplitudes may be further reduced if the Bose statistics is used for the two photons. The terms A_1 and A_2 in (3.28) are antisymmetric under the interchange $(k_1 \varepsilon_1) \leftrightarrow (k_2 \varepsilon_2)$. The decay of spin 1 particles in to two photons has been first considered by LANDAU and YANG [19].

4. Soft Photons and Cluster Decomposition of the *S*-matrix

We have seen that in the case of scattering of charged particles we have to use in the initial and final states products of reducible representations of massive and massless particles (soft photons). This leads us to consider from the beginning larger S-matrix elements. In the example of Fig. 2 we assume, for simplicity, a single soft photon and separate from the S-matrix the disconnected parts.

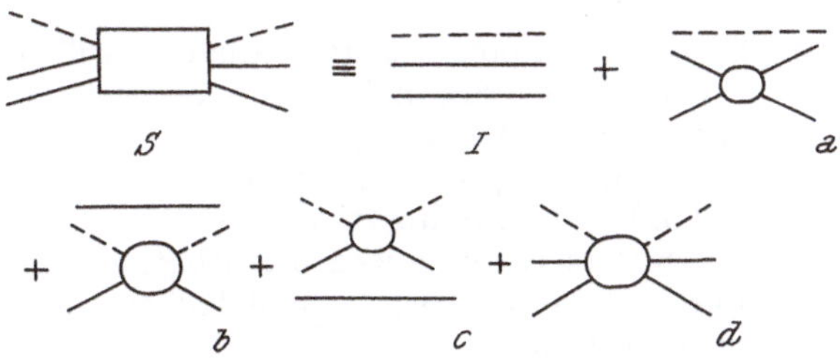

Fig. 2

In Fig. 2, a straight line indicates an energy-momentum δ-function factor in the corresponding amplitude. The part (d) represents the connected part of the amplitude. Thus, all the terms (a), (b), (c) and (d) must be separately evaluated. In the next section we explicitly carry an approximation method in which every amplitude occurring in the unitarity equation is approximated by its "pole" terms in all channels. We assume thereby the e^2 is small and collect all terms in Fig. 2 which are of the same order in e^2. We do this to show the connection with the renormalized perturbation theory.

In the lowest order we get a contribution from (a), Fig. 2, and we shall write the unitarity for the connected part of (a) in the next Section. The connected parts of (b) and (c) may at first be neglected because of the very small momenta of the soft photons. Hence the lowest order term is essentially the pole terms of (a). The higher order terms of (a) must now be combined with the terms of the same order in (b), (c) and (d). It is interesting that unitarity condition and (a) alone, in the higher orders, would give divergent results (Section 7) and that the cluster expansion of Fig. 2 is essential to obtain final results. This so-called infra-red problem occurs, of course, almost in the same form in field theory.

If the initial charged particles are well prepared and one considers scattering from neutral targets then soft photons only in the final states need to be considered, namely those in the range of experimental detection of charged particles; energy range ΔE and solid angle range $\Delta \Omega$ [20].

5. Successive Pole Approximation

In this Section we discuss how unitarity and analyticity are used to evaluate the scattering amplitude. If we write the S-matrix as

$$S = I + R = I + i(2\pi)^4\, \delta(P_f - P_i)\, G, \tag{3.30}$$

then the unitarity condition is given by

$$G_{fi} - G_{fi}^{\dagger} = i \sum_j G_{fj}\, G_{ji}^{\dagger} (2\pi)^4\, \delta(P_f - P_i). \tag{3.31}$$

The sum over states will be taken to be

$$\sum_{\text{states of } j} = \sum_{\text{spins}} \int \frac{dk_j}{(2\pi)^4} 4\pi\, \delta(k_j^2 - m_j^2)\, \theta(k_j^0). \tag{3.32}$$

As we have explained in Chapter I. 9, this equation will be used as an analytic equation even below threshold where we pick up the analytically continued processes. We shall show that in quantum electrodynamics it is a very good approximation to consider only the one-particle states in (3.31) and their iterations. At least, this is sufficient to reproduce the results of renormalized perturbation theory. It is of course also possible to treat the two — and higher particle states — different from the perturbation theory.

We shall introduce the analyticity assumptions, that is, the nature of singularities in the complex planes of the variables to every order of unitarity as we go along.

It is best to illustrate the procedure in terms of an example; it can easily be applied to any other case.

Example: Compton Scattering of a Fermion [21].

Let k_1 and k_3 be the momenta of the fermions f and f', and (k_2, ε_2), (k_4, ε_4) the momenta and the polarization vectors of the photons. There are, in general, four related antiparticle processes: s-channel[10]: $f\gamma \to f'\gamma$ t-channel: $f f' \to \gamma\gamma$: u-channel: $f\gamma \to f'\gamma$, and the decay channel $f' \to f + \gamma + \gamma$. [Note that a reaction like $\mu \to e + \gamma + \gamma$ is not forbidden by Lorentz invariance, but by an additional quantum number, the μ-ness quantum number.]

There are 36 amplitudes to begin with, starting with massive vector particles, but this number reduces to six under (3.7), (3.12), plus the parity and time reversal invariance. One could write down these

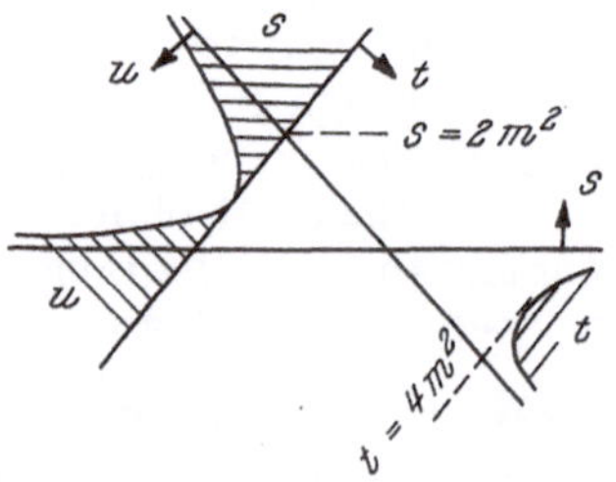

Fig. 3

amplitudes. It is however most convenient to choose the basis functions Y_i in such a way that the unitarity condition is diagonalized, at least for one-particle states in all channels. We therefore let the unitarity

[10] We use the standard variables $s = (k_1 + k_2)^2$, $t = (k_1 - k_3)^2$, $u = (k_1 - k_4)^2$, with $s + t + u = \Sigma\, m_i^2$.

condition choose the basis functions. The physical regions of the three scattering channels in the variables s, t, u (for equal masses of the fermions) and, diagrammatically, the three unitarity conditions are shown in the following figures:

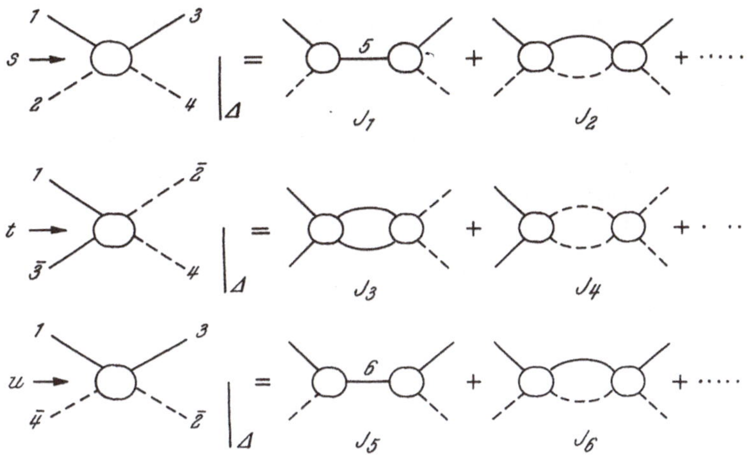

Fig. 4

Because we are interested primarily to reproduce the field theory results here we shall use on the right-hand side of the unitarity equation the minimal coupling (3.21). Even without a fundamental principle of minimal coupling we would be justified to take the e-term alone on the right-hand side of unitarity: This is because the unitarity is a non-linear equation, and because $\mu \ll e$, we would get terms like $e\mu$ and μ^2 which are lower order than e^2. (See also next Section.)

The contribution of J_1 is given by, with (3.21),

$$\sum_j A_j Y_j - \left(\sum_j A_j Y_j\right)^\dagger = i \sum_{\text{spins}} \varrho_1 \mathscr{Y}_e^\dagger \mathscr{D}^{01}\left(\frac{k_5 \cdot \tilde{\sigma}}{m_5}\right) \mathscr{Y}_e =$$

$$= i\,\varrho_1 |e|^2 \left[\left(\varepsilon_4 \cdot \sigma + \frac{k_3 \cdot \sigma}{m_3}\, \varepsilon \cdot \tilde{\sigma} \frac{k_5 \cdot \sigma}{m_5}\right) \frac{k_5 \cdot \tilde{\sigma}}{m_5} \left(\varepsilon_2 \cdot \sigma + \frac{k_5 \cdot \sigma}{m_5}\, \varepsilon_2 \cdot \tilde{\sigma} \frac{k_1 \cdot \sigma}{m_1}\right)\right], \tag{3.33}$$

where we can put symmetrically

$$k_5 = \frac{1}{2}(k_1 + k_2 + k_3 + k_4).$$

We now choose the basis Y_1 to be just the term in the square bracket in (3.33). This term has positive parity[11] and satisfies

$$Y_1^\dagger = Y_1. \tag{3.34}$$

[11] Note that because unitarity is a non-linear equation we must use all positive parity terms throughout.

Consequently the term J_1 contributes only to the amplitude A_1 so defined and we have immediately

$$
\begin{aligned}
A_1 - A_1^* &= i\,\varrho_1\,|e|^2 = i\,4\pi\,|e|^2\,\delta(s - m_5^2) \\
A_j - A_j^* &= 0, \qquad j \neq 1,
\end{aligned}
\tag{3.35}
$$

where we have used the one particle phase space factor

$$
\varrho_1 = 4\pi\,\delta(s - m^2). \tag{3.36}
$$

In exactly the same way, the term J_5 in the unitarity equation of the u-channel, leads to a second basis function

$$
\begin{aligned}
Y_2 = \Bigg[& \varepsilon_2\cdot\sigma\, k_6\cdot\tilde{\sigma}\,\varepsilon_4\cdot\sigma + \frac{k_3\cdot\sigma}{m_3}\,\varepsilon_2\cdot\tilde{\sigma}\,\frac{k_6\cdot\sigma}{m_6}\,\varepsilon_4\cdot\tilde{\sigma}\,\frac{k_1\cdot\sigma}{m_1} + \\
& + \varepsilon_2\cdot\sigma\,\varepsilon_4\cdot\tilde{\sigma}\,\frac{k_1\cdot\sigma_1}{m_1} + \frac{k_3\cdot\sigma}{m_3}\,k_2\cdot\tilde{\sigma}\,\varepsilon_4\cdot\sigma \Bigg]
\end{aligned}
\tag{3.37}
$$

$$
k_6 = \frac{1}{2}\,(k_1 - k_4 + k_3 - k_2)
$$

$$
Y_2^\dagger = Y_2;
$$

i.e. Y_2 is obtained from Y_1 by the interchange $\varepsilon_2 \leftrightarrow \varepsilon_4$ and $k_2 \leftrightarrow -k_2$, $k_4 \leftrightarrow -k_4$. [The arguments of ε_2 and ε_4 in (3.37) are $(-k)$.] Thus, J_5 contributes only to the amplitude A_2 so defined:

$$
\begin{aligned}
A_2 - A_2^* &= i\,4\,\pi\,|e|^2\,\delta(u - m_6^2) \\
A_j - A_j^* &= 0, \qquad j \neq 2.
\end{aligned}
\tag{3.38}
$$

In the t-channel, as well as for the higher order terms in s- and u-channels, we have again two-particle amplitudes on the right-hand side. We have to write a unitarity condition for these amplitudes as well. It is clear that if these amplitudes themselves are approximated by their one-particle contributions the results will be of higher order in $|e|^2$. Thus to order $|e|^2$ the terms (3.35) and (3.38) are the only terms.

Now we consider the equations (3.35) and (3.38) as analytic equations with only singularities indicated by the right-hand side. It is true that many particle intermediate states in the unitarity give branch points and other singularities. But the discontinuities of these singularities are of higher order in $|e|^2$ than the one-particle states so that to this order we can neglect these higher singularities. Thus the approximate appropriate analyticity assumptions to order $|e|^2$ can be expressed by the dispersion relations

$$
A_i = \frac{1}{\pi}\int_{-\infty}^{+\infty} \frac{\operatorname{Im} A_i(s_i')\,ds_i'}{(s_i' - s_i)}, \qquad i = 1, 2. \qquad s_1 = s, \qquad s_2 = u. \tag{3.39}
$$

No subtractions are necessary to this order because Im A is just a δ-function, and we obtain

$$A_i = -\frac{2\,e^2}{s_i - m_i^2}, \qquad i = 1, 2. \tag{3.40}$$

Hence the full amplitude to this order is given by

$$M = -2\,e^2\left[\frac{Y_1}{s - m_s^2} + \frac{Y_2}{u - m_u^2}\right]. \tag{3.41}$$

We can now easily pass from this spinorial amplitude to the actual R- or G-amplitudes by Eq. (2.30). Denoting the corresponding basis functions by R_1 and R_2 we have finally

$$G = -2\,e^2\left[\frac{R_1}{s - m_s^2} + \frac{R_2}{u - m_u^2}\right],$$

where (3.42)

$$R_i = \sqrt{\frac{k_3 \cdot \tilde{\sigma}}{m_3}}\, Y_i \sqrt{\frac{k_1 \cdot \tilde{\sigma}}{m_1}}, \qquad i = 1, 2,$$

with Y_1 and Y_2 given by (3.33) and (3.37). Note that in the t-channel G is invariant under the interchange of the two photons: $-k_2 \leftrightarrow k_4$; $-\varepsilon_2 \leftrightarrow \varepsilon_4$.

The remaining trace calculations are easier to do in the above two-component form rather than in four-component form. But because these calculations already exist in four-component form it is sufficient to give the four-component form of (3.42). By means of Appendix I we find

$$\begin{aligned} R_1 &= \bar{u}_3(k_3)\,\varepsilon_4\cdot\gamma(k_1\cdot\gamma + k_2\cdot\gamma + m)\,\varepsilon_2\cdot\gamma\,u_1(k_1)\\ R_2 &= \bar{u}_3(k_3)\,\varepsilon_2\cdot\gamma(k_1\cdot\gamma - k_4\cdot\gamma + m)\,\varepsilon_4\cdot\gamma\,u_1(k_1). \end{aligned} \tag{3.43}$$

With these (3.42) is precisely the field theoretical result [22]. By crossing symmetry, (3.42) holds for all three channels in their respective physical regions.

Remark: Clearly the spin summation in the unitarity contains only physical polarization states, because the particles are on the mass shell. In particular, for the one-photon intermediate state in $e - e$ scattering,

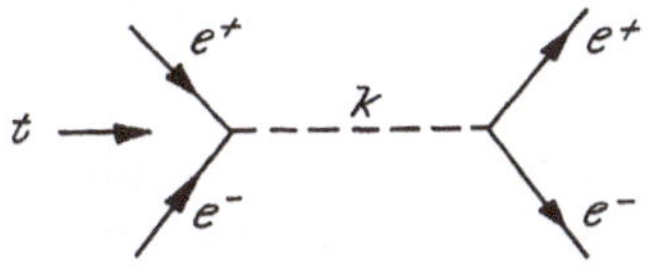

Fig. 5

for example, (Fig. 5), the two vertex amplitudes are already gauge invariant, then in the unitarity of the M-functions we use the relation

$$\varepsilon_\mu \mathscr{D}^{01}(k \cdot \tilde{\sigma})\, \varepsilon_\nu = g_{\mu\nu}. \tag{3.44}$$

Although $\mathscr{D}^{01}$ appears in the spin sum of the intermediate state, only the physical states of the photon contribute. We give here, for further reference, the first order result for $e - e$ scattering which can be derived in a similar way [21]

$$G_{ee\to ee} = -2e^2\left[\frac{\bar{u}_4 \gamma^\mu u_2 \otimes \bar{u}_3 \gamma_\mu u_1}{t} - \frac{\bar{u}_3 \gamma^\mu u_2 \otimes \bar{u}_4 \gamma_\mu u_1}{u}\right]. \tag{3.45}$$

The relative sign between the two terms is such that G is antisymmetric under the exchange of two identical fermions in s-channel:

$$G(1 \leftrightarrow 2) = G(3 \leftrightarrow 4) = -G.$$

Before we consider higher order scattering terms J_2, J_3, J_4, J_6 terms in (Fig. 4) we discuss in the next Section the anomalous magnetic moment of the leptons.

6. Anomalous Magnetic Moment of the Leptons

The vertex amplitude, with all the particles on the mass shell, cannot, of course, be realized with real momenta and there are no variables to

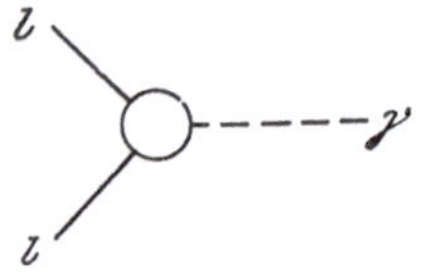

Fig. 6

write dispersion relations. The magnetic moment of the lepton l is measured by the interaction with an external field or by a scattering experiment. We can represent this complicated interaction (e.g. the external field) by introducing a fictitious spin 0 particle α carrying energy-momentum. Because we are interested in the limiting case of

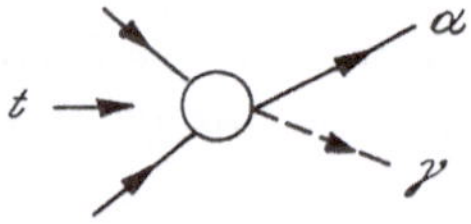

Fig. 7

Fig. 6, it is a sufficient approximation to couple the sum of the momenta of α and the photon to leptons. This is equivalent to giving the photon a mass.

We therefore need only forward scattering dispersion relations in t (for $s = 0$) in Fig. 7. Both forward scattering dispersion relation and the dispersion relation for the vertex amplitude in the mass of one of the particles, are proved to all order of perturbation theory, whereas dispersion relations in both variables s, t (Mandelstam representation) are at present not proved. We need in our approximations only the forward scattering dispersion relations [23].

In Fig. 7 we have then one variable t and we can write unitarity and dispersion relations (forward scattering). The unitarity in all three channels is shown in Fig. 8:

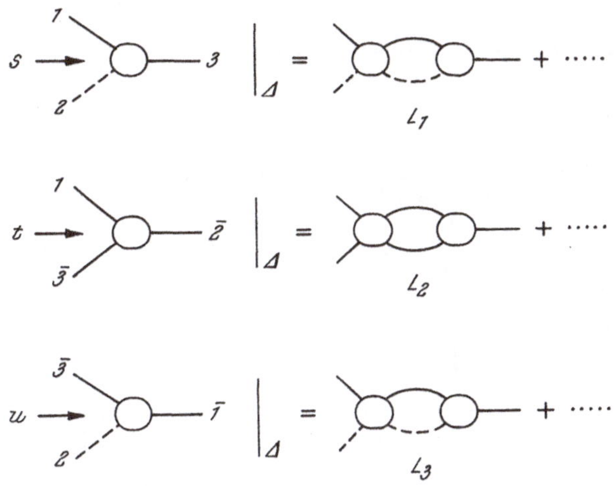

Fig. 8

Now we approximate the two-body amplitudes in L_1, L_2 and L_3 by their pole terms in *all* channels (Fig. 8′).

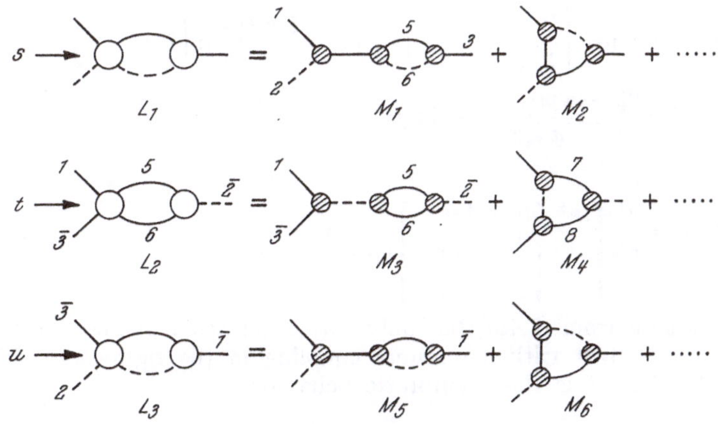

Fig. 8′

In the interesting case of t-channel we find [21]

$$M_3 = -\int \varrho_2 \, d\Omega \left[\varepsilon_2 \cdot \sigma + \frac{k_5 \cdot \sigma}{m_5} \varepsilon_2 \cdot \tilde{\sigma} \frac{k_6 \cdot \sigma}{m_6} \right] \frac{k_6 \cdot \tilde{\sigma}}{m_6} \otimes \frac{k_5 \cdot \tilde{\sigma}}{m_5} \cdot$$

$$\cdot \frac{-2e^2}{m_2{}^2} \left[\left(\sigma^\mu + \frac{k_6 \cdot \sigma}{m_6} \tilde{\sigma}^\mu \frac{k_5 \cdot \sigma}{m_5} \right) \otimes \left(\sigma_\mu + \frac{k_3 \cdot \sigma}{m_3} \tilde{\sigma}_\mu \frac{k_1 \cdot \sigma}{m_1} \right) \right] \cdot$$

$$\cdot (2\pi)^4 \, \delta(P_{56} - P_{13}) = 2\, h(t) \left[\varepsilon_2 \cdot \sigma + \frac{k_3 \cdot \sigma}{m_3} \varepsilon_2 \cdot \tilde{\sigma} \frac{k_1 \cdot \sigma}{m_1} \right], \quad (3.46)$$

where ϱ_2 is the two-body phase space,

$$\varrho_2 = \frac{1}{4\pi^2} \frac{1}{2\,t} \left[(t - (m_5 + m_6)^2) \, (t - (m_5 - m_6)^2) \right]^{1/2} \quad (3.47)$$

and we have used t-channel Born approximation to Møller Scattering given in (3.45) and the relation (3.44). Similarly,

$$M_4 = -\int \varrho_2 \, d\Omega \, e \left(\varepsilon_2 \cdot \sigma + \frac{k_7 \cdot \sigma}{m_7} \varepsilon_2 \cdot \tilde{\sigma} \frac{k_8 \cdot \sigma}{m_8} \right) \frac{k_8 \cdot \tilde{\sigma}}{m_8} \otimes \frac{k_7 \cdot \tilde{\sigma}}{m_7} \frac{2\,e^2}{(k_7 - k_8)^2} \cdot$$

$$\cdot \left(\sigma^\mu + \frac{k_3 \cdot \sigma}{m_3} \tilde{\sigma}^\mu \frac{k_7 \cdot \sigma}{m_7} \right) \otimes \left(\sigma_\mu + \frac{k_8 \cdot \sigma}{m_8} \tilde{\sigma}_\mu \frac{k_1 \cdot \sigma}{m_1} \right) (2\pi)^4 \, \delta(P_{78} - P_{15}) =$$

$$= 2\, f_1(t) \left[\varepsilon_2 \cdot \sigma + \frac{k_3 \cdot \sigma}{m_3} \varepsilon_2 \cdot \tilde{\sigma} \frac{k_1 \cdot \sigma}{m_1} \right] + 2\, f_2(t) \left[\left(\frac{k_1}{m_1} + \frac{k_3}{m_3} \right) \cdot \sigma (k_1 + k_3) \cdot \varepsilon \right]. \quad (3.48)$$

Integrals of this type have been already evaluated by KÄLLEN [15] and we find

$$h(t) = \text{const.} \sqrt{1 - \frac{4\,m^2}{t}} \left(1 + \frac{2\,m^2}{t} \right) \theta(t - 4\,m^2) \underset{|t| \to \infty}{\to} \text{const.}$$

$$f_1(t) = \text{const.} \left[\frac{3}{2} \left(1 - \frac{4\,m^2}{t} \right) - \left(1 - \frac{2\,m^2}{t} \right) \right] \ln \left(1 + \frac{t - 4\,m^2}{\lambda} \right) \cdot$$

$$\cdot \frac{\theta(t - 4\,m^2)}{\sqrt{1 - \dfrac{4\,m^2}{t}}} \underset{|t| \to \infty}{\to} \ln |t|$$

$$f_2(t) = \frac{e^2}{4\pi} \left[\frac{m^2}{t} \frac{\theta(t - 4\,m^2)}{\sqrt{1 - \dfrac{4\,m^2}{t}}} \right] \underset{|t| \to \infty}{\to} 0. \quad (3.49)$$

It is seen from (3.48) that only $f_2(t)$ generates a magnetic moment term, if we start with a minimal coupling in the right-hand side of Fig. 8′. And it is the asymptotic behavior

$$f_2(t) \underset{|t| \to \infty}{\to} 0$$

that allows us to calculate μ; whereas due to the logarithmic and constant asymptotic behavior of $h(t)$ and $f_1(t)$ we cannot calculate the electric coupling constant e, as we shall see. In the integration of $f_1(t)$ we have taken the lower limit of integration to be λ to avoid infra-red divergences; $f_2(t)$ is free of infra-red terms.

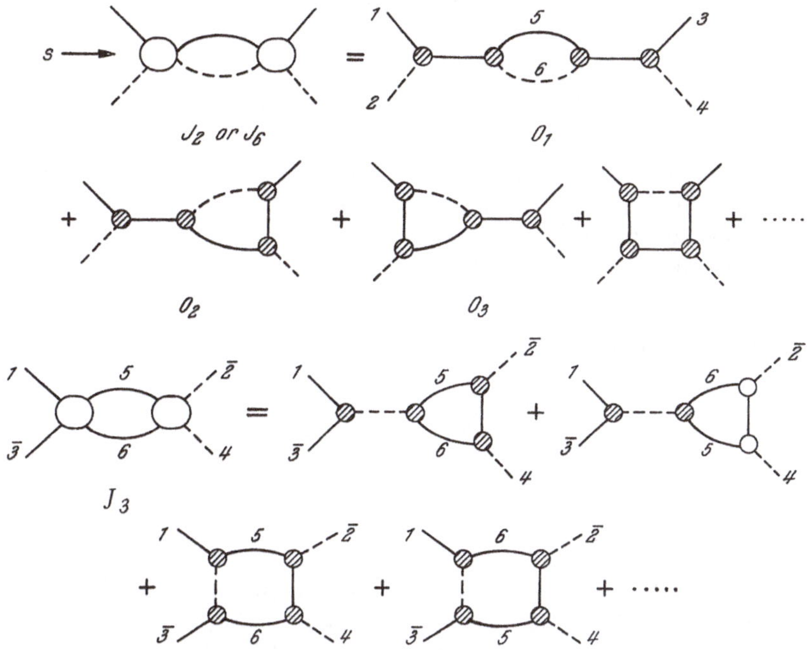

Fig. 9

Now we write dispersion relations for the amplitude $G(t)$ and because of the asymptotic behavior of $h(t)$ and $f_1(t)$ we must use one subtraction in these terms, whereas $f_2(t)$ term does not need any subtraction:

$$G(t) = \left\{ \frac{t}{\pi} \int_{-\infty}^{\infty} \frac{dt'}{t'(t'-t)} [h(t') + f_1(t') + 2 f_2(t')] + [h(0) + f_1(0) + 2 f_2(0)] \right\} \times$$

$$\times \mathcal{Y}_e(K) + \frac{1}{\pi} \int_{-\infty}^{\infty} \frac{dt'}{t'-t} f_2(t') \mathcal{Y}_\mu(K), \tag{3.50}$$

where the spinor functions $\mathcal{Y}_e(K)$ and $\mathcal{Y}_\mu(K)$ are those given in Eq. (3.18). For $t = 0$, (3.50) and (3.18) must coincide, hence

$$h(0) + f_1(0) + 2 f_2(0) = e \tag{3.51}$$

and

$$\frac{1}{\pi}\int_{-\infty}^{\infty}\frac{dt'}{t'}f_2(t') = \mu(2\,m\,\pi) \tag{3.52}$$

or

$$\mu = \frac{2}{\pi}\int_{4m^2}^{\infty}\frac{e^3\,m}{16\pi t'}\frac{dt'}{\sqrt{t'(t'-4\,m^2)}} = \frac{e^3}{16\pi^2 m} \tag{3.53}$$

which is the known lowest order value of the anomalous magnetic moment.

7. Higher Order Terms and Relation to Feynman Graphs

To evaluate the second order corrections J_2, J_3 and J_6 to Compton scattering shown in Fig. 4, we approximate the amplitudes in these terms by their lowest order terms in all channels. We have thus the situation shown in Fig. 9.

For comparison the Feynman diagrams corresponding to $O_1 - O_4$ are shown in Fig. 10.

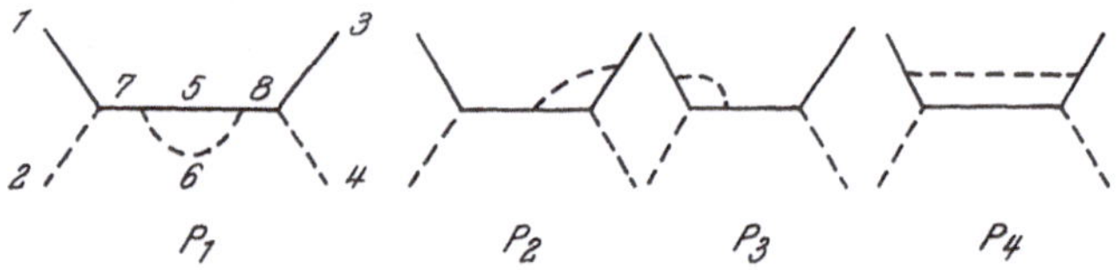

Fig. 10

The relation of the perturbation terms to the unitarity diagrams is the following. In Fig. 9, we calculate the contribution of terms $O_1, \ldots$ to the imaginary part of the amplitude. In Fig. 10, we calculate the contribution of $P_1, \ldots$ to the full amplitude. It will be shown that the imaginary part of the Feynman diagram, say for P_1, is equal to the contribution of O_1, and so on. The contribution of P_1 is infinite by the renormalization terms and can be "renormalized" by evaluating its imaginary part and recovering its real part through the dispersion integral, which is precisely the prescription of the S-matrix theory from the beginning. The contribution of O_1 (Fig. 9) is given by (in four-component form, for comparison)

$$O_1 = i\sum_{\substack{\text{spins}\\5,6}}\int\frac{d^4k_5}{(2\pi)^4}\frac{d^4k}{(2\pi)^4}(4\pi)^2\,\delta(k_5{}^2-m_5{}^2)\,\theta(k_5{}^0)\,\delta(k_6{}^2)\,\theta(k_6{}^0)\,\times$$

$$\times\left[-2\,e^2\,\bar{u}_3\,\underline{\varepsilon_4}\cdot\gamma\,\frac{(\underline{p}_{12}\cdot\gamma+m)}{s-m^2}\,\underline{\varepsilon_6}\cdot\gamma\,u_5\right]\cdot$$

$$\cdot\left[-2\,e^2\,\bar{u}_5\,\underline{\varepsilon_6}\cdot\gamma\,\frac{(\underline{p}_{12}\cdot\gamma+m)}{s-m^2}\,\underline{\varepsilon_2}\cdot\gamma\,u_1\right]\times(2\pi)^4\,\delta(P_{56}-P_{12}). \tag{3.54}$$

The corresponding Feynman amplitude (Fig. 10) is

$$P_1 = i\,(\text{const.}) \sum \int \frac{dk_5}{k_5^2 - m^2 - i\,\varepsilon_5} \cdot$$
$$\cdot \frac{dk_6}{k_6^2 - i\,\varepsilon_6}\,\frac{dk_7}{k_7^2 - m^2 - i\,\varepsilon_7}\,\frac{dk_8}{k_8^2 - m^2 - i\,\varepsilon_8} \times$$
$$\times\, G(8 \to 31)\,\delta(k_{34} - k_8)\,G(56 \to 8)\,\delta(k_8 - k_{56})\,G(7 \to 56) \times$$
$$\times\, \delta(k_{56} - k_7)\,G(12 \to 7)\,\delta(k_7 - k_{12}). \tag{3.55}$$

We can take the imaginary part of (3.51) by the rule

$$\frac{1}{k_j^2 - m_j^2 - i\,\varepsilon_j} = P\,\frac{1}{k_j^2 - m_j^2} + i\pi\,\delta(k_j^2 - m_j^2). \tag{3.56}$$

It is seen then that the integrals involving the principal parts vanish; the cross terms also vanish for any $s = k_{12} \geqslant m^2 + \lambda^2$. The resulting imaginary part is equivalent to O_1, except for $\delta(k_{34} - k_{12})$ [which is due to the fact that P_1 is actually $G\,\delta(k_{34} - k_{12})$] and for $\theta(k_5^0)\,\theta(k_6^0)$ in O_1. These θ-functions serve to eliminate the poles in the unphysical sheets of the s-plane.

In general there are more Feynman diagrams than unitarity diagrams. Those Feynman diagrams which have no corresponding unitarity diagrams contribute only to renormalization constants which are of course absent in S-matrix theory.

8. On the Derivation of Charge Conservation

As we have seen, the condition of masslessness considerably restricts the form of the amplitude. As a further application we consider a process involving n spinless particles and a photon of momentum k. The spinorial amplitude can be written as

$$M = \sum_i A_i(s, t, \ldots)\, k_i^\mu\, \varepsilon_\mu. \tag{3.57}$$

The gauge condition implies

$$\sum_i A_i(s, t, \ldots)\, k_i^\mu\, k_\mu = 0. \tag{3.58}$$

Let us draw all particles as ingoing; the conservation of momenta, $k + \Sigma\, k_j = 0$, gives

$$\sum_{j=1}^{n} k_j^\mu\, k_\mu = 0. \tag{3.59}$$

If we introduce new form factors $E_i(s, t, \ldots)$ defined by

$$E_i = A_i(s, t \ldots)\,(k_i \cdot k), \tag{3.60}$$

then (3.58) gives immediately

$$\sum_{i=1}^{n} E_i(s, t, \ldots) = 0. \qquad (3.61)$$

Thus the form factors E_i so introduced satisfy a conservation law. If we go to crossed channels by changing some k_j into $-k_j$, then it follows from (3.58)—(3.60) that E_j must change signs as well; i.e. these form factor behave like additive quantum numbers; they have opposite signs for antiparticles. The amplitude has now the form

$$M = \sum_{i=1}^{n} E_i(s, t, \ldots) \frac{k_i \cdot \varepsilon}{k_i \cdot k}. \qquad (3.62)$$

Because there is only *one* unknown transition amplitude under the proper Lorentz group, (3.62) is of the form

$$M = E(s, t, \ldots) \sum_{i=1}^{n} \eta_j \frac{k_i \cdot \varepsilon}{k_i \cdot k} \qquad (3.63)$$

with constants η_i such that

$$\sum_{j=1}^{n} \eta_j = 0. \qquad (3.64)$$

Clearly there are many ways of introducing form factors. The form factors E_i have been so introduced that they go in the limit $k \to 0$ (soft photon) to the charge coupling constant e_i. For if we approximate for $k \to 0$ the amplitude we are considering as a sum of pole terms (Fig. 11).

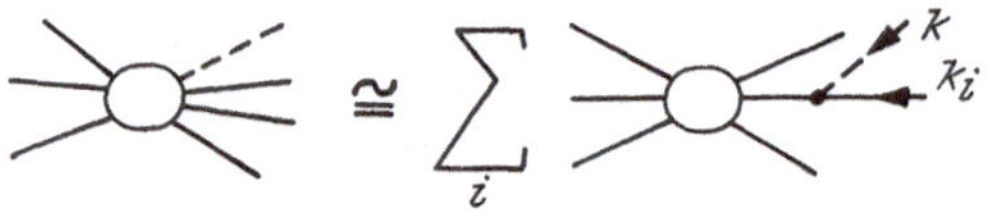

Fig. 11

We have

$$M \cong \sum_{i} e_i \frac{k_i \cdot \varepsilon}{(k + k_i)^2 - m_i^2} = \sum_{i} e_i \frac{k_i \cdot \varepsilon}{2\, k \cdot k_i}, \qquad (3.65)$$

which is the same equation (3.62) or (3.63) for constant form factors. Thus the approximation Fig. 11 implies charge conservation [24].

IV. On Gravitational Interactions

We discuss in this chapter some consequences of the interaction of a massless particle of spin 2 (and higher). We deviate from the general relativity insofar as the particles interacting via the graviton are assumed to have definite spin and mass, i.e. are defined again by the unitary representations of the Lorentz group.

1. A spin 2 massless particle of momentum k will be described by the symmetric *polarization tensor*

$$\varrho_m^{\mu\nu}(k) = [112]_m^{\alpha\beta}\, \varepsilon_\alpha^\mu(k)\, \varepsilon_\beta^\nu(k), \tag{4.1}$$

which then transforms according to

$$\mathcal{D}^{02}\, \varrho^{\mu\nu}(\Lambda^{-1} k) = \Lambda_\sigma^\mu\, \Lambda_\lambda^\nu\, \varrho^{\sigma\lambda}(k). \tag{4.2}$$

The ε^μ's have been given in Eq. (3.6). We have now the restrictions

$$\varrho^{\mu\nu}(k)\, k_\mu = \varrho^{\mu\nu}(k)\, k_\nu = 0 \tag{4.3}$$

and

$$\varrho_\mu{}^\mu(k) = 0. \tag{4.4}$$

The condition of masslessness is expressed again by the invariance of the spinorial amplitudes under the replacement $\varepsilon \to \varepsilon + \lambda\, k$, for all λ, or, by

$$\varrho^{\mu\nu} \to \varrho^{\mu\nu} + \lambda\, [112]_m^{\alpha\beta}\, k^\mu\, \varepsilon^\nu + \lambda\, [112]\, \varepsilon^\nu\, k^\mu + \lambda^2\, [112]\, k^\mu\, k^\nu,$$
$$\lambda \text{ arbitrary}. \tag{4.5}$$

2. Consider first the vertex $\langle 002 \rangle$ of two spin 0 particles of momenta k_1 and k_2 with $k = k_1 + k_2$. We have for this case the single amplitude

$$M = f(k_1 - k_2)^\mu\, (k_1 - k_2)^\nu\, \varrho_{\mu\nu}. \tag{4.6}$$

Gauge invariance (4.5) implies

$$f\,\{2\,\lambda(k_1 - k_2)\cdot(k_1 + k_2)\, [112]\, (k_1 - k_2)\cdot\varepsilon + \lambda^2(k_1^2 - k_2^2)^2\, [112]\} = 0.$$

Hence

$$f = 0, \qquad \text{unless} \qquad k_1^2 = k_2^2. \tag{4.7}$$

Thus, a spinless particle cannot decay with one graviton emission as in the case of photon.

Next we consider the vertex $\left\langle \frac{1}{2}\, \frac{1}{2}\, 2 \right\rangle$. Using the "currents" of definite parity, Eq. (3.15), we have

$$\begin{aligned} M = {} & A_1 \left(\frac{k_1}{m_1} + \frac{k_2}{m_2}\right) \cdot \sigma (k_1 - k_2)^\mu\, (k_1 - k_2)^\nu\, \varrho_{\mu\nu} + \\ & + A_2 \left(\frac{k_1}{m_1} - \frac{k_2}{m_2}\right) \cdot \sigma (k_1 - k_2)^\mu\, (k_1 - k_2)^\nu\, \varrho_{\mu\nu} + \\ & + A_3 \left(\sigma^\mu + \frac{k_2 \cdot \sigma}{m_2}\, \tilde{\sigma}^\mu \frac{k_1 \cdot \sigma}{m_1}\right) (k_1 - k_2)^\nu\, \varrho_{\mu\nu} + \\ & + A_4 \left(\sigma^\mu - \frac{k_2 \cdot \sigma}{m_2}\, \tilde{\sigma}^\mu \frac{k_1 \cdot \sigma}{m_1}\right) (k_1 - k_2)^\nu\, \varrho_{\mu\nu} + \end{aligned}$$

$$+ A_5 \left[\frac{k_2 \cdot \sigma}{m_2} (\tilde{\sigma}^\mu \sigma^\nu - \tilde{\sigma}^\nu \sigma^\mu) + (\sigma^\mu \tilde{\sigma}^\nu - \sigma^\nu \tilde{\sigma}^\mu) \frac{k_1 \cdot \sigma}{m_1} \right] \varrho_{\mu\nu} +$$

$$+ A_6 \left[\frac{k_2 \cdot \sigma}{m_2} (\tilde{\sigma}^\mu \sigma^\nu - \tilde{\sigma}^\nu \sigma^\mu) - (\sigma^\mu \tilde{\sigma}^\nu - \sigma^\nu \tilde{\sigma}^\mu) \frac{k_1 \cdot \sigma}{m_1} \right] \varrho_{\mu\nu}. \quad (4.8)$$

The three form factors A_1, A_3, A_5 have positive parity, the other three negative parity. It has been pointed out recently [25] that the experimental results so far cannot tell us whether P, C or T are conserved in gravitational interactions.

3. We now consider the scattering of two spinless particles via the exchange of a graviton in pole approximation of the unitarity condition. In t-channel the extended unitarity gives the following single particle contribution

$$f^2 \delta(t)\, \varrho_{\mu\nu} \mathcal{D}^{02}(k \cdot \tilde{\sigma})\, \varrho_{\lambda\sigma} (k_1 - k_3)^\mu (k_1 - k_3)^\nu (k_2 - k_4)^\lambda (k_2 - k_4)^\sigma. \quad (4.9)$$

It follows from (4.1) and (3.40) that

$$\varrho_{\mu\nu} \mathcal{D}^{02} \varrho_{\lambda\sigma} = g_{\mu\lambda} g_{\nu\sigma}.$$

Hence (4.9) becomes

$$f^2\, \delta(t)\, [(k_1 - k_3) \cdot (k_2 - k_4)]^2$$

and with the same analyticity assumption as in (3.35) we find

$$R = f^2 \left\{ \frac{[(k_1 - k_3) \cdot (k_2 - k_4)]^2}{t} + \frac{[(k_1 + k_4) \cdot (k_2 + k_3)]^2}{u} \right\}. \quad (4.10)$$

4. Finally we consider the analog considerations leading to the conservation of charge (III.8). For a process involving n spinless particles and a graviton of momentum k we write

$$M = \sum A_i(s, t, \ldots)\, k_i^\mu k_i^\nu \varrho_{\mu\nu}(k). \quad (4.11)$$

The gauge condition (4.5) gives

$$\sum_i A_i(s, t, \ldots)\, (k_i \cdot k)\, (k_i \cdot \varepsilon) = 0$$

and

$$\sum_i A_i(s, t, \ldots)\, (k_i \cdot k)^2 = 0.$$

If we now define new form factors $F_i(s, t, \ldots)$ by

$$F_i(s, t, \ldots) = A_i(s, t, \ldots)\, (k_i \cdot k) \quad (4.12)$$

we obtain

$$M = \sum F_i(s, t, \ldots) \frac{k_i^\mu k_i^\nu \varrho_{\mu\nu}}{k_i \cdot k} \quad (4.13)$$

with

$$\sum_i F_i(k_i \cdot \varepsilon) = 0, \qquad \text{and} \qquad \sum F_i(k_i \cdot k) = 0. \tag{4.14}$$

These equations must be satisfied together with the momentum conservation equations

$$\sum_i (k_i \cdot k) = 0$$

and

$$\sum_i (k_i \cdot \varepsilon) = 0. \tag{4.15}$$

The solution of these equations is

$$F_i(s, t, \ldots) = F(s, t, \ldots), \qquad \text{for all } i. \tag{4.16}$$

On the other hand, (4.13) is a sum of pole terms as in Fig. 11. Thus, if the approximation of Fig. 11 holds in the limit $k \to 0$ the single form factor F goes over to the coupling constant f which is then universal.

In general, if we have a spin j, mass zero particle, the condition (4.14) modifies into

$$\sum F_i(k_i \cdot k)^{j-1} = 0 \tag{4.17}$$

which cannot be satisfied for $j \geqslant 3$, *if*, there is a universal coupling constant.

V. Description of Neutrino Processes

If a particle of half integer spin is massless we get restrictions on the form of amplitudes just as the restrictions we obtained for photons. The spinor index of a massless half integer spin particle must transform like a *null spinor*. Analogous to the definition of polarization vector ε_μ, Eq. (3.2), we define (and give explicitly) a *polarization spinor* $\xi(k)$ by

$$\mathscr{D}^{0(1/2)}(A)\, \xi(\Lambda^{-1} k) = \xi(k), \tag{5.1}$$

which is a null spinor, i.e.

$$\mathscr{D}^{0(1/2)}\left(\frac{k \cdot \sigma}{2\,\lambda}\right) \xi(k) = \xi(k). \tag{5.2}$$

Indeed, we have the factorization property,

$$[k \cdot \sigma]_{\alpha\dot{\alpha}} = \xi_\alpha(k) \xi_{\dot{\alpha}}(k), \qquad \text{if} \qquad k^2 = 0, \tag{5.3}$$

hence,

$$A\, k \cdot \sigma\, A^\dagger = A\, \xi_\alpha(k)\, \xi_{\dot{\alpha}}(k)\, A^\dagger = \xi_\beta(\Lambda\, k)\, \xi_{\dot{\beta}}(\Lambda\, k) =$$
$$= k' \cdot \sigma = \xi_\beta(k')\, \xi_{\dot{\beta}}(k').$$

Explicitly ξ_α is given by

$$\xi_\alpha = \begin{pmatrix} \sqrt{k^0 + k^3} \\ \dfrac{k' + i\,k^2}{\sqrt{k^0 + k^3}} \end{pmatrix}, \qquad \xi_{\dot\alpha} = \xi_\alpha{}^* \tag{5.4}$$

and satisfies

$$\left(\frac{k\cdot\sigma}{2\,k^0}\right)_{\alpha\dot\alpha} \xi^{\dot\alpha} = \xi_\alpha; \qquad (\xi^{\dot\alpha} \equiv \xi_\alpha). \tag{5.5}$$

Note that null spinors behave like rotational spinors: an upper dotted index is the same as a lower undotted index. Because of (5.2) we have also the condition

$$(k\cdot\tilde\sigma)\,\xi(k) = 0. \tag{5.6}$$

There is also a parity conjugate null spinor $\eta(k)$ satisfying

$$\begin{aligned} \left(\frac{k\cdot\tilde\sigma}{2\,k^0}\right)\eta &= \eta \\ (k\cdot\tilde\sigma)^{\dot\alpha\alpha} &= \eta^{\dot\alpha}\,\eta^\alpha \\ (k\cdot\sigma)\,\eta &= 0; \end{aligned} \tag{5.7}$$

parity conjugate because

$$\eta(k) = \xi(\tilde k); \qquad \tilde k: (k^0, -\boldsymbol{k}). \tag{5.8}$$

For massive particles, $k^2 > 0$, there is *no* factorization of the type (5.3). The simplest spinorial object that one can construct in the massive case involves two particles and is of the form $k\cdot\sigma$.

If we now have a general spinorial amplitude M, then for every neutrino in the final state of momentum k_f we must impose the condition

$$(k_f\cdot\tilde\sigma)\,M = 0, \qquad k_f{}^2 = 0, \tag{5.9}$$

and for every neutrino in the initial state the condition

$$M(k_i\cdot\tilde\sigma) = 0, \qquad k_i{}^2 = 0. \tag{5.10}$$

These conditions considerably reduce the number of amplitudes, even in the presence of parity non-conservation. For example, processes like $\pi\pi \to \nu\,\bar\nu$ are described by one amplitude, $K \to \pi\,e\,\nu$ by two, processes like $n \to p\,e\,\bar\nu$ by six (without isotopic spin parity invariance), etc.

The conditions (5.9), (5.10) are equivalent to the $V - A$ theory in the sense that the number of amplitudes obtained in both cases are the same.

Appendix I

Connection Between Two and Four Component Form of the Amplitudes

A. Introduction of Four Component Spinors

Given a two component (dotted) spinor $\phi_{\dot\alpha}$, for example, which in a particular frame has the components

$$\phi_{i/2} = \frac{1}{\sqrt2}\begin{pmatrix}1\\0\end{pmatrix}, \qquad \phi_{-(i/2)} = \frac{1}{\sqrt2}\begin{pmatrix}0\\1\end{pmatrix} \tag{A. 1}$$

we define a four component spinor by

$$u(k)_{\dot{a}} = \begin{pmatrix} B\,\phi_{\dot{a}} \\ B^{-1\dagger}\,\phi_{\dot{a}} \end{pmatrix}, \tag{A. 2}$$

where B is the 2-by-2 matrix corresponding to a Lorentz fransformation from rest frame to an arbitrary momentum k

$$B \equiv B_{k\leftarrow p} = \sqrt{\frac{k\cdot\sigma}{m}}\,U$$

$$B^{-1} = U^{-1}\sqrt{\frac{k\cdot\tilde{\sigma}}{m}}, \qquad B^{-1\dagger} = \sqrt{\frac{k\cdot\tilde{\sigma}}{m}}\,U; \tag{A. 3}$$

then the four components spinor $u(k)$ satisfies the Dirac equation:

$$(\gamma\cdot k - m)\,u(k) = 0,$$
$$\gamma\cdot k = \gamma^{\mu}\,k_{\mu}. \tag{A. 4}$$

Proof: We can choose

$$\gamma_{\mu} = \begin{pmatrix} 0 & \sigma_{\mu} \\ \tilde{\sigma}_{\mu} & 0 \end{pmatrix} \tag{A. 5}$$

then (A. 4) becomes

$$\gamma\cdot k\,u = \begin{pmatrix} 0 & \sigma\cdot k \\ \tilde{\sigma}\cdot k & 0 \end{pmatrix}\begin{pmatrix} B\,\phi \\ B^{-1\dagger}\,\phi \end{pmatrix} = \begin{pmatrix} k\cdot\sigma\,B^{-1\dagger}\,\phi \\ k\cdot\tilde{\sigma}\,B\,\phi \end{pmatrix} = m\begin{pmatrix} \sqrt{k\cdot\sigma/m}\,U\,\phi \\ \sqrt{k\cdot\tilde{\sigma}/m}\,U\,\phi \end{pmatrix} = m\,u(k)$$

which proves the statement.

We also *define* the conjugate spinor

$$\bar{u}(k) = u(k)^{\dagger}\,\gamma_0,$$

or

$$\bar{u}(k) = \left(B^{\dagger}\,\phi^*\;\; B^{-1}\phi^*\right)\begin{pmatrix} 0 & 1 \\ 1 & 0 \end{pmatrix} = \begin{pmatrix} B^{-1}\,\phi^* \\ B^{\dagger}\,\phi^* \end{pmatrix}, \tag{A. 6}$$

and the spinor

$$v(k) \equiv C\,\bar{u}(k)^T = \begin{pmatrix} C\,B^{-1T}\,\phi^* \\ C\,B^{\dagger T}\,\phi^* \end{pmatrix} \tag{A. 7}$$

then $v(k)$ satisfies the Dirac equation

$$(\gamma\cdot k + m)\,v(k) = 0. \tag{A. 8}$$

Proof:

$$\gamma\cdot k\,v = \begin{pmatrix} 0 & k\cdot\sigma \\ k\cdot\tilde{\sigma} & 0 \end{pmatrix}\begin{pmatrix} C\,B^{-1\,T}\,\phi^* \\ C\,B^{\dagger\,T}\,\phi^* \end{pmatrix} = \begin{pmatrix} k\cdot\sigma\,C\,B^{\dagger T}\,\phi^* \\ k\cdot\tilde{\sigma}\,C\,B^{-1\,T}\,\phi^* \end{pmatrix} =$$

$$= \begin{pmatrix} k\cdot\sigma\,C\,\sqrt{k\cdot\sigma^T/m}\,U^*\,\phi^* \\ k\cdot\tilde{\sigma}\,C\sqrt{k\cdot\tilde{\sigma}^T/m}\,U^*\phi^* \end{pmatrix} = -\begin{pmatrix} k\cdot\sigma\,C^{-1}\sqrt{k\cdot\sigma^T/m}\,C\,C^{-1}\,U^*\,C\,C^{-1}\phi^* \\ k\cdot\tilde{\sigma}\,C^{-1}\sqrt{k\cdot\tilde{\sigma}^T/m}\,C\,C^{-1}\,U^*\,C\,C^{-1}\phi^* \end{pmatrix} =$$

$$= -\begin{pmatrix} k\cdot\sigma\,\sqrt{k\cdot\tilde{\sigma}/m}\,U\,C^{-1}\,\phi^* \\ k\cdot\tilde{\sigma}\,\sqrt{k\cdot\sigma/m}\,U\,C^{-1}\phi^* \end{pmatrix} = -\,m\begin{pmatrix} B\,C^{-1}\phi^* \\ B^{-1\dagger}\,C^{-1}\phi^* \end{pmatrix} = -\,m\begin{pmatrix} C\,B^{-1T}\phi^* \\ C\,B^{\dagger T}\,\phi^* \end{pmatrix} =$$

$$= -\,m\,v(k).$$

Where we have used the relations

$$C M^{-1} C^{-1} = M^T$$

for any 2×2, det $M = 1$;

$$C U^* C^{-1} = U,$$

$$\tilde{\sigma}_\mu = C^{-1} \sigma_\mu{}^T C; \qquad \sigma_\mu = C^{-1} \tilde{\sigma}_\mu{}^T C; \qquad C = \begin{pmatrix} 0 & -1 \\ 1 & 0 \end{pmatrix}.$$

B. *Amplitudes*

We consider now the amplitudes in the four component form

$$R = \bar{u}(k')_\alpha \, T^{\alpha \dot{\alpha}} \, u(k)_{\dot{\alpha}}$$

i.e.

$$R = \widehat{\phi_\alpha{}^* B^{-1} \phi_\alpha{}^\dagger B^\dagger} \begin{pmatrix} T_{11} & T_{12} \\ T_{21} & T_{22} \end{pmatrix} \begin{pmatrix} B \phi_{\dot{\alpha}} \\ B^{-1\dagger} \phi_{\dot{\alpha}} \end{pmatrix} \tag{A. 9}$$

or

$$R = \phi_\alpha{}^* B^{-1}(k') \, T_{11} \, B(k) \, \phi_{\dot{\alpha}}(k) + \phi_\alpha{}^* B^\dagger(k') \, T_{21} \, B(k) \, \phi_{\dot{\alpha}} +$$
$$+ \phi_\alpha{}^* B^{-1}(k') \, T_{12} \, B^{-1\dagger}(k) \, \phi_{\dot{\alpha}} + \phi_\alpha{}^* B^\dagger(k') \, T_{22} \, B^{-1\dagger}(k) \, \phi_{\dot{\alpha}}.$$

In the frame where (A. 1) holds the quantities $\phi_\alpha{}^* B^{-1}$, $B\phi_\alpha$, ... act as follows

$$\phi_{\alpha, i}{}^* B^{-1ij} = \frac{1}{\sqrt{2}} \begin{pmatrix} 1 \\ 0 \end{pmatrix} \begin{pmatrix} \cdot & \cdot \\ \cdot & \cdot \end{pmatrix} = B^{-1}(k')_\alpha{}^j,$$

i.e. ϕ, ϕ^* just pick up definite elements of B. Thus we have the two component form

$$R = \frac{1}{2} \cdot \tag{A. 10}$$

$$\cdot \, [B^{-1}(k') \, T_{11} \, B(k) + B^\dagger(k') \, T_{21} \, B(k) + B^{-1}(k') \, T_{12} \, B^{-1\dagger}(k) + B(k') \, T_{22} \, B^{-1\dagger}(k)].$$

Here a 2-by-2 matrix multiplication is understood.

Finally using the connection between the R and the spinerial amplitudes,

$$R = B^{-1}(k') \, M \, B^{\dagger^{-1}}(k),$$

we obtain the connection between M and the T matrix in (A. 9).

$$M = \frac{1}{2} [T_{11} \, B(k) \, B^\dagger(k) + B(k') \, B^\dagger(k') \, T_{21} \, B(k) \, B^\dagger(k) + T_{12} + B(k') \, B^\dagger(k') \, T_{22}]. \tag{A. 11}$$

The matrix T is made up of Dirac Matrices. The following table shows the values of T_{ij} in the representation (A. 5)

T	T_{11}	T_{12}	T_{21}	T_{22}
I	I	0	0	I
$\gamma_\mu\gamma_\nu - \gamma_\nu\gamma_\mu$	$\sigma_\mu\tilde{\sigma}_\nu - \sigma_\nu\tilde{\sigma}_\mu$	0	0	$\tilde{\sigma}_\mu\sigma_\nu - \tilde{\sigma}_\nu\sigma_\mu$
$\gamma_5\gamma_\mu$	0	$-\sigma_\mu$	$\tilde{\sigma}_\mu$	0
γ_μ	0	σ_μ	$\tilde{\sigma}_\mu$	0
γ_5	$-I$	0	0	I

(A. 12)

Given now a four-component T-matrix we can easily evaluate M in two component form using (A. 12) and (A. 11), and *vice versa*.

References

1. This question has often been answered in the negative; see for example, G. F. CHEW, Science Progress, **51**, 529 (1963), P. V. LANDSHOFF, The S-Matrix Theory without Field Theory, Cambridge Lecture notes, **1964** (unpublished).
2. We follow here a group theoretical specification of the S-matrix; the Hilbert space being explicitly given by the direct sum of the representation spaces of the symmetry groups. The kinematical framework is thus the same as in quantum field theory. All properties of the S-matrix, including C, P, T and spin-statistics will be obtained from the corresponding unitary representations. See also A. O. BARUT, The Framework of S-Matrix Theory, in *Strong Interactions and High Energy Physics*, Oliver and Boyd, Edinburgh **1964**, and references therein. See also H. JOOS, Fortschritte der Physik **10**, **3** (**1962**).
3. G. C. WICK, A. S. WIGHTMAN, E. P. WIGNER, Phys. Rev. **88**, **101** (**1952**).
4. M. L. GOLDBERGER and K. M. WATSON, Phys. Rev. **127**, **2264** (**1962**); M. FROISSART, M. L. GOLDBERGER and K. M. WATSON, Phys. Rev. **131**, 2820 (**1963**).
5. E. P. WIGNER, Ann. Math. **40**, 149 (1939). A. S. WIGHTMAN in *Dispersion Relations and Elementary Particle Physics*, ed. by C. DE WITT (John Wiley Sons, Inc., New York, 1960).
6. There are also infinite dimensional representations of E_2 (E. P. WIGNER, in *Theoretical Physics*, International Atomic Energy Agency, Vienna, **1963**). But these are not realized for free particles.
7. D. ZWANZIGER, Phys. Rev. **133**, B 1056 (**1964**).
8. Unstable particles have also been represented by *non-unitary* representations of the Poincaré group with a complex energy momentum vector. D. ZWANZIGER, Phys. Rev. **131**, 2818 (**1963**); E. G. BELTRAMETTI and G. LUZZATTO (preprint).
9. This proof was given in A. O. BARUT, I. MUZINICH, and D. N. WILLIAMS, Phys. Rev. **130**, **442** (**1963**).
10. For a detailed discussion of invariant amplitudes and in particular the problem of kinematical singularities see K. HEPP, Helv. Physica Acta **36**, **355** (**1963**); **37**, 55 (**1964**); D. N. WILLIAMS, Lawrence Radiation Lab. Report UCRL-11113, Berkeley **1965**; H. JOOS, Forts. d. Phys. **10**, **3** (**1962**) and reference [9].
11. R. JOST, Helv. Physica Acta **30**, **409** (**1947**).
12. H. P. STAPP, Phys. Rev. **128**, **2139** (**1962**).
13. For a recent discussion of extended unitarity see J. B. BOYLING, Nuovo Cimento **33**, **1356** (**1964**) and references therein.

14. D. I. OLIVE, Phys. Rev. **135**, B 745 (1964); H. P. STAPP, Lectures on *S*-Matrix Theory, to be published by the International Centre for Theoretical Physics, in 1965.
15. G. A. O. KÄLLEN, in *Elementary Particle Physics and Field Theory*, K. W. FORD, editor (W. A. Benjamin, New York 1961), and references therein.
16. K. NISHIJIMA, *Fundamental Particles* (W. A. Benjamin, New York 1963).
17. A. PETERMANN, Helv. Phys. Acta **36**, 942 (1963).
18. D. ZWANZIGER, in *Proceedings of the Symposium on the Lorentz Group*, University of Colorado Press, Boulder, 1965.
19. L. LANDAU, Doklady **60**, 207 (1948); C. N. YANG, Phys. Rev. **77**, 242 (1950).
20. For a recent discussion of infrared corrections see, for example, D. R. YENNIE, S. C. FRAUTSCHI and H. SUURA, Ann. of Physics **13**, 379 (1961).
21. We follow here essentially A. O. BARUT and R. BLADE, Nuovo Cimento **39**, 331 (1965).
22. See, for example, J. M. JAUCH and F. ROHRLICH, *The Theory of Electrons and Photons*, Addison-Wesley Publishing Co., Cambridge, 1953, p. 229.
23. In fact if one starts from the dispersion relation for a scattering process (for all s and t) one can derive a dispersion relation for the form factor only if one neglects the crossed channels [see H. P. STAPP, UCRL 11766, part IV]. This seems then to pose a problem how the double dispersion relations are compatible with the dispersion relations for form factors.
24. The derivation based on the pole approximation of Fig. 11 was first given by S. WEINBERG, Physics Letters **9**, 357 (1964). For the derivation of the conservation of form factors (3.57), see A. O. BARUT, Physics Letters **10**, 356 (1964).
25. J. LEITNER and S. OKUBO, Phys. Rev. **136**, B 1542 (1964); K. HIIDA and Y. YAMAGUCHI, Prog. Theor. Physics (1965).

Electromagnetic Properties of Hadrons in the Static SU_6 Model*

By

W. Thirring

Institute for Theoretical Physics, University of Vienna

In the current thinking about strongly interacting particles one usually imagines that they are mixtures of certain bare particles e.g. the proton being partly a bare proton and partly a bare proton with a virtual p-wave pion around it. This picture was never very successful in explaining the electromagnetic properties of the baryons. The reason is that the well-explored outer part of the pion cloud does not contain much charge and magnetic moment and the core is rather complicated. It is the more remarkable that an alternative model where the baryons consist of three spin 1/2 particles called quarks is numerically very successful in predicting the electromagnetic properties. In this model one assumes that the quark interaction is spin- and unitary spin-independent and that there are no orbital contributions to the magnetic moments. The following table lists the quantum numbers of the three fundamental entities:

	Q	Y	N	T_3	μ
p	2/3	1/3	1/3	1/2	2/3
n	$-1/3$	1/3	1/3	$-1/2$	$-1/3$
λ	$-1/3$	$-2/3$	1/3	0	$-1/3$

μ is the magnetic moment measured in quark magnetons.

For the antiparticles they all change their sign. In the following we shall draw the state vectors of the physical particles regarding their spin and quark content. These vectors can be found by simple symmetry considerations. In the four columns we list the following expectation values

---

* Lecture given at the IV. Internationale Universitätswochen für Kernphysik, Schladming, 25 February—10 March 1965.

$$\mu = \left\langle \sum_i Q_i\, \sigma_i^{(z)} \right\rangle$$

$$\delta m_0 = \left\langle \sum 3\, N\, Q_i \right\rangle$$

$$\delta m_e = \left\langle \sum_{i>k} Q_i Q_k \right\rangle$$

$$-\,\delta m_m = \left\langle \sum_{i>k} Q_i Q_k (\vec{\sigma}_i \cdot \vec{\sigma}_k) \right\rangle .$$

δm_0 corresponds to a mass difference of the quarks with $|Q| = 1/3$ and $2/3$. δm_e is the electrostatic interaction and should be multiplied with e^2/R where R is a mean distance of the quarks. δm_m is the magnetic interaction and for s-states one expects a factor $2/3\ (\mu_0{}^2/R^3)$.

In this model all particles are constructed from 3 2-component fields which we unite in one entity ψ_α, $\alpha = 1 - 6$. In a more physical notation we shall give the spin direction and the quark content, e.g.

$$\psi_\alpha = \underset{p}{\uparrow},\ \underset{p}{\downarrow},\ \underset{n}{\uparrow},\ \underset{n}{\downarrow},\ \underset{\lambda}{\uparrow},\ \underset{\lambda}{\downarrow}.$$

The 9 vector and 8 pseudoscalar bosons are in a $3 \times 9 + 8 = 35$ representation of the quark-antiquark states $\psi_\alpha\, \psi_\beta{}^+$. They are orthogonal to the ninth pseudoscalar meson.

$$x = \frac{1}{\sqrt{6}}\, \psi_\alpha\, \psi_\alpha{}^+ = \frac{1}{\sqrt{6}} \left(\underset{\bar{p}\ p}{\uparrow\downarrow} - \underset{\bar{p}\ p}{\downarrow\uparrow} + \underset{\bar{n}\ n}{\uparrow\downarrow} - \underset{\bar{n}\ n}{\downarrow\uparrow} + \underset{\bar{\lambda}\ \lambda}{\uparrow\downarrow} - \underset{\bar{\lambda}\ \lambda}{\downarrow\uparrow} \right)$$

which is an $S\,U_6$ singulet. Assuming that the space wave function corresponds to an s-state we note that the particles have the correct parity and the neutral vector and pseudoscalar particles have charge conjugation -1 and $+1$ resp.

	μ	δm_0	δm_e	δm_m
$\varrho^+ = \underset{\bar{n}\ p}{\uparrow\uparrow}$	1	$\frac{1}{3}$	$\frac{2}{9}$	$-\frac{2}{9}$
$\varrho^- = \underset{\bar{p}\ n}{\uparrow\uparrow}$	-1	$\frac{1}{3}$	$\frac{2}{9}$	$-\frac{2}{9}$
$\varrho^0 = \frac{1}{\sqrt{2}} \left(\underset{\bar{p}\ p}{\uparrow\uparrow} - \underset{\bar{n}\ n}{\uparrow\uparrow} \right)$	0	$\frac{1}{3}$	$-\frac{5}{18}$	$\frac{5}{18}$
$K^{*+} = \underset{\bar{\lambda}\ p}{\uparrow\uparrow}$	1	$\frac{1}{3}$	$\frac{2}{9}$	$-\frac{2}{9}$

	μ	δm_0	δm_e	δm_m
$K^{*-} = \underset{\bar p}{\uparrow}\underset{\lambda}{\uparrow}$	-1	$\frac{1}{3}$	$\frac{2}{9}$	$-\frac{2}{9}$
$K^{*0} = \underset{\bar\lambda}{\uparrow}\underset{n}{\uparrow}$	0	$-\frac{2}{3}$	$-\frac{1}{9}$	$+\frac{1}{9}$
$\bar K^{*0} = \underset{\bar n}{\uparrow}\underset{\lambda}{\uparrow}$	0	$-\frac{2}{3}$	$-\frac{1}{9}$	$+\frac{1}{9}$
$\phi_8 = \frac{1}{\sqrt{6}}(2\underset{\bar\lambda}{\uparrow}\underset{\lambda}{\uparrow} - \underset{\bar p}{\uparrow}\underset{p}{\uparrow} - \underset{\bar n}{\uparrow}\underset{n}{\uparrow})$	0	$-\frac{1}{3}$	$-\frac{1}{6}$	$\frac{1}{6}$
$\omega_1 = \frac{1}{\sqrt{3}}(\underset{\bar\lambda}{\uparrow}\underset{\lambda}{\uparrow} + \underset{\bar p}{\uparrow}\underset{p}{\uparrow} + \underset{\bar n}{\uparrow}\underset{n}{\uparrow})$	0	0	$-\frac{5}{27}$	$\frac{5}{27}$
$\phi = \underset{\bar\lambda}{\uparrow}\underset{\lambda}{\uparrow}$	0	$-\frac{2}{3}$	$-\frac{1}{9}$	$+\frac{1}{9}$
$\omega = \frac{1}{\sqrt{2}}(\underset{\bar p}{\uparrow}\underset{p}{\uparrow} + \underset{\bar n}{\uparrow}\underset{n}{\uparrow})$	0	$\frac{1}{3}$	$-\frac{5}{18}$	$\frac{5}{18}$
$\pi^+ = \frac{1}{\sqrt{2}}(\underset{\bar n}{\uparrow}\underset{p}{\downarrow} - \underset{\bar n}{\downarrow}\underset{p}{\uparrow})$	0	$\frac{1}{3}$	$\frac{2}{9}$	$\frac{2}{3}$
$\pi^- = \frac{1}{\sqrt{2}}(\underset{\bar n}{\uparrow}\underset{p}{\downarrow} - \underset{\bar n}{\downarrow}\underset{p}{\uparrow})$	0	$\frac{1}{3}$	$\frac{2}{9}$	$\frac{2}{3}$
$\pi^0 = \frac{1}{2}(\underset{\bar p}{\uparrow}\underset{p}{\downarrow} - \underset{\bar p}{\downarrow}\underset{p}{\uparrow} - \underset{\bar n}{\uparrow}\underset{n}{\downarrow} + \underset{\bar n}{\downarrow}\underset{n}{\uparrow})$	0	$\frac{1}{3}$	$-\frac{5}{18}$	$-\frac{5}{6}$
$K^+ = \frac{1}{\sqrt{2}}(\underset{\bar\lambda}{\uparrow}\underset{p}{\downarrow} - \underset{\bar\lambda}{\downarrow}\underset{p}{\uparrow})$	0	$\frac{1}{3}$	$\frac{2}{9}$	$\frac{2}{3}$
$K^- = \frac{1}{\sqrt{2}}(\underset{\bar p}{\uparrow}\underset{\lambda}{\downarrow} - \underset{\bar p}{\downarrow}\underset{\lambda}{\uparrow})$	0	$\frac{1}{3}$	$\frac{2}{9}$	$\frac{2}{3}$
$K^0 = \frac{1}{\sqrt{2}}(\underset{\bar\lambda}{\uparrow}\underset{n}{\downarrow} - \underset{\bar\lambda}{\downarrow}\underset{n}{\uparrow})$	0	$-\frac{2}{3}$	$-\frac{1}{9}$	$-\frac{1}{3}$
$\bar K^0 = \frac{1}{\sqrt{2}}(\underset{\bar n}{\uparrow}\underset{\lambda}{\downarrow} - \underset{\bar n}{\downarrow}\underset{\lambda}{\uparrow})$	0	$-\frac{2}{3}$	$-\frac{1}{9}$	$-\frac{1}{3}$
$\eta = \frac{1}{\sqrt{12}}(2\underset{\bar\lambda}{\uparrow}\underset{\lambda}{\downarrow} - 2\underset{\bar\lambda}{\downarrow}\underset{\lambda}{\uparrow} - \underset{\bar p}{\uparrow}\underset{p}{\downarrow} + \underset{\bar p}{\downarrow}\underset{p}{\uparrow} - \underset{\bar n}{\uparrow}\underset{n}{\downarrow} + \underset{\bar n}{\downarrow}\underset{n}{\uparrow})$	0	$-\frac{3}{9}$	$-\frac{1}{6}$	$-\frac{1}{2}$

For spin 0 mesons $\langle\mu\rangle$ is, of course, zero but they have a sort of inner magnetic moment which manifests itself in transitions to the spin 1 states. Thus for the transition moment $\langle I = 0, I_z = 0 \,|\mu_Z|\, I = 1, I_z = 0\rangle$ one finds

$$\langle\pi^+|\mu|\varrho^+\rangle = \langle\pi^0|\mu|\varrho^0\rangle = \langle\pi^-|\mu|\varrho^-\rangle = -1/3$$
$$\langle\pi^0|\mu|\omega\rangle = -1, \langle\pi^0|\mu|\phi\rangle = 0$$
$$\langle\eta|\mu|\omega\rangle = -\frac{1}{3\sqrt{3}}, \langle\eta|\mu|\phi\rangle = -\frac{2\sqrt{2}}{3\sqrt{3}}, \langle\eta|\mu|\varrho^0\rangle = -\frac{1}{\sqrt{3}}$$
$$\langle K^+|\mu|K^{*+}\rangle = \langle K^-|\mu|K^{*-}\rangle = \frac{1}{3}, \langle K^0|\mu|K^{*0}\rangle = \langle \bar{K}^0|\mu|\bar{K}^{*0}\rangle = \frac{2}{3}.$$

The baryons are composed of three quarks $\psi_\alpha\,\psi_\beta\,\psi_\gamma$ in an $S\,U_6$ symmetric state. Thus the radial wave-function must be rather complicated if they are in an s-state and obey Fermi statistics.

	μ	δm_0	δm_e	δm_m
$\Omega^- = \uparrow_\lambda\uparrow_\lambda\uparrow_\lambda$	-1	-1	$\frac{1}{3}$	$-\frac{1}{3}$
$\Xi^{*-} = \frac{1}{\sqrt{3}}(\uparrow_\lambda\uparrow_\lambda\uparrow_n + \uparrow_\lambda\uparrow_n\uparrow_\lambda + \uparrow_n\uparrow_\lambda\uparrow_\lambda)$	-1	-1	$\frac{1}{3}$	$-\frac{1}{3}$
$\Xi^{*0} = \frac{1}{\sqrt{3}}(\uparrow_\lambda\uparrow_\lambda\uparrow_p + \uparrow_\lambda\uparrow_p\uparrow_\lambda + \uparrow_p\uparrow_\lambda\uparrow_\lambda)$	0	0	$-\frac{1}{3}$	$+\frac{1}{3}$
$Y^{*-} = \frac{1}{\sqrt{3}}(\uparrow_\lambda\uparrow_n\uparrow_n + \uparrow_n\uparrow_\lambda\uparrow_n + \uparrow_n\uparrow_n\uparrow_\lambda)$	-1	-1	$\frac{1}{3}$	$-\frac{1}{3}$
$Y^{*0} = \frac{1}{\sqrt{6}}(\uparrow_\lambda\uparrow_n\uparrow_p + \uparrow_\lambda\uparrow_p\uparrow_n + \uparrow_n\uparrow_\lambda\uparrow_p + \uparrow_p\uparrow_\lambda\uparrow_n + \uparrow_p\uparrow_n\uparrow_\lambda + \uparrow_n\uparrow_p\uparrow_\lambda)$	0	0	$-\frac{1}{3}$	$+\frac{1}{3}$
$Y^{*+} = \frac{1}{\sqrt{3}}(\uparrow_\lambda\uparrow_p\uparrow_p + \uparrow_p\uparrow_\lambda\uparrow_p + \uparrow_p\uparrow_p\uparrow_\lambda)$	1	1	0	0
$N^{*-} = \uparrow_n\uparrow_n\uparrow_n$	-1	-1	$\frac{1}{3}$	$-\frac{1}{3}$
$N^{*0} = \frac{1}{\sqrt{3}}(\uparrow_n\uparrow_n\uparrow_p + \uparrow_n\uparrow_p\uparrow_n + \uparrow_p\uparrow_n\uparrow_n)$	0	0	$-\frac{1}{3}$	$+\frac{1}{3}$
$N^{*+} = \frac{1}{\sqrt{3}}(\uparrow_p\uparrow_p\uparrow_n + \uparrow_p\uparrow_n\uparrow_p + \uparrow_n\uparrow_p\uparrow_p)$	1	1	0	0

	μ	δm_0	δm_e	δm_m
$N^{*++} = \uparrow_p \uparrow_p \uparrow_p$	2	2	$\frac{4}{3}$	$-\frac{4}{3}$
$P = \frac{1}{\sqrt{18}} (2\uparrow_p\downarrow_n\uparrow_p + 2\uparrow_p\uparrow_p\downarrow_n + 2\downarrow_n\uparrow_p\uparrow_p - \uparrow_p\downarrow_p\uparrow_n - \uparrow_p\uparrow_n\downarrow_p - \downarrow_p\uparrow_n\uparrow_p - \uparrow_n\downarrow_p\uparrow_p - \uparrow_n\uparrow_p\downarrow_p - \downarrow_p\uparrow_p\uparrow_n)$	1	1	0	$-\frac{4}{3}$
$N = \frac{1}{\sqrt{18}} (-2\uparrow_n\downarrow_p\uparrow_n - 2\uparrow_n\uparrow_n\downarrow_p - 2\downarrow_p\uparrow_n\uparrow_n + \uparrow_p\downarrow_n\uparrow_n + \uparrow_n\uparrow_p\downarrow_n + \downarrow_n\uparrow_p\uparrow_n + \uparrow_n\downarrow_n\uparrow_p + \uparrow_p\uparrow_n\downarrow_n + \downarrow_n\uparrow_n\uparrow_p)$	$-\frac{2}{3}$	0	$-\frac{1}{3}$	-1
$\Sigma^+ = \frac{1}{\sqrt{18}} (2\uparrow_p\downarrow_\lambda\uparrow_p + 2\uparrow_p\uparrow_p\downarrow_\lambda + 2\downarrow_\lambda\uparrow_p\uparrow_p - \uparrow_p\downarrow_p\uparrow_\lambda - \uparrow_p\uparrow_\lambda\downarrow_p - \downarrow_p\uparrow_\lambda\uparrow_p - \uparrow_\lambda\downarrow_p\uparrow_p - \uparrow_\lambda\uparrow_p\downarrow_p - \downarrow_p\uparrow_p\uparrow_\lambda)$	1	1	0	$-\frac{4}{3}$
$\Sigma^- = \frac{1}{\sqrt{18}} (-2\uparrow_n\downarrow_\lambda\uparrow_n - 2\uparrow_n\uparrow_n\downarrow_\lambda - 2\downarrow_\lambda\uparrow_n\uparrow_n + \uparrow_\lambda\downarrow_n\uparrow_n + \uparrow_n\uparrow_\lambda\downarrow_n + \downarrow_n\uparrow_\lambda\uparrow_n + \uparrow_n\downarrow_n\uparrow_\lambda + \uparrow_\lambda\uparrow_n\downarrow_n + \downarrow_n\uparrow_n\uparrow_\lambda)$	$-\frac{1}{3}$	-1	$\frac{1}{3}$	$\frac{1}{3}$
$\Sigma^0 = \frac{1}{\sqrt{36}} (+2\uparrow_n\downarrow_\lambda\uparrow_p - \downarrow_n\uparrow_\lambda\uparrow_p - \uparrow_\lambda\downarrow_n\uparrow_p + 2\downarrow_\lambda\uparrow_n\uparrow_p + 2\uparrow_p\uparrow_n\downarrow_\lambda -$	$\frac{1}{3}$	0	$-\frac{1}{3}$	0

	μ	δm_0	δm_e	δm_m
$-\underset{p\,n\,\lambda}{\uparrow\downarrow\uparrow}-\underset{p\,\lambda\,n}{\uparrow\uparrow\downarrow}+2\underset{p\,\lambda\,n}{\uparrow\downarrow\uparrow}+$ $+2\underset{\lambda\,p\,n}{\downarrow\uparrow\uparrow}-\underset{\lambda\,p\,n}{\uparrow\uparrow\downarrow}-\underset{n\,p\,\lambda}{\downarrow\uparrow\uparrow}+$ $+2\underset{n\,p\,\lambda}{\uparrow\uparrow\downarrow}-\underset{\lambda\,p\,n}{\uparrow\downarrow\uparrow}-\underset{p\,\lambda\,n}{\downarrow\uparrow\uparrow}-$ $-\underset{n\,\lambda\,p}{\uparrow\uparrow\downarrow}-\underset{n\,p\,\lambda}{\uparrow\downarrow\uparrow}-\underset{p\,n\,\lambda}{\downarrow\uparrow\uparrow}-$ $-\underset{\lambda\,n\,p}{\uparrow\uparrow\downarrow})$				
$\Lambda=\frac{1}{\sqrt{12}}(\underset{p\,n\,\lambda}{\uparrow\downarrow\uparrow}-\underset{p\,n\,\lambda}{\downarrow\uparrow\uparrow}-\underset{n\,p\,\lambda}{\uparrow\downarrow\uparrow}+$ $+\underset{n\,p\,\lambda}{\downarrow\uparrow\uparrow}+\underset{\lambda\,p\,n}{\uparrow\uparrow\downarrow}-\underset{\lambda\,p\,n}{\uparrow\downarrow\uparrow}-$ $-\underset{\lambda\,n\,p}{\uparrow\uparrow\downarrow}+\underset{\lambda\,n\,p}{\uparrow\downarrow\uparrow}+\underset{n\,\lambda\,p}{\downarrow\uparrow\uparrow}-$ $-\underset{n\,\lambda\,p}{\uparrow\uparrow\downarrow}-\underset{p\,\lambda\,n}{\downarrow\uparrow\uparrow}+\underset{p\,\lambda\,n}{\uparrow\uparrow\downarrow})$	$-\frac{1}{3}$	0	$-\frac{1}{3}$	$-\frac{2}{3}$
$\Xi^0=\frac{1}{\sqrt{18}}(-2\underset{\lambda\,p\,\lambda}{\uparrow\downarrow\uparrow}-2\underset{\lambda\,\lambda\,p}{\uparrow\uparrow\downarrow}-$ $-2\underset{p\,\lambda\,\lambda}{\downarrow\uparrow\uparrow}+\underset{p\,\lambda\,\lambda}{\uparrow\downarrow\uparrow}+\underset{\lambda\,p\,\lambda}{\uparrow\uparrow\downarrow}+$ $+\underset{\lambda\,p\,\lambda}{\downarrow\uparrow\uparrow}+\underset{\lambda\,\lambda\,p}{\uparrow\downarrow\uparrow}+\underset{p\,\lambda\,\lambda}{\uparrow\uparrow\downarrow}+$ $+\underset{\lambda\,\lambda\,p}{\downarrow\uparrow\uparrow})$	$-\frac{2}{3}$	0	$-\frac{1}{3}$	-1
$\Xi^-=\frac{1}{\sqrt{18}}(2\underset{\lambda\,n\,\lambda}{\uparrow\downarrow\uparrow}+2\underset{\lambda\,\lambda\,n}{\uparrow\uparrow\downarrow}+$ $+2\underset{n\,\lambda\,\lambda}{\downarrow\uparrow\uparrow}-\underset{\lambda\,\lambda\,n}{\uparrow\downarrow\uparrow}-\underset{\lambda\,n\,\lambda}{\uparrow\uparrow\downarrow}-$ $-\underset{\lambda\,n\,\lambda}{\downarrow\uparrow\uparrow}-\underset{n\,\lambda\,\lambda}{\uparrow\downarrow\uparrow}-\underset{n\,\lambda\,\lambda}{\uparrow\uparrow\downarrow}-$ $-\underset{\lambda\,\lambda\,n}{\downarrow\uparrow\uparrow})$	$-\frac{1}{3}$	-1	$\frac{1}{3}$	$\frac{1}{3}$

The transition moments are found to be (in absolute value)

$$\langle P|\mu|N^{*+}\rangle=\langle\Sigma^+|\mu|Y^{*+}\rangle=\langle N|\mu|N^{*0}\rangle=2\,\langle\Sigma^0|\mu|Y^{*0}\rangle=$$
$$=\frac{\sqrt{3}}{2}\,\langle\Lambda|\mu|Y^{*0}\rangle=\langle\Xi^0|\mu|\Xi^{*0}\rangle=\frac{2\sqrt{2}}{3}.$$

For the electromagnetic mass splittings the following relations follow from the table.

$$Y_1{}^{*0} - Y_1{}^{*0} = N^{*0} - N^{*+} = \frac{1}{3}(N^{*-} - N^{*++})$$

$$N^{*-} - N^{*0} = Y_1{}^{*-} - Y_1{}^{*0} = \Xi^{*-} - \Xi^{*0} = N - P + \Sigma^- + \Sigma^+ - 2\Sigma_0.$$

The following experimental data are available on that.

Magnetic moments in nuclear magnetons:

$$\mu_p = 2.78, \qquad \mu_N = -1.91, \qquad \mu_\Lambda = -0.7 \pm 0.3, \qquad \mu_{\Sigma^+} = 2.2 \pm ?$$

Thus if the quark magneton equals the proton moment we get excellent agreement with the known moments; the transition moments $N \to N^*$ are experimentally somewhat bigger than the predictions. Regarding the mass differences we can calculate from $N - P, \Sigma^+ - \Sigma^0, \Sigma^0 - \Sigma^-$ the three parameters and find

$$\delta m = -(1.9 \pm 0.2)\,\delta m_0 + (3.5 \pm 0.1)\,\delta m_e + (1.6 \pm 0.3)\,\delta m_m.$$

Giving experimental numbers in parenthesis this compars as follows with the data.

$$\Xi^- - \Xi^0 = 6.3 \pm 0.3 \quad (6.5 \pm 1)$$
$$\Xi^{*-} - \Xi^{*0} = 3.1 \pm 0.3 \quad (6 \pm 3)$$
$$N^{*0} - N^{*++} = 0.7 \pm 0.6 \quad (0.4 \pm 0.8)$$
$$N^{*-} - N^{*++} = 3.6 \pm 0.6 \quad (0.6 \pm 5)$$
$$Y_1{}^{*-} - Y_1{}^{*+} = 4.4 \pm 0.5 \quad (4.3 \pm 2).$$

Thus the theoretically expected sign of δm_e and δm_m reproduces the data within the errors. They correspond to a mean quark distance $R \sim 0.4 \cdot 10^{-13}$ cm in reasonable agreement with the expected order of the Compton wavelength of the vector mesons.

For the Bosons only the $\pi^+ - \pi^0, K^+ - K^0$ mass splitting are experimentally known. They do not determine 3 parameters but we will try and see what happens if we use the numbers from the baryons. We find

$$\pi^+ - \pi^0 = 4.2 \pm 0.4 \quad (4.6 \pm 0.07)$$
$$K^0 - K^+ = -0.9 \pm 0.2 \quad (4.2 \pm 0.5).$$

Thus it works for the pion but for the kaon we get a wrong number.

Finally from the transition moments for the bosons we find for the partial width

$$\Gamma_{\omega \to \pi_0 + \gamma} = 1.2\,\text{MeV}$$

assuming the quark moment for Bosons is the same as for Baryons.

This is in excellent agreement with the experimental branching ratio of 11% for this decay. For decays like $\varrho \to \pi + \gamma$ are theoretically only 1/10% well below the experimental limit.

References

Beg, M., e. a. Phys. Rev. Lett. **13**, 514 (1964).
Dolgov, A., e. a. Phys. Lett. **15**, 84 (1965).

Theories with Gauge Groups* **

By

Bruno Zumino

Department of Physics
New York University, New York, New York 10003

(*Received April 26, 1965*)

1. Introduction

The success of quantum electrodynamics suggests the investigation of theories constructed in analogy with it. The gauge invariance of second kind with variable phase functions can be viewed as a generalization of the gauge invariance of first kind (with constant phase) which forces the introduction of the electromagnetic potential, so that the gauge transformation on the matter field can be cancelled by the gauge transformation of the potential. In this way, the generalization of the gauge transformation to a coordinate dependent transformation can be taken as the principle generating the electromagnetic field. This point of view has been applied (YANG-MILLS [1]) to the group of isospin transformations of a nucleon field. When these transformations are made coordinate dependent, one is led to introduce a vector field, the b field, which is both a Lorentz vector and an isospin vector, and which is the analogue of the electromagnetic potential. The theory of YANG and MILLS is described in Section 3 after a brief review of electrodynamics, given in Section 2. The analogies are stressed in the formalism, as well as the differences, which arise mainly from the non-abelian nature of the isospin group, as contrasted with the abelian nature of the group of ordinary phase transformations.

Special problems arise when one attempts to quantize the YANG-MILLS theory. In quantum electrodynamics the natural way to quantize the theory is to remove the arbitrariness of the field variables by imposing a gauge condition. In this way one eliminates the redundant variables and one succeeds in formulating the theory in terms of only physical degrees of freedom which are then quantized. Two examples of gauges are considered; the Coulomb gauge and the axial gauge. The

* Lecture given at the IV. Internationale Universitätswochen für Kernphysik, Schladming, 25 February—10 March 1965.

** This work was supported in part by the National Science Foundation.

same procedure of imposing gauge conditions can be used as a preliminary to the quantization of the Yang-Mills theory and again one can use either the Coulomb gauge or the axial gauge. The Coulomb gauge method, however, is much more complicated for the Yang-Mills theory than in the case of electrodynamics. The axial gauge method is not so complicated, but there are certain problems peculiar to the axial gauge also. At any rate the quantization can be performed [2, 3] even though some problems appear to remain, as explained in Section 4.

The generalization of the Yang-Mills approach to other internal symmetry groups (for instance, SU_3) can be carried out without too much difficulty, especially in the case of groups of unitary matrices, in which case it is almost obvious. Quite generally it has been described first by Utiyama [4]. We shall not go into it explicitly in these notes, since the developments are purely formal. We prefer to inquire about the applicability of the Yang-Mills type of theories to physical reality.

Particles with the quantum numbers of the Yang-Mills field are now known (the ϱ-meson), as well as particles with the quantum numbers of the other vector mesons required by the Yang-Mills approach as applied to SU_3 (the K^* and the ϕ-meson). Their existence would seem to represent a success of the Yang-Mills idea and was anticipated especially by Sakurai [5]. As a matter of fact, in deference to this point of view, it is customary to attribute the vector mesons to the regular (adjoint) representation of whatever internal symmetry group one is using (the representation according to which the infinitesimal generators of the group themselves transform). This procedure of attributing the vector mesons to the regular representation is considered by some physicists as a sort of basic postulate which one has to keep at all costs. However, the interpretation of the physical vector mesons as Yang-Mills particles presents some difficulties. Furthermore, alternative interpretations are possible.

The difficulties are well known. They are related to the question of the mass of the b field. The bare mass of the b field is zero, while the masses of the physical vector mesons are quite large. Arguments have been given [6] to show that the physical mass of the Yang-Mills field can be different from zero, even though the bare mass is zero and the theory is gauge invariant. The situation is still anything but clear and will be discussed briefly in Section 5, where also an alternative approach to the mass problem will be described.

There are alternative interpretations of the physical vector mesons. First of all, it is possible to make theories different from the Yang-Mills theory which are invariant only under gauge transformations with constant coefficients and not under the coordinate dependent gauge group. Such theories have vector particles with the right quantum numbers; they also have conserved currents and, furthermore, they are not afflicted by the mass problem. In the second place, it is possible and may be very plausible that the physical vector mesons should be described as composite particles, with no fundamental fields associated with them. This is done, for instance, in the "triplet" theories of SU_3 and related formulations [7, 8, 9, 10]. Finally, one must still keep in mind that isospin invariance (and, of course, even more the higher symmetries) is only an approximate invariance. It is possibly slightly exaggerated to make an approximate invariance into a very basic generalized gauge invariance (this criticism does not apply, of course, to the introduction of a vector gauge field associated to an exact symmetry, such as barion number conservation; however, those are abelian groups).

After these critical remarks, one may say that possibly the strongest reason for studying in detail gauge theories of the Yang-Mills type is their aesthetic appeal and their inner consistency. They represent useful preliminary models for the study of more complicated theories like Einstein's theory of gravitation. Their quantization is a good preliminary exercise for solving the problem of quantization of such more complicated theories and has been looked at in this light by various people (however, it is not at all clear why one should want to quantize Einstein's theory!).

2. Quantum Electrodynamics

In this section we review briefly quantum electrodynamics itself. One can take the first order Lagrangian density

$$L = L_M + L_D \tag{1}$$

$$L_M = \frac{1}{4} F_{\mu\nu} F^{\mu\nu} - \frac{1}{2} F_{\mu\nu}(\partial^\mu A^\nu - \partial^\nu A^\mu) \tag{2}$$

$$L_D = i\,\bar{\psi}\,[\gamma^\mu(\overleftrightarrow{\partial}_\mu - i\,e\,A_\mu) + m]\,\psi, \tag{3}$$

where

$$\gamma^\mu \gamma^\nu + \gamma^\nu \gamma^\mu = 2\,g^{\mu\nu} \qquad (g^{00} = -1)$$

$$\overleftrightarrow{\partial}_\mu = \frac{1}{2}(\overrightarrow{\partial}_\mu - \overleftarrow{\partial}_\mu)$$

$$\bar{\psi} = \psi^* \gamma_0,$$

and the corresponding action

$$W = \int L\,\underset{4}{dx} = W_M + W_D. \tag{4}$$

It gives rise to the equations

$$\frac{\delta W}{\delta F^{\mu\nu}} \equiv F_{\mu\nu} - (\partial_\mu A_\nu - \partial_\nu A_\mu) = 0 \tag{5}$$

$$\frac{\delta W}{\delta A^\nu} \equiv \partial^\mu F_{\mu\nu} + e\,\bar{\psi}\,\gamma_\nu\,\psi = 0 \tag{6}$$

$$\frac{\delta W}{\delta \bar{\psi}} \equiv [\gamma^\mu(\partial_\mu - i\,e\,A_\mu) + m]\,\psi = 0, \qquad \frac{\delta W}{\delta \psi} = 0. \tag{7}$$

The theory is invariant under the gauge transformation of the second kind

$$A_\mu \to A_\mu + \partial_\mu \Lambda \qquad \psi \to \psi \exp\,[i\,e\,\Lambda]. \tag{8}$$

Taking an infinitesimal gauge transformation, we obtain

$$\delta W = \int \left(\frac{\delta W}{\delta \psi}\,\delta\psi + \delta\bar{\psi}\,\frac{\delta W}{\delta\bar{\psi}} + \frac{\delta W}{\delta A^\mu}\,\delta A^\mu + \frac{\delta W}{\delta F_{\mu\nu}}\,\delta F_{\mu\nu} \right) d\underset{4}{x} = 0 \tag{9}$$

with

$$\delta A_\mu = \partial_\mu \Lambda, \qquad \delta\psi = i\,e\,\Lambda\,\psi, \qquad \delta\bar{\psi} = -\,i\,e\,\Lambda\,\bar{\psi}, \qquad \delta F_{\mu\nu} = 0. \tag{10}$$

Integrating by parts and using the fact that Λ is arbitrary, we obtain

$$\partial_\mu \frac{\delta W}{\delta A_\mu} + i\,e\,\bar{\psi}\,\frac{\delta W}{\delta\bar{\psi}} - i\,e\,\frac{\delta W}{\delta\psi}\,\psi = 0 \tag{11}$$

which is an identity satisfied by the left hand sides of the field equations (5), (6), and (7). This is a special case of a theorem of E. Noether. The field equations are not independent.

We can make use of the more stringent condition that W_M and W_D are separately gauge invariant. We then have, for instance,

$$\partial_\mu \frac{\delta W_D}{\delta A_\mu} + i\,e\,\bar{\psi}\,\frac{\delta W_D}{\delta\bar{\psi}} - i\,e\,\frac{\delta W_D}{\delta\psi}\,\psi = 0. \tag{12}$$

Here only two of the functional derivatives correspond to field equations (Dirac equation). Assuming the Dirac equation to be satisfied, we obtain

$$\partial_\mu \frac{\delta W_D}{\delta A_\mu} = 0 \tag{13}$$

which gives us the conservation of current. The current is

$$\frac{\delta W_D}{\delta A_\mu} = e\,\bar{\psi}\,\gamma^\mu\,\psi. \tag{14}$$

In order to quantize the theory it is convenient to impose gauge conditions which remove the gauge arbitrariness of the field quantities. In this way only those degrees of freedom are left in the theory which actually have physical meaning and which must be quantized. Two gauges have been used mostly, the Coulomb (or radiation) gauge and the axial gauge. The Coulomb gauge is obtained by restricting the space part of the four potential through

$$\partial_r A_r = 0 \tag{15}$$

(latin indices run from 1 to 3). It is quite well known and we shall not discuss it here. The axial gauge is characterized by the condition

$$A_3 = 0. \tag{16}$$

It will be discussed in Section 4.

3. The Yang-Mills Theory

The Lagrangian of the Yang-Mills theory can be taken to be

$$L = L_1 + L_2 \tag{17}$$

$$L_1 = \frac{1}{4}\vec{F}_{\mu\nu}\cdot\vec{F}^{\mu\nu} - \frac{1}{2}\vec{F}_{\mu\nu}\cdot(\partial^\mu\vec{b}^\nu - \partial^\nu\vec{b}^\mu + 2g\,\vec{b}^\mu\times\vec{b}^\nu) \tag{18}$$

$$L_2 = i\,\bar{\psi}\,[\gamma^\mu(\overleftrightarrow{\partial_\mu} - i\,g\,\vec{b}_\mu\cdot\vec{\tau}) + m]\,\psi. \tag{19}$$

The vector indices refer to isospin space and $\vec{\tau} = (\tau_1, \tau_2, \tau_3)$ are the Pauli matrices; ψ is a spinor in isospace as well as a spinor in space time. The corresponding field equations are

$$\frac{\delta W}{\delta\vec{F}_{\mu\nu}} \equiv \vec{F}^{\mu\nu} - (\partial^\mu\vec{b}^\nu - \partial^\nu\vec{b}^\mu + 2g\,\vec{b}^\mu\times\vec{b}^\nu) = 0 \tag{20}$$

$$\frac{\delta W}{\delta\vec{b}^\nu} \equiv \partial^\mu\vec{F}_{\mu\nu} + 2g\,\vec{b}^\mu\times\vec{F}_{\mu\nu} + g\,\bar{\psi}\gamma_\nu\vec{\tau}\,\psi = 0 \tag{21}$$

$$\frac{\delta W}{\delta\psi} \equiv [\gamma^\mu(\partial_\mu - i\,g\,\vec{b}_\mu\cdot\vec{\tau}) + m]\,\psi = 0, \qquad \frac{\delta W}{\delta\bar{\psi}} = 0. \tag{22}$$

The theory is invariant under a gauge transformation which, in the infinitesimal form, can be written as

$$\begin{aligned}\delta\vec{b}_\mu &= \partial_\mu\vec{\lambda} - 2g\,\vec{\lambda}\times\vec{b}_\mu\\ \delta\vec{F}_{\mu\nu} &= -2g\,\vec{\lambda}\times\vec{F}_{\mu\nu}\\ \delta\psi &= i\,g\,\vec{\lambda}\cdot\vec{\tau}\,\psi.\end{aligned} \tag{23}$$

To write this transformation in finite form it is convenient to introduce the 2 by 2 matrices

$$b_\mu = \vec{b}_\mu\cdot\vec{\tau} \qquad F_{\mu\nu} = \vec{F}_{\mu\nu}\cdot\vec{\tau} \qquad U = e^{i g\vec{\lambda}\cdot\vec{\tau}}. \tag{24}$$

The gauge transformation is then

$$\begin{aligned} b_\mu &\to U\,b_\mu\,U^{-1} - \frac{i}{g}\,U\,\partial_\mu U^{-1} = U\,b_\mu\,U^{-1} + \frac{i}{g}\,\partial_\mu U\,U^{-1}\\ F_{\mu\nu} &\to U\,F_{\mu\nu}\,U^{-1}\\ \psi &\to U\,\psi.\end{aligned} \tag{25}$$

Observe that, with 2 by 2 matrices,

$$F_{\mu\nu} = (\partial_\mu - i\,g\,b_\mu)\,b_\nu - (\partial_\nu - i\,g\,b_\nu)\,b_\mu \tag{26}$$

and the other equations could also be written in this notation.

The invariance of the action under an infinitesimal gauge transformation is stated by

$$\delta W = \int\left(\frac{\delta W}{\delta\psi}\,\delta\psi + \delta\bar{\psi}\,\frac{\delta W}{\delta\bar{\psi}} + \frac{\delta W}{\delta\vec{b}_\mu}\,\delta\vec{b}_\mu + \frac{\delta W}{\delta\vec{F}_{\mu\nu}}\,\delta\vec{F}_{\mu\nu}\right)\underset{4}{dx} = 0, \tag{27}$$

where the infinitesimal changes in the field quantities are given by (23). Using the arbitrariness of $\vec{\lambda}$ this gives

$$(\partial_\mu + 2g\vec{b}_\mu \times)\frac{\delta W}{\delta \vec{b}_\mu} + ig\bar{\psi}\vec{\tau}\frac{\delta W}{\delta \bar{\psi}} - ig\frac{\delta W}{\delta \psi}\vec{\tau}\psi + 2g\frac{\delta W}{\delta \vec{F}_{\mu\nu}} \times \vec{F}_{\mu\nu} = 0 \tag{28}$$

which is a Noether type identity between the field equations (20), (21) and (22).

We can again make use of the more stringent requirement that W_1 and W_2 are separately gauge invariant. This gives, since W_2 does not contain $\vec{F}_{\mu\nu}$,

$$(\partial_\mu + 2g\vec{b}_\mu \times)\frac{\delta W_2}{\delta \vec{b}_\mu} + ig\bar{\psi}\vec{\tau}\frac{\delta W_2}{\delta \bar{\psi}} - ig\frac{\delta W_2}{\delta \psi}\vec{\tau}\psi = 0 \tag{29}$$

and, using the field equations (22),

$$(\partial_\mu + 2g\vec{b}_\mu \times)\frac{\delta W_2}{\delta \vec{b}_\mu} = 0. \tag{30}$$

The isospin current of the ψ field is given by

$$\frac{\delta W_2}{\delta \vec{b}_\mu} = g\bar{\psi}\gamma^\mu\vec{\tau}\psi. \tag{31}$$

We see from (30) that it is not truly conserved, there is an additional term; rather the "covariant" divergence of the ψ current vanishes. This is physically quite reasonable since the b field also carries isospin. From the field equation (21), since $\vec{F}_{\mu\nu}$ is skewsymmetric, it follows that the total isospin current

$$\vec{S}_\nu = g\bar{\psi}\gamma_\nu\vec{\tau}\psi + 2g\vec{b}^\mu \times \vec{F}_{\mu\nu} \tag{32}$$

satisfies a true conservation equation

$$\partial^\nu\vec{S}_\nu = 0. \tag{33}$$

In $\vec{S}_\nu$ we see, in addition to the current of the ψ field, a contribution due to the b field.

The validity of (33) implies that the additional term in (30) can also be written as a divergence

$$2g\vec{b}_\nu \times \frac{\delta W_2}{\delta \vec{b}_\nu} = 2g\,\partial^\nu(\vec{b}^\mu \times \vec{F}_{\mu\nu}). \tag{34}$$

Using the field equation (21) one can eliminate the ψ field from (34) and write

$$-\vec{b}_\nu \times \frac{\delta W_1}{\delta \vec{b}_\nu} = \partial^\nu(\vec{b}^\mu \times \vec{F}_{\mu\nu}). \tag{35}$$

This equation contains only the fields $\vec{b}_\mu$ and $\vec{F}_{\mu\nu}$. It is very satisfactory for the consistency of the theory that (35) follows from the field equation which we have not used yet, namely (20). To see this, we can use the fact that the Lagrangian density L_1 is invariant under isospin transformations with *constant* $\vec{\lambda}$. This means that we have the identity

$$\frac{\partial L_1}{\partial \vec{F}_{\mu\nu}} \times \vec{F}_{\mu\nu} + \frac{\partial L_1}{\partial \vec{b}_{\mu,\nu}} \times \vec{b}_{\mu,\nu} + \frac{\partial L_1}{\partial \vec{b}_\mu} \times \vec{b}_\mu = 0, \tag{36}$$

as one can see by using (23), with constant $\vec{\lambda}$, and remembering that the vector $\vec{\lambda}$ is otherwise arbitrary. Now (36) can be transformed into

$$\frac{\partial L_1}{\partial \vec{F}_{\mu\nu}} \times \vec{F}_{\mu\nu} + \partial_\nu \left(\frac{\partial L_1}{\partial \vec{b}_{\mu,\nu}} \times \vec{b}_\mu \right) + \left(\frac{\partial L_1}{\partial \vec{b}_\mu} - \partial_\nu \frac{\partial L_1}{\partial \vec{b}_{\mu,\nu}} \right) \times \vec{b}_\mu = 0 \tag{37}$$

which is still an identity. If we now use the field equation (20), the first term in (37) vanishes and we obtain (35), since

$$\frac{\partial L_1}{\partial \vec{b}_{\mu,\nu}} = \vec{F}_{\mu\nu} \tag{38}$$

and

$$\frac{\delta W_1}{\delta \vec{b}_\mu} = \frac{\partial L_1}{\partial \vec{b}_\mu} - \partial_\nu \frac{\partial L_1}{\partial \vec{b}_{\mu,\nu}}. \tag{39}$$

We could apply the invariance under constant isospin transformations to the part L_2 of the Lagrangian. This, however, would give nothing essentially new; we would get first the identity

$$\frac{\partial L_2}{\partial \vec{b}_\mu} \times \vec{b}_\mu + \left(\frac{\partial L_2}{\partial \psi} - \partial_\mu \frac{\partial L_2}{\partial \psi_{,\mu}} \right) i g \vec{\tau} \psi - i g \bar{\psi} \vec{\tau} \left(\frac{\partial L_2}{\partial \bar{\psi}} - \partial_\mu \frac{\partial L_2}{\partial \bar{\psi}_{,\mu}} \right) + \\ + \partial_\mu \left(\frac{\partial L_2}{\partial \psi_{,\mu}} i g \vec{\tau} \psi - i g \bar{\psi} \vec{\tau} \frac{\partial L_2}{\partial \bar{\psi}_{,\mu}} \right) = 0 \tag{40}$$

and finally, using the field equations (22), we would effectively obtain (30), written in a slightly different form. The sum of (40) and (37) gives an identity which, by use of the field equations, produces the conservation equation for the total isospin current. This local conservation law can therefore be derived completely within the framework of coordinate independent isospin invariance, with the additional restrictions on the form of the action functional, namely that it be the integral of a Lagrangian density depending only upon the fields and their derivatives. This derivation of current conservation from the invariance under constant isospin transformations can also be described in the following way. Let the field variables undergo a transformation

$$\begin{aligned} \delta \vec{b}_\mu &= -2 g \vec{\lambda} \times \vec{b}_\mu \\ \delta \vec{F}_{\mu\nu} &= -2 g \vec{\lambda} \times \vec{F}_{\mu\nu} \\ \delta \psi &= i g \vec{\lambda} \cdot \vec{\tau} \psi. \end{aligned} \tag{41}$$

This transformation has the form of a constant isospin transformation, but we shall allow $\vec{\lambda}$ to be coordinate dependent. The total Lagrangian is clearly not invariant under (41) when $\vec{\lambda}$ is not a constant. This is due to the presence in the Lagrangian of the first derivatives of the field variables and the change induced by (41) is obviously proportional to $\partial_\mu \vec{\lambda}$. It is actually quite easy to see that one has

$$\delta L = -\vec{S}^\mu \cdot \partial_\mu \vec{\lambda}, \tag{42}$$

where

$$\vec{S}^\mu = -\frac{\partial L}{\partial \vec{b}_{\nu,\mu}} \vec{b}_\nu - \frac{\partial L}{\partial \psi_{,\mu}} i g \vec{\tau} \psi + i g \bar{\psi} \vec{\tau} \frac{\partial L}{\partial \bar{\psi}_{,\mu}} \tag{43}$$

is also the same as given by (32). Equation (42) can be taken as the definition of the conserved current $\vec{S}^\mu$. The absence in (42) of a term proportional to $\vec{\lambda}$ with no differentiation is due to the invariance of the Lagrangian under constant isospin transformations. The absence of derivatives of $\vec{\lambda}$ higher than the first is due to the fact that only first order derivatives of the field variables occur in the Lagrangian. From (42), by integration, one obtains the change in the action. If we choose $\vec{\lambda}$ different from zero only in a small domain, the change in the action must vanish as a consequence of the field equations, since (41) are admissible variations. Integrating by parts, the conservation of current (33) follows. The preceding shows that, if one would define the current (in analogy with (14)) by (31), one would not obtain the total conserved current. This can be instead defined by (42) or more explicitly by (43).

The conserved current $\vec{S}_\mu$ has transformation properties under the gauge group which can be best derived from the field equation (21) or

$$\vec{S}_\mu = -\partial^\lambda \vec{F}_{\lambda\mu}. \tag{44}$$

We see that the infinitesimal change of $\vec{S}_\mu$ is

$$\delta \vec{S}_\mu = -\partial^\lambda (\delta \vec{F}_{\lambda\mu}) \tag{45}$$

with $\delta \vec{F}_{\lambda\mu}$ given by (23). In particular the isospin density $\vec{S}^0$ changes by

$$\delta \vec{S}^0 = 2g\, \partial_r (\vec{\lambda} \times \vec{F}^{r0}). \tag{46}$$

If the gauge function $\vec{\lambda}$ tends to a constant $\vec{\lambda}_c$ at infinity in spatial directions in a sufficiently uniform way, we can evaluate the change in the integral

$$\delta \int_3 \vec{S}^0\, dx = 2g \int_3 \partial_r (\vec{\lambda} \times \vec{F}^{r0})\, dx = 2g\, \vec{\lambda}_c \times \int_3 \partial_r F^{r0}\, dx =$$

$$= -2g\, \vec{\lambda}_c \times \int_3 \vec{S}^0\, dx. \tag{47}$$

We see that the total conserved isospin of the system transforms like an isovector, with an isospin rotation which is given by the asymptotically constant value of the coordinate dependent isospin rotation.

4. Gauge Conditions and Quantization

As a preliminary to the study of gauge conditions for the Yang-Mills field, let us consider the axial gauge in electrodynamics (the Coulomb gauge method in electrodynamics is very well known [11] and shall not be described here).

The axial gauge, first used by J. Anderson, is characterized by the gauge condition

$$A_3 = 0. \tag{48}$$

It is convenient to separate explicitly the value 3 of the indices and to introduce indices i and j which take only the values 1 and 2. The field equations (5) and (6) split into equations of motion (containing a time derivative)

$$\partial_0 A_i = \partial_i A_0 - E_i \tag{49}$$

$$-\partial_0 E_i = e\,\bar{\psi}\gamma_i\psi + \partial_j F_{ji} + \partial_3 F_{3i} \tag{50}$$

and equations of constraint (with no time derivatives)

$$E_3 = \partial_3 A_0 \tag{51}$$

$$F_{i3} = -\partial_3 A_i \tag{52}$$

$$F_{ij} = \partial_i A_j - \partial_j A_i \tag{53}$$

$$\partial_i E_i + \partial_3 E_3 + e\,\bar{\psi}\gamma_0\psi = 0 \tag{54}$$

($E_i = -F_{0i}$ is the electric field). The Dirac equation (7) is also an equation of motion. We have not included among the equations of motion the remaining equation

$$-\partial_0 E_3 = e\,\bar{\psi}\gamma_3\psi + \partial_i F_{i3} \tag{55}$$

because it can be regarded as a consequence of the other equations. Indeed, combining (50), (54), and the current conservation equation, one obtains

$$\partial_3(\partial_0 E_3 + e\,\bar{\psi}\gamma_3\psi + \partial_i F_{i3}) = 0 \tag{56}$$

which, with suitable boundary conditions at infinity, should imply (55).

In the axial gauge the independent dynamical variables are A_i and E_i and the matter field ψ, $\bar{\psi}$. In the quantized version of the theory they should satisfy the equal time commutation relations

$$[A_i, A_j] = [E_i, E_j] = 0 \tag{57}$$

and

$$[A_i, E_j] = -i\,\delta_{ij}\,\underset{3}{\delta}(x - x') \tag{58}$$

plus the obvious ones for the ψ field. The other field quantities can be expressed as functions of these independent variables by solving for them in the equations of constraint (51) to (54).

A difficulty arises, however, in connection with equation (54). Let us write its three-dimensional Fourier transform as

$$- i k_3 E_3(k) = J_0(k) + i k_i E_i(k). \tag{59}$$

For $k_3 = 0$ we obtain a relation between E_1 and E_2, which therefore no longer appear as two independent degrees of freedom. In coordinate space the same difficulty can be seen if one solves (54) for E_3 obtaining

$$E_3(x) = -\frac{1}{2}\left(\int\limits_{-\infty}^{x_3} - \int\limits_{x_3}^{\infty} \right) [J_0(x') + \partial_i' E_i(x')]\, dx_3'. \tag{60}$$

As $x_3 \to +\infty$ or $x_3 \to -\infty$, (60) gives a non-vanishing result, while we would expect the field quantity E_3 to tend to zero.

A further objection is that the axial gauge condition does not determine the gauge uniquely. One can make a gauge transformation with a gauge function independent of x_3, but otherwise arbitrary, and the condition (48) will still be preserved. The fact that, as we shall see below, the axial gauge presents considerable advantages in the case of the Yang-Mills field, gives us an incentive to try to overcome these difficulties. A possible solution may be the modified axial gauge suggested by Y. P. YAO [12]. It consists in quantizing the system in a finite volume and then giving different gauge conditions for $k_3 \neq 0$ andfor $k_3 = 0$. More precisely, for $k_3 \neq 0$ one retains (48) while for $k_3 = 0$ (68) is replaced by

$$k_i A_i(k) = 0 \qquad (k_3 = 0). \tag{61}$$

Although this problem is probably not completely solved, in the following we shall ignore this difficulty and shall proceed to apply the axial gauge method to the Yang-Mills theory.

The gauge condition is

$$\vec{b}_3 = 0 \tag{62}$$

and the equations of motion are ($\vec{F}_{0r} \equiv -\vec{E}_r$)

$$\partial_0 \vec{b}_i = \partial_i \vec{b}_0 - \vec{E}_i - 2g\, \vec{b}_0 \times \vec{b}_i \tag{63}$$

$$\begin{aligned} - \partial_0 \vec{E}_i = g\, \bar{\psi} \gamma_i \vec{\tau}\, \psi + 2g\, \vec{b}_0 \times \vec{E}_i + 2g\, \vec{b}_j \times \vec{F}_{ji} + \\ + 2g\, \vec{b}_3 \times \vec{F}_{3i} + \partial_j \vec{F}_{ji} + \partial_3 \vec{F}_{3i}, \end{aligned} \tag{64}$$

with the addition of the Dirac equation (22), while the equations of constraint are

$$\vec{E}_3 = \partial_3 \vec{b}_0 \tag{65}$$

$$\vec{F}_{i3} = - \partial_3 \vec{b}_i \tag{66}$$

$$\vec{F}_{ij} = \partial_i \vec{b}_j - \partial_j \vec{b}_i + 2g\, \vec{b}_i \times \vec{b}_j \tag{67}$$

$$\partial_i \vec{E}_i + \partial_3 \vec{E}_3 + 2g\, \vec{b}_i \times \vec{E}_i + g\, \bar{\psi} \gamma_0 \vec{\tau}\, \psi = 0. \tag{68}$$

The remaining equation

$$-\partial_0 \vec{E}_3 = g\, \bar{\psi} \gamma_3 \vec{\tau}\, \psi + 2g\, \vec{b}_0 \times \vec{E}_3 + 2g\, \vec{b}_j \times \vec{F}_{j3} + \partial_j \vec{F}_{j3} \tag{69}$$

is not independent, in the sense that its ∂_3 derivative follows from (64), (68), and the conservation equation (33). Observe that the validity of (33) is ensured, without having to require (69). It is true that we have derived it (in Section 3) by using the Dirac equation (22), the identity (37), and the field equations (20) *and* (21). However, it is very easy to see that, due to the condition $\vec{b}_3 = 0$, the equation (21) for $\nu = 3$ is actually not needed in the argument.

The independent variables are $\vec{b}_i$, $\vec{E}_i$, and ψ, $\bar{\psi}$. The simplicity of the constraint equations (65) to (68) allows immediate solution for the other variables in terms of the independent variables. However, the same difficulties arise here as for electrodynamics. Ignoring them, we write the quantization conditions in the form

$$[b_i^\alpha, b_j^\beta] = [E_i^\alpha, E_j^\beta] = 0 \tag{70}$$

$$[b_i^\alpha, E_j^\beta] = -i\, \delta_{ij}\, \delta^{\alpha\beta}\, \underset{3}{\delta}(x - x') \tag{71}$$

(α, β are isospin indices).

We proceed next to apply the Coulomb gauge method. The gauge condition is, in analogy with (15),

$$\partial_r \vec{b}_r = 0. \tag{72}$$

To proceed it is convenient to introduce a notation. For a vector in three-dimensional physical space, say V_r, we introduce the essentially unique decomposition into its longitudinal and its transverse parts

$$V_r = V_r^T + V_r^L, \tag{73}$$

where V^T is solenoidal and V^L is a gradient, so that

$$\partial_r V_r^T = 0$$
$$\partial_r V_s^L - \partial_s V_r^L = 0. \tag{74}$$

With these definitions, (72) can be written equivalently as

$$\vec{b}_r^L = 0, \qquad \vec{b}_r = \vec{b}_r^T. \tag{75}$$

Similarly, we shall write

$$\vec{E}_r = \vec{E}_r^T + \vec{E}_r^L. \tag{76}$$

The field equations (20) and (21) separate again into equations of motion and equations of constraint. In the Coulomb gauge the equations of motion are (besibes (22))

$$-\partial_0 \vec{b}_r = \vec{E}_r^T + 2g(\vec{b}_0 \times \vec{b}_r)^T \tag{77}$$

$$-\partial_0 \vec{E}_r^T = (g\, \bar{\psi} \gamma_r \vec{\tau}\, \psi + 2g\, \vec{b}_0 \times \vec{E}_r + 2g\, \vec{b}_s \times \vec{F}_{sr})^T + \partial_s \vec{F}_{sr} \tag{78}$$

while the constraint equations become

$$\vec{E}_r{}^L - \partial_r \vec{b}_0 + 2g(\vec{b}_0 \times \vec{b}_r)^L = 0 \tag{79}$$

$$\vec{F}_{rs} = \partial_r \vec{b}_s - \partial_s \vec{b}_r + 2g\,\vec{b}_r \times \vec{b}_s \tag{80}$$

$$\partial_r \vec{E}_r{}^L + 2g\,\vec{b}_r \times (\vec{E}_r{}^L + \vec{E}_r{}^T) + g\,\bar{\psi}\gamma_0\,\vec{\tau}\,\psi = 0. \tag{81}$$

The remaining equation

$$-\,\partial_0 \vec{E}_r{}^L = (g\,\bar{\psi}\gamma_r\,\vec{\tau}\,\psi + 2g\,\vec{b}_0 \times \vec{E}_r + 2g\,\vec{b}_s \times \vec{F}_{sr})^L \tag{82}$$

is again not independent, this time in the sense that its space divergence follows from the other equations, and a longitudinal vector is essentially determined by its divergence. We shall not give the derivation, but only note that it is based on the uniqueness of the solution of (85) below.

The basic dynamical variables are now, besides ψ and $\bar{\psi}$, $\vec{b}_r$ and $\vec{E}_r{}^T$. One can impose the commutation relations

$$[b_r{}^\alpha, b_s{}^\beta] = [E_r{}^{T\alpha}, E_s{}^{T\beta}] = 0 \tag{83}$$

$$[b_r{}^\alpha, E_s{}^{T\beta}] = -\,i\,\delta^{\alpha\beta}\left(\delta_{rs} - \partial_r\,\partial_s\,\frac{1}{\nabla^2}\right)\underset{3}{\delta}(x - x'). \tag{84}$$

The problem of determining the remaining field variables in terms of the basic ones through the constraint equations appears to be quite complicated. However, the solution of both (81) for $\vec{E}_r{}^L$ and (79) for $\vec{b}_0$ is clearly reducible to the solution of the inhomogeneous differential equation for the isovector $\vec{\phi}$

$$\nabla^2\vec{\phi} + 2g\,\vec{b}_r \times \partial_r\vec{\phi} = \vec{\chi} \tag{85}$$

which, in turn, depends upon finding its Green's function. There is no difficulty in principle in doing this and we shall assume that it can be done. In particular, we shall assume that the solution of (85) is unique, with suitable boundary conditions (this is needed, e.g., to derive the divergence of (82) from the other equations).

The use of the above described gauge conditions is only the first step towards the quantization of the theory. We have treated the fields as classical quantities in our manipulations until now. When they become operators, the order in which they appear in product expressions becomes important and should be specified. As it turns out, this can be done rather simply by suitable symmetrizations or antisymmetrizations. A more difficult problem is that of verifying the relativistic covariance of the theory, since the gauge conditions spoil the manifest covariance of the original Lagrangian. Finally, one may ask whether quantization in different gauges gives equivalent results and, in particular, whether one is entitled to work in relativistic gauges, with the consequent simplifications in the handling of calculations. We shall not go here into a discussion of these problems which have been dealt with in detail and with his usual skill by SCHWINGER [3]. Let us only say that some questions appear to remain open especially in connection with the use of relativistic gauges and their consistency [13]. These difficulties are probably related to the fact that in the quantized version of the Yang-Mills theory one has not yet applied the same sophistication of argument which has shown itself necessary in the case of quantum electrodynamics in order to avoid certain paradoxes, and one uses still products of field operators at the same space-time point without specifying the particular limiting process involved in letting the space-time points coincide.

5. The Mass Problem

Can the physical mass of a gauge field be different from its bare mass, which is zero? This question is very relevant to attempts to interpret all current conservation equations as deriving from the invariance of the theory under gauge transformations of the second kind. It has been suggested, for instance, that the conservation of barion number is of this type. On the other hand, theories of the Yang-Mills type originate from the assumption that even approximate conservation laws like isospin conservation can be connected to a gauge invariance of the second kind.

LEE and YANG [14] have observed that, if the mass of these vector gauge fields is indeed zero, they would generate long range forces which would add to the known long range force of gravity. This added force would give rise to deviations from the equality between the observed gravitational mass and the inertial mass. From the very precise experiments which show the equality of these two masses, one can set very stringent limits to the strength with which matter can be coupled to a zero mass vector meson. For instance, if the "barion number charge" is denoted by η, LEE and YANG show that available experimental evidence requires

$$\frac{\eta^2}{G m_p^2} < 10^{-5},$$

where G is Newton's gravitational constant and m_p the proton mass. Furthermore, the known physical vector mesons all have masses comparable to the nucleon mass.

The question therefore arises if the *physical* mass of a vector gauge field can possibly be different from zero. This problem is already so difficult for theories with an abelian gauge group (like electrodynamics) that hardly any attempts have been made to study it for the Yang-Mills case. The abelian case is of interest because it is relevant to the exact conservation laws (like barion number conservation). Furthermore, the non-abelian case can probably be reduced to the abelian case: for instance, the Yang-Mills theory with vanishing nucleon field is isomorphic to a form of the theory of a charged mass-less vector field in interaction with the electromagnetic field. We shall therefore limit ourselves to a brief discussion of the case of quantum electrodynamics.

To understand better what is meant by the mass problem we must first clarify a few points which may cause confusion. First, we must agree that gauge invariance of the second kind implies that the *bare* mass of the photon vanishes. This statement has been challenged in the past few years in a number of papers [15] which point out that it is possible to formulate a gauge invariant theory with a non-vanishing photon bare mass. These papers describe, with small modifications, a theory given many years ago by STÜCKELBERG [16] (for a different purpose) and we shall therefore consider only STÜCKELBERGs formulation. In order to describe real massive vector mesons in interaction with a spinor field, STÜCKELBERG introduced, in addition to the vector field A_μ, a scalar field B. The field equations can then be taken to be the following

$$\begin{gathered} F_{\mu\nu} = \partial_\mu A_\nu - \partial_\nu A_\mu \qquad \phi_\mu = A_\mu + \frac{1}{\varkappa}\,\partial_\mu B \\ \partial^\lambda F_{\lambda\mu} - \varkappa^2\,\phi_\mu + e\,\bar{\psi}\gamma_\mu\psi = 0 \\ \partial^\mu\phi_\mu = 0 \\ [\gamma^\mu(\partial_\mu - i\,e\,A_\mu) + m]\,\psi = 0. \end{gathered} \tag{86}$$

These equations are clearly invariant under the gauge transformation

$$\psi \to \psi \exp\,[i\,e\,\Lambda], \quad A_\mu \to A_\mu + \partial_\mu \Lambda, \quad B \to B - \varkappa\,\Lambda. \tag{87}$$

On the other hand, they clearly describe a vector particle of mass $\varkappa$ in interaction with the spinor field ψ. However it is quite clear that the gauge group of this theory is of a rather spurious nature. It is very easy to introduce completely gauge invariant fields, namely ϕ_μ itself, and

$$\Psi = \psi \exp\left[i\,\frac{e}{\varkappa} B\right] \tag{88}$$

and to verify that the equations (86) can be written directly in terms of these gauge invariant fields alone. One could say that one has taken the usual equations of the vector meson theory with mass (which do not admit a gauge group) and one has artifically introduced into the theory a gauge group by adding redundant degrees of freedom, those of the field B. We shall say that the gauge group of the STÜCKELBERG theory is a "trivial" gauge group as opposed to the "essential" gauge group of electrodynamics, where no such immediate elimination of the gauge dependent quantities is possible [17]. We shall restrict ourselves to the consideration of theories with essential gauge groups. The invariance of the theory under such a group implies the vanishing of the bare mass of the vector particle. This would obviously also be the case in equations (86) if we wanted to submit ϕ_μ itself to a gauge transformation.

The second point to clarify has to do with the transverse nature of the polarization tensor $\Pi_{\mu\nu}$. This tensor, which corresponds to the sum of all proper photon self energy graphs, must be of the form

$$\Pi_{\mu\nu}(k) = (g_{\mu\nu}\,k^2 - k_\mu\,k_\nu)\,\Pi(k^2) \tag{89}$$

in a correctly gauge invariant theory. In perturbation theory it does actually have the transverse form (89) provided care is taken to ensure that the expansion is gauge invariant, for instance by introduction of suitable line integrals of the type

$$\exp\left[i e \int A_\mu \, d\xi^\mu\right] \tag{90}$$

in the definition of the current. A non-transverse form such as

$$\Pi_{\mu\nu}(k) = (g_{\mu\nu} k^2 - k_\mu k_\nu) \Pi(k^2) + C g_{\mu\nu} \tag{91}$$

would give rise to a non-vanishing photon self energy through the term $C g_{\mu\nu}$. Such a spurious photon self energy (first considered by Wentzel) can only originate from lack of care in preserving gauge invariance in the perturbation expansion. We shall assume that such spurious self energy terms are in fact not present.

Now that we have dispelled these two points of confusion, we can formulate the mass problem correctly. We shall use formally the unrenormalized quantities, but it would be just as easy to work with the renormalized propagators. The photon propagator $\mathcal{G}_{\mu\nu}$ can be obtained from DYSONs equation

$$\mathcal{G}_{\mu\nu} = \mathcal{G}_{\mu\nu}^{(0)} - \mathcal{G}_{\mu\lambda}^{(0)} \Pi^{\lambda\sigma} \mathcal{G}_{\sigma\nu}, \tag{92}$$

where (if we choose e.g. the Landau gauge) the unperturbed propagator can be written as

$$\mathcal{G}_{\mu\nu}^{(0)} = \left(g_{\mu\nu} - \frac{k_\mu k_\nu}{k^2}\right) \frac{1}{k^2}. \tag{93}$$

Solving DYSONs equation we obtain

$$\mathcal{G}_{\mu\nu} = \left(g_{\mu\nu} - \frac{k_\mu k_\nu}{k^2}\right) \mathcal{G}(k^2), \tag{94}$$

where the function $\mathcal{G}(k^2)$ is gauge invariant and is connected with the photon self energy by

$$\mathcal{G}(k^2) = \frac{1}{k^2 + k^2 \Pi(k^2)}. \tag{95}$$

The position of the pole of the function $\mathcal{G}(k^2)$ determines the physical photon mass. In perturbation theory the function $\Pi(k^2)$ is regular for $k^2 = 0$ and therefore the physical photon mass vanishes. A non-vanishing photon mass can only arise if the function $\Pi(k^2)$ behaves like $1/k^2$ for $k^2 = 0$ and this in turn can only come from a breakdown of perturbation theory.

SCHWINGER [6] has argued that, if we allow the strength of the coupling constant to increase, values may be reached where the residue of the zero mass photon pole goes to zero. When this happens the nature of the energy spectrum suddenly changes. In a suitable range of values for the coupling constant states like the triplet state of positronium may become truly bound and appear like a stable vector particle of finite mass. A non-relativistic situation where something rather similar happens is that of the electron gas: here the long range Coulomb forces become screened through the interaction. To study the problem further, Schwinger has considered a model, quantum electrodynamics in two-dimensional space-time with vanishing spinor mass, and has solved it exactly in the Landau gauge showing that something like a non-vanishing photon mass renormalization does actually occur. The exact solution is possible only for vanishing spinor mass. However, one can learn a few things from the second order photon self energy calculated leaving the spinor mass m finite. In two dimensions the calculation done by means of the usual Feynman rules gives rise to a convergent result. However, the $\Pi_{\mu\nu}$ one obtains is *not* transverse.

$$\Pi_{\mu\nu} = (g_{\mu\nu} k^2 - k_\mu k_\nu)\, \Pi(k^2) + \frac{e^2}{2\pi} g_{\mu\nu} \tag{96}$$

with

$$\Pi(k^2) = \frac{e^2}{2\pi} \int_{-1}^{1} d\eta\, \frac{1-\eta^2}{4m^2 + k^2(1-\eta^2)}\,. \tag{97}$$

The unwanted (Wentzel type) term is clearly due to a failure of the Feynman rules. It could be removed by means of a Pauli-Villars regularization, since this kind of regularization amounts to a differentiation with respect to the spinor mass followed by an integration and a limiting process on the cut off M

$$\lim_{M^2 \to \infty} \int_{M^2}^{m^2} \frac{\partial}{\partial m^2}\,, \tag{98}$$

and the unwanted term does not depend upon the spinor mass m. The Pauli-Villars regularization shows itself here for what it is, a "brute force" method for eliminating unwanted terms, since clearly, when a theory is finite, regularization should not change the result. Alternatively, one can use a gauge invariant perturbation expansion (based on a correct gauge invariant definition for the current) and the unwanted term will cancel automatically. We shall therefore take for $\Pi_{\mu\nu}$ the expression (89) with $\Pi(k^2)$ given by (97). From (97) one can see that, if $m \neq 0$, $\Pi(k^2)$ is regular for $k^2 = 0$, and the propagator (95) has a pole at the origin. However, if $m = 0$, one obtains

$$\Pi(k^2) = \frac{e^2}{\pi k^2} \tag{99}$$

and the propagator has a pole for

$$k^2 + \frac{e^2}{\pi} = 0 \tag{100}$$

which corresponds to a finite photon self energy (observe that in two-dimensional space-time (100) is dimensionally correct). We see here how a "dynamical" difference can change the spectral properties of the photon propagator. In this case it is the value of the spinor mass; in a more realistic four-dimensional case it is hoped that it might be the value of the coupling constant.

Taken as a mathematical example of the expected analytic behaviour of propagators Schwinger's model is quite satisfactory. If, however, one tries to take it more seriously it loses must of its appeal. For instance, if one formulates the model in the Coulomb gauge, where only physical degrees of freedom occur, one must impose the transversality condition on the potential, which in one space dimension becomes

$$\partial_1 A_1 = 0. \tag{101}$$

With suitable boundary conditions at infinity in space this equation gives

$$A_1 = 0. \tag{102}$$

We see that there are no quantized degrees of freedom associated with the photon field and one is left with a direct Coulomb interaction of the spinor field with itself. The relevance of the model to the mass problem is therefore somewhat reduced. One can write the Hamiltonian of the model as

$$\mathscr{H} = \int \psi^*(-i\,\alpha^1 \overleftrightarrow{\partial_1})\,\psi\,dx_1 + \frac{e^2}{2}\int\int J^0(x_1)\,v(x_1 - x_1')\,J^0(x_1')\,dx_1\,dx_1'. \tag{103}$$

In two-dimensional space time the Dirac spinors have two components. We take the matrix α^1 to be diagonal and set

$$\alpha^0 = \begin{vmatrix} 1 & 0 \\ 0 & 1 \end{vmatrix}, \quad \alpha^1 = \begin{vmatrix} 1 & 0 \\ 0 & -1 \end{vmatrix}.$$

The current is defined as

$$J^\mu(x) = \lim_{\varepsilon \to 0} \frac{1}{2}\,[\psi^*(x+\varepsilon), \alpha^\mu\,\psi(x-\varepsilon)], \tag{104}$$

where the limit is taken symmetrically for positive and negative (space-like) ε; $v(x_1)$ is the "Coulomb potential" in one space dimension. In the unperturbed part of the Hamiltonian one should also understand a suitable antisymmetrization of the product of spinor fields.

BROWN [18] and the author [19] have given an exact solution of the model in the Coulomb gauge and have investigated the question of Lorentz covariance. We want to describe here only the first step in the solution, in order to point out certain difficulties which arise. The Hamiltonian (103) gives rise to the conservation equation

$$\partial_0 J^0 + \partial_1 J^1 = 0 \tag{105}$$

as well as to the equation

$$\partial_0 J_1 - \partial_1 J_0 = \frac{e}{\pi} F_{10}, \tag{106}$$

where the "electric field" F_{10} satisfies

$$\begin{aligned} \partial_0 F^{01} &= -e J^1 \\ \partial_1 F^{10} &= -e J^0. \end{aligned} \tag{107}$$

Equation (106) can be understood as follows. If we had no interaction, $e = 0$, we would expect conservation of the γ_5 current, since the spinor mass vanishes. In two-dimensional space-time the γ_5 current has the same components as the vector current, except for the fact that the components are exchanged. So, for $e = 0$, (106) is identical with the γ_5 current conservation, which we would expect from the free-part of the Hamiltonian (103). The term on the right hand side of (106) comes from the interaction, if one uses the equal time commutation relation

$$[J^0(x), J^1(x')] = -\frac{i}{\pi} \partial_1 \delta(x - x') \tag{108}$$

which in turn follows from the definition (104).

Combining (106) and (107) we obtain the wave equation

$$\left(\Box - \frac{e^2}{\pi}\right) F_{01} = 0 \tag{109}$$

for the electric field. This equation is exact and shows the appearance of a massive vector "bound state" in the theory confirming our previous result. However, we encounter a difficulty. From (107) we see that the total charge is given by

$$Q = \int_{-\infty}^{\infty} e J^0 \, dx' = F_{10}(+\infty) - F_{10}(-\infty). \tag{110}$$

If we do not want the total charge to vanish we must allow that F_{10} does not vanish at infinity in space, but rather admit that it tends to a constant. However, from (109) we see then immediately that the total charge satisfies the harmonic equation

$$\ddot{Q} + \frac{e^2}{\pi} Q = 0. \tag{111}$$

Therefore if we want Q to be a constant in time we must restrict ourselves to states of zero charge (doing this, however, would defeat the very purpose of the whole approach). The origin of this difficulty can be understood if one examines carefully the propagators in the Coulomb gauge or the matrix elements of the field operators. One discovers that, due to the one-dimensional nature of time *and* to the vanishing of the spinor mass, the asymptotic behaviour of the propagators for large space distance is such that differential conservation laws, like (105), do not imply the usual integral conservation laws. The usual proofs of Lorentz covariance fail for the same reason: when one attempts to verify the commutation relations between the infinitesimal generators of the Lorentz group, one is stopped by the impossibility to perform the necessary partial integrations which presuppose an asymptotic behaviour not verified by the field operators of the model. It seems very likely that the difficulty described here for the two-dimensional model will not appear in a more realistic theory. However, their presence leaves the situation somewhat confused. One may say that no serious objection has been raised to Schwinger's conjecture, but also that no really satisfactory example is known where the desired non-vanishing photon mass actually arises.

If a photon mass arises in a gauge invariant theory, the structure of the singularities of the propagators must undergo a very drastic rearrangement. We want to show, in particular, that the Dyson vertex part $\Gamma_\mu(p, q)$ with both momenta on the mass shell must acquire a pole for $(p-q)^2 = 0$. We begin with the definition of Γ_μ in terms of a time ordered vacuum expectation value of Heisenberg field operators

$$\langle(\psi(x)\, A_\mu(z)\, \bar{\psi}(y))_+\rangle \Leftrightarrow \mathcal{G}_{\mu\lambda}(k)\, G(p)\, \Gamma^\lambda(p, q)\, G(q). \tag{112}$$

Here $k = p - q$; the $\Leftrightarrow$ sign indicates that the right hand side is, up to factors, the Fourier transform of the left hand side with respect to the coordinate differences and

$$\mathcal{G}_{\mu\nu} \Leftrightarrow \langle(A_\mu A_\nu)_+\rangle \tag{113}$$

and

$$G \Leftrightarrow \langle(\psi\, \bar{\psi})_+\rangle \tag{114}$$

are, respectively, the exact photon and electron propagators. If we go in (112) to the asymptotic limit, replacing ψ by $\psi^{(\mathrm{out})}$ and $\bar{\psi}$ by $\bar{\psi}^{(\mathrm{in})}$, we must put the two momenta p and q in the right hand side on the mass shell, which we indicate by using bold face type. We obtain in this way the matrix element of the A_μ between one particle states

$$\langle \mathbf{p}|\, A_\mu\, |\mathbf{q}\rangle \Leftrightarrow \bar{u}(\mathbf{p})\, \mathcal{G}_{\mu\lambda}(k)\, \Gamma^\lambda(\mathbf{p}, \mathbf{q})\, v(\mathbf{q}), \tag{115}$$

where u and v are one particle wave functions. Observe that the application here made of the asymptotic condition is perfectly justified if the photon has actually acquired a mass because then none of the usual difficulties associated with the infrared problem can arise. If we now go over to the current, by means of Maxwell's equations, and also write explicitly the space-time dependence of the operators, we obtain

$$\langle \mathbf{p}|\, J_\mu(z)\, |\mathbf{p}\rangle \cong \bar{u}(\mathbf{p})\, k^2\, \mathcal{G}(k)\, \Gamma_\mu(\mathbf{p}, \mathbf{q})\, v(\mathbf{q})\, e^{ik\cdot z}, \tag{116}$$

where $\mathcal{G}(k)$ is the gauge invariant part of the photon propagator, defined in (95). We can use the time component of (116), integrating over all space to calculate the matrix elements of the total charge between charge one, one-particle states. We obtain upon integration

$$\langle \mathbf{p}|\, Q\, \mathbf{q}\rangle \cong \underset{3}{\delta}(\mathbf{p}-\mathbf{q})\, \bar{u}(\mathbf{p})\, k^2\, \mathcal{G}(k)\, \Gamma_0(\mathbf{p}, \mathbf{q})\, v(\mathbf{q}). \tag{117}$$

The left hand side is also proportional to a δ-function

$$\langle p | Q | q \rangle \cong \underset{3}{\delta}(p - q)\, \bar{u}(p)\, \gamma_0\, v(q). \tag{118}$$

Going back to four-dimensional expressions we obtain finally the relation (for renormalized quantities)

$$\lim_{k^2 \to 0} k^2\, \mathcal{G}(k)\, \Gamma_\mu(p, q) = \gamma_\mu. \tag{119}$$

It is clear from our derivation that (119) should be valid independently of perturbation theory. In the conventional type of solution in quantum electrodynamics we know that as $k^2 \to 0$

$$\mathcal{G}(k) \sim \frac{1}{k^2} \qquad \Gamma_\mu \sim \gamma_\mu \tag{120}$$

and (119) is clearly satisfied. If solutions exist, for suitably strong coupling, where $\mathcal{G}$ has no pole at the origin, then (119) shows that Γ_μ must have a pole at the origin. Approximation methods suitable for finding this new kind of solutions must be devised keeping in mind the singularity at the origin in Γ_μ. JOHNSON [20] has made an attempt to develop such an approximation method.

A much less ambitious (and less fundamental) approach to the mass problem consists in simply adding a mass term to the gauge invariant Lagrangian. For instance, to (1) one would add simply

$$\frac{1}{2} \varkappa^2 A_\mu A^\mu. \tag{121}$$

Such a term clearly destroys the gauge invariance of the theory. However, just because the theory was gauge invariant before the inclusion of the mass term, the interacting field satisfies the same condition

$$\partial^\mu A_\mu = 0 \tag{122}$$

as the free vector field. This follows simply by using the obvious generalization of (11) to the present case. Similarly, adding a mass term

$$\frac{1}{2} \varkappa^2 \vec{b}_\mu \cdot \vec{b}^\mu \tag{123}$$

to the Yang-Mills Lagrangian (17), one obtains field equations which imply

$$\partial^\mu \vec{b}_\mu = 0. \tag{124}$$

If the original Lagrangian had not been invariant under the coordinate dependent gauge group but, e.g., only under the constant isospin group, one would have instead of (124) a more complicated relation between the four components of the vector field. The conservation of current woulde be still valid, however. A theory in which (124) is valid appears to have a special place among all those invariant under constant isospin rotations in the sense that it satisfies special requirements of mathematical consistency and possibly also in the sense that it leads itself to a particularly simple physical interpretation. This point has been considered by OGIEVETSKIJ and POLUBARINOV [21] but it deserves further investigation.

We would like to contribute here a few simple ideas in this connection. Let us consider a Yang-Mills field with mass in interaction with the nucleon field. Furthermore, let us assume that the coupling constant describing the self-interaction of the Yang-Mills field can be different from that describing the coupling to the nucleon field. We have then the field equations

$$\vec{F}_{\mu\nu} - (\partial_\mu \vec{b}_\nu - \partial_\nu \vec{b}_\mu + 2\,f\,\vec{b}_\mu \times \vec{b}_\nu) = 0 \tag{125}$$

$$\partial^\mu \vec{F}_{\mu\nu} + 2\,f\,\vec{b}^\mu \times \vec{F}_{\mu\nu} + g\,\bar{\psi}\gamma_\nu \vec{\tau}\,\psi - \varkappa^2 \vec{b}_\nu = 0 \tag{126}$$

$$[\gamma^\mu(\partial_\mu - i\,g\,\vec{\tau}\cdot\vec{b}_\mu) + m]\,\psi = 0, \tag{127}$$

where the coupling constants f and g can be different. Clearly these equations do not admit the gauge group (23), although they are invariant under constant isospin rotations. The coordinate dependent group is violated because of the presence of the mass term and also because the two coupling constants are taken to be different. For $f = 0$, one obtains the "naive" theory describing the coupling of an isovector and Lorentz vector to the nucleon field. We want to describe some consequences of the field equations (125), (126) and (127).

We first observe that, since the Dirac type equation (127) has not been changed, we can still derive from it the "covariant" divergence equation

$$(\partial^\mu + 2\,g\,\vec{b}^\mu \times)\,g\,\bar{\psi}\gamma_\mu\vec{\tau}\,\psi = 0. \tag{128}$$

Furthermore, since the equations are invariant under constant isospin rotations, it is possible, just as in Section 3, to rewrite (128) as a true conservation equation

$$\partial^\mu(g\,\bar{\psi}\gamma_\mu\vec{\tau}\,\psi + 2\,g\,\vec{b}^\lambda \times \vec{F}_{\lambda\mu}) = 0. \tag{129}$$

On the other hand, from the field equation (126) we obtain immediately

$$\varkappa^2\,\partial^\mu \vec{b}_\mu = \partial^\lambda(2\,f\,\vec{b}^\mu \times \vec{F}_{\mu\lambda} + g\,\bar{\psi}\gamma_\lambda\vec{\tau}\,\psi). \tag{130}$$

We see that there are two independent conservation equations. Combining (129) and (130) we obtain the relation

$$\varkappa^2\,\partial^\mu \vec{b}_\mu = 2(f - g)\,g\,\vec{b}^\mu \times (\bar{\psi}\gamma_\mu\vec{\tau}\,\psi). \tag{131}$$

We can now discuss to a certain extent the consistency of the theory. First, let us assume that $\varkappa = 0$. Then we must have at least one of the following

$$f = g \tag{132}$$

$$g = 0 \tag{133}$$

$$\vec{b}^\mu \times (\bar{\psi}\gamma_\mu\vec{\tau}\,\psi) = 0. \tag{134}$$

Equation (132) gives the Yang-Mills theory. Equation (133) gives the Yang-Mills theory with zero coupling to the nucleon field. In both cases the coordinate dependent gauge group for the vector field $\vec{b}$ is valid. Equation (134) is an algebraic constraint equation between the various field components involved. It would seem that one cannot impose it on top of the field equations without incurring some sort of contradiction. Let us now take $\varkappa \neq 0$. Then, if either $f = g$ or $g = 0$, we obtain the simple relation (124). Otherwise, the full equation (131) remains. Equation (124) is identical to that which in the free field case ensures that the field describes particles of spin one and has no spin zero components. The full equation (131) also gives a connection between the four components of the vector field but its physical interpretation is certainly not very simple.

References

1. C. N. Yang and R. L. Mills, Phys. Rev. **96**, 191 (1954).
2. R. L. Arnowitt and S. I. Fickler, Phys. Rev. **127**, 1821 (1954).
3. J. Schwinger, Phys. Rev. **125**, 1043 (1962); **127**, 324 (1962); **130**, 402 (1962).
4. R. Utiyama, Phys. Rev. **101**, 1597 (1956). See also M. Gell-Mann and S. Glashow, Ann. Phys. (New York) **15**, 437 (1961).
5. J. J. Sakurai, Ann. Phys. (New York) **11**, 1 (1960).
6. J. Schwinger, Phys. Rev. **125**, 397 (1962); **128**, 2425 (1962).
7. M. Gell-Mann, Physics Letters **8**, 214 (1964).
8. G. Zweig, CERN Report (1964).
9. J. Schwinger, Phys. Rev. **135**, B 816 (1964); **136**, B 1821 (1964).
10. T. D. Lee, F. Gursey, M. Nauenberg, Phys. Rev. **135**, B 467 (1964).
11. B. Zumino, J. Math. Phys. **1**, 1 (1960).
12. Y. P. Yao, J. Math. Phys. **5**, 1319 (1964).
13. Unpublished remarks by R. P. Feynman. See also B. DeWitt, Phys. Rev. Letters **12**, 742 (1964).
14. C. N. Yang and T. D. Lee, Phys. Rev. **98**, 1501 (1955).
15. See, e.g., G. Feldman and P. T. Matthews, Phys. Rev. **130**, 1633 (1963); ibid **132**, 823 (1963); Y. Fujii and S. Kamefuchi, Nuovo Cimento **33**, 1639 (1964).
16. E. C. G. Stückelberg, Helv. Phys. Acta **11**, 299 (1938).
17. It must be admitted that the difference between essential and spurious gauge groups is not defined above in a completely satisfactory manner. Nevertheless, such a distinction clearly exists.
18. L. S. Brown, Nuovo Cimento **29**, 617 (1963).
19. B. Zumino, 1962 Eastern Theoretical Physics Conference, edited by M. E. Rose, Gordon and Breach Publishers, New York (1963); Phys. Lett. **10**, 224 (1964).
20. K. Johnson, Theoretical Physics (Trieste Seminar directed by A. Salam), International Atomic Energy Agency, Vienna 1963.
21. V. I. Ogievetskij and I. V. Polubarinov, Ann Phys. (New York) **25**, 358 (1963).

The Zero Mass Limit of a Charged Particle in Quantum Electrodynamics*

By

M. Nauenberg**

Stanford University, Physics Department

In the renormalized perturbation theory in quantum electrodynamics, the mass m and the charge e of a particle are parameters which are in general determined experimentally. I would like to make a few remarks about this theory in the limit $m \to 0$. Since m characterizes the energy scale, we may obtain some information on the high energy limit of the theory. We are also interested in the question whether the theory exists for zero mass charged particles. The work I will discuss has been done in collaboration with T. D. LEE [1]. A detailed analysis of the zero mass problem was first given by KINOSHITA [2].

As is well known, higher order terms in the perturbation expansion of the transition amplitudes in powers of e are divergent in the limit $m \to 0$. This divergence has the same origin as the infrared divergence which arises because the photon has zero mass; in both cases we are dealing with a breakdown of the perturbation expansion because the Hamiltonian has degenerate eigenstates. Consider, for example, an electron state with momentum $\boldsymbol{p}$ and energy $\varepsilon = \sqrt{\boldsymbol{p}^2 + m^2}$, and an electron plus photon state with momentum $\boldsymbol{p} - \boldsymbol{k}$ and $\boldsymbol{k}$ respectively, and total energy $E = \sqrt{(\boldsymbol{p} - \boldsymbol{k})^2 + m^2} + |\boldsymbol{k}|$. In the limit $m \to 0$, $\varepsilon = |\boldsymbol{p}|$ and $E = |\boldsymbol{p} - \boldsymbol{k}| + |\boldsymbol{k}|$, and we see that if the electron and photon momentum are parallel we have degeneracy.

We want to show that for transition probabilities suitable averaged over degenerate states, a formal perturbation series in e can be developed in which each term is finite when $m \to 0$. This is well known in the case of the infrared problem, although, as we will see, the problem is more complicated in our case.

Let us begin by considering an example due to KINOSHITA and SIRLIN [3]. We examine the lowest order radiative corrections to

---

* Lecture given at the IV. Internationale Universitätswochen für Kernphysik, Schladming, 25 February—10 March 1965.

** ALFRED, P. SLOAN Fellow.

μ-decay in the limit that the electron mass $m_e \to 0$. The diagram for the process is

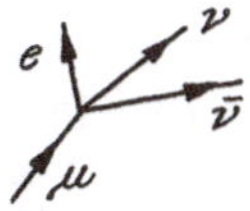

For our purpose, the $\nu - \bar{\nu}$ pair will be replaced by a vertex V that absorbs energy and momentum. The radiative corrections are then given by the diagrams

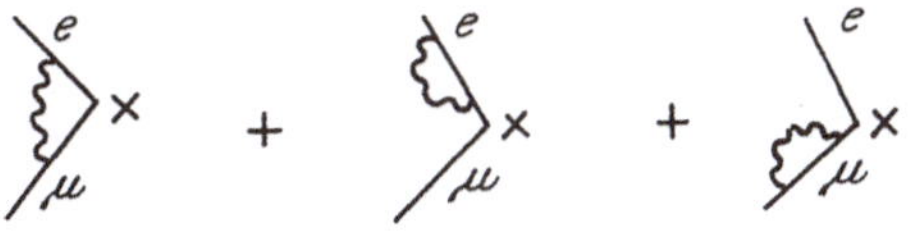

In addition we have to consider also the emmission of real photons,

Real soft photons are required to eliminate the infrared divergences in the transition probability. In the case $m_e \to 0$ the radiative corrections have also a $\log(m_e/\varepsilon)$ singularity. Correspondingly, such a singularity occurs in the transition probability for the emission of a real photon along the direction of the outgoing electron. We will sketch this calculation here.

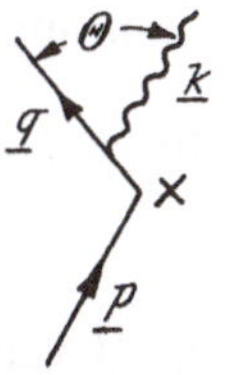

$$\text{Ampl.} = \bar{u}_e(q)\, \varepsilon \cdot \gamma \frac{(q+k) \cdot \gamma + m_e}{(q+k)^2 - m_e^2} V\, u_\mu(p)$$

where V is the matrix which corresponds to the $\bar{\nu}\,\nu$ states. The denominator $(q+k)^2 - m_e^2 = 2\, q \cdot k$, which for small m_e/ε and θ becomes equal to $\varepsilon\, \omega\, [\theta^2 + (m_e^2/\varepsilon^2)]$, while the numerator is proportional to θ. (Helicity is preserved in the numerator if $m_e = 0$ (γ_5-invariance) and angular conservation forbids emission or absorption of photons along the electron momentum).

The transition probability for the emission of a photon along the electron momentum is then proportional to

$$\int_0^{\theta_0} \theta\, d\theta \frac{\theta^2}{\left[\theta^2 + \frac{m^2}{\varepsilon^2}\right]^2} \sim \ln \frac{m}{\varepsilon}$$

The interesting point is that the $\ln(m/\varepsilon)$ singularity which is due to the *virtual* photon contribution to the transition probability to order e^2 (which we did not calculate here) is exactly equal in magnitude but opposite in sign to the singularity in the transition probability corresponding to the emission of a *real* photon. Hence the sum of these transition probabilities contains no singularities in the limit $m_e \to 0$.

Suppose now that we let also the mass of the muon, m_μ, vanish, i.e. we consider the scattering of a massless electron from an external potential. We then find that there are additional $\ln(m/\varepsilon)$ singularities and in order to cancel these we have to consider the transition probability for the process

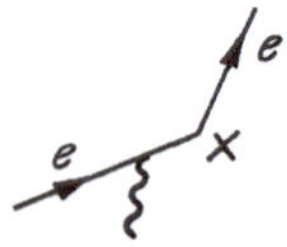

in which a photon is absorbed in the initial state. This is a new feature which appears in the theory of massless charged particles, namely (that such particles must always be accompanied by a "cloud" of real photons. The cancellation of the $\ln(m/\varepsilon)$ singularity is achieved if the momentum distribution of these photons along the electron momentum is given by phase space. That this must be the case becomes clear after considering the process of production of such particles, as in the example of the decay of the muon into a massless electron.

Let me now state a general theorem concerning perturbation expansions for a Hamiltonian which has degenerate eigenstates which shows why there is a cancellation of mass singularities in the processes which we have been discussing. It also explains the cancellation of infrared divergences.

We consider a Hamiltonian $H_0 + g\,H_1$ which can be diagonalized by a unitary matrix U

$$U^\dagger (H_0 + g\,H_1)\,U = E$$

where g is the interaction coupling constant (e in QED). If there is a continuum of eigenstates we have $U = U_+$ or U_- for outgoing or ingoing scattered waves. The S-matrix is then given by

$$S = U_-^\dagger\,U_+$$

The transition probability from a set of states D_a to a set of states D_b can then be written in the form

$$\sum_{D_b, D_a} |S_{ba}|^2 = \mathrm{Sp}\; T_-(D_b)\; T_+(D_a)$$

where

$$T_{ij}(D_a) = \sum_{D_a} U_{ia}\,U_{aj}^\dagger$$

Suppose now that we have a parameter m and that the degeneracy of states in the total Hamiltonian occurs in the limit $m \to 0$. Then the renormalized perturbation series of U in powers g becomes divergent term by term. This becomes clear already from the familiar first order contribution.

$$U_{ij}^{(1)} = \frac{1}{(E_j - E_i)} H_{ij}(1 - \delta_{ij})$$

which becomes infinite if $E_i = E_j$ for $i \neq j$ and $H_{ij} \neq 0$.

It is an elementary exercise to show that in the corresponding order, $T_{ij}^{(1)}(D_a)$ is finite provided the set D_a consists of all degenerate states of a given energy E_a.

The theorem which is proved in reference 1 states that under certain general conditions the formal expansion of $T(D_a)$ in powers of g exists in the limit $m \to 0$ provided D_a is the subset of degenerate states of energy E_a. For the quantum electrodynamics of a zero mass charged particle we have discussed an example of such a subset of degenerate states D_a consisting of an electron state and the electron + photon states with parallel momenta. Note in particular that this theorem requires in general that we average transition probabilities over degenerate *initial* as well as *final* states. The reason why it is sufficient to average only over final degenerate states to eliminate the infrared divergences is that the amplitudes for the emission and absorption of infrared photons are the same.

I would like to conclude with the application of this theorem to bremsstrahlung, because it illustrates some more peculiar features of massless charged particles. The process we have in mind is described by the diagram

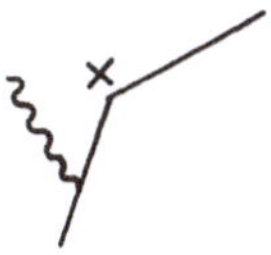

in which the photon is emitted by the incoming electron parallel to its momentum. We again encounter a $\log(m/\varepsilon)$ singularity which is now completely cancelled by the interference of the following two processes

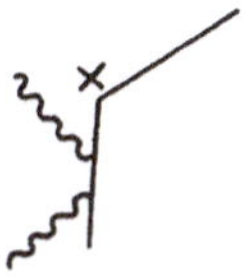

(i)

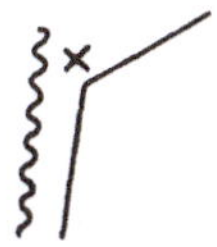

(ii)

In diagram (ii) we are dealing with a disconnected Feynman diagram, which depends on the volume of quantization. It turns out that care must be taken with the epsilontics due to the poles in (i) to obtain the correct contribution from the interference of (i) and (ii).

It appears therefore that a formal renormalized perturbation theory for a massless charged particle which is finite in each order can be constructed for the transition probabilities, provided an average is taken over initial as well as final degenerate states.

References

1. T. D. LEE and M. NAUENBERG, Phys. Rev. **133** B, 1549 (1964).
2. T. KINOSHITA, J. Math. Phys. **3**, 650 (1962).
3. T. KINOSHITA and A. SIRLIN, Phys. Rev. **113**, 1652 (1959).

Summary*

By

J. D. Björken

Stanford University, Stanford Linear Accelerator Center

With 6 Figures

There is a standard paragraph which opens conference summary talks, a paragraph born out of a profound feeling of futility on the part of the speaker. I will not repeat this paragraph, but refer you to the extensive literature [1].

I do want, however, to take the time to express on behalf of the participants our thanks and gratitude to Prof. URBAN and his many collaborators for organizing so well this conference, even to the extent of providing good snow and good weather. And special thanks go to the many students and staff who have worked so diligently in providing lecture notes on the spot.

Now let us turn to the physics. I will restrict most of these comments to the main theme of the conference — high energy behavior — and, for lack of time, not review the interesting but more pedagogical topics on functional methods, lasers, SU_6, and so on. We start with the experimental status.

In a word, it is good, with theory tested to distances $\sim 10^{-14}$ cm for e and μ except for one recent experiment at CEA on $e^+ e^-$ pair production which indicates a deviation at virtual electron momenta ~ 400 MeV/c, the dominant effect being that the observed cross section is too big at the highest energy ($E_\gamma \sim 5$ BeV) by a factor ~ 1.5. We have also heard of the proposal of Low to search systematically for resonant states of leptons and photons at very high energies. This is very recent, and we will hear more of this in the future.

Let us turn now to the theoretical status. On one point there seems to be agreement: Renormalized perturbation theory works at present energies. The interesting questions come when we ask what happens at higher energies, and here we may conveniently use the photon propagator

* Lecture given at the IV. Internationale Universitätswochen für Kernphysik, Schladming, 25 February—10 March 1965.

as an illustration of some of the points of view discussed. We may distinguish at least three different viewpoints, illustrated in Fig. 1.

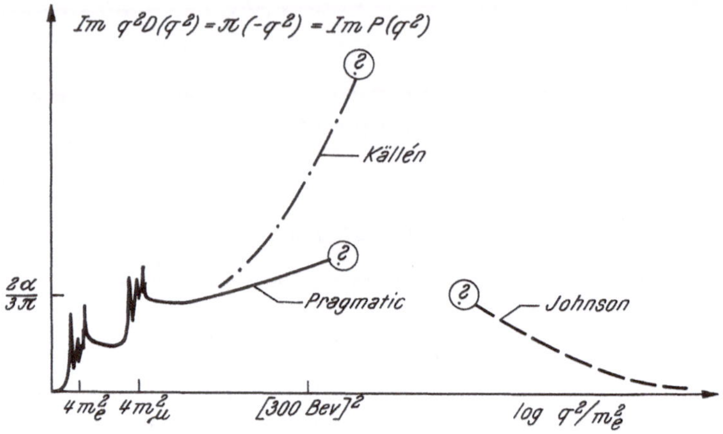

Fig. 1

I. Pragmatic

In the pragmatic, or conventional, point of view the theory works until it becomes inapplicable, directly, because of weak and possibly strong interaction processes. For example, the graph of Fig. 2 (with point couplings assumed)

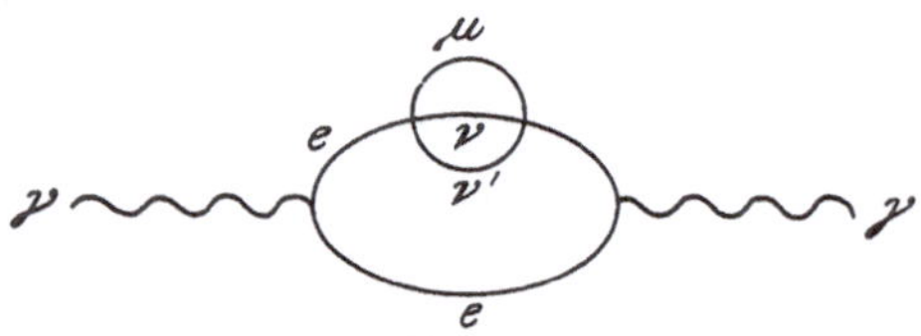

Fig. 2

grows like $\pi \sim \alpha G^2 q^4$, and for $q^2 \gtrsim (300 \text{ BeV})^2$ is a significant modification to the purely electrodynamic contribution. This is not too strong an argument because we don't understand weak interactions at high energies at all. Accepting this point of view, it might be argued that consideration of very high energy electrodynamics is uninteresting, inasmuch as *direct* application to physics is unlikely. However, interesting questions will come if ideas developed here are applicable to field theories of strong or weak interactions.

II. Johnson

Putting aside pragmatic objections, we then turn to an optimistic point of view on the high energy behavior put forward by Johnson. This starts with the *hypothesis* that all is well with high energy behavior, renormalizations are finite if one is careful to choose the right gauge, and that canonical field theory works. This means that $\pi(q^2) \to 0$ as $q^2 \to \infty$ in order that the dispersion integral defining Z_3 converges. This hypothesis is supported by a proposed calculational scheme, carried out in practice one or two orders.

The method is a non-perturbative expansion in terms of α_0, the bare charge, about the values of the unrenormalized Green's functions at infinite energies. The main points are the following:

1. $Z_1 = Z_2 = N^2$ is gauge dependent. The gauge dependence is calculable. The gauge is then chosen such that Z_1 is finite. This is the Landau gauge, within corrections $0(\alpha_0)$. Using renormalization group results Johnson argues that in higher orders it is always possible to find a "finite" gauge and verifies this explicitly through fourth order.

2. δm is to be made finite by setting the bare mass $m_0 = 0$ and obtaining m by a "spontaneous breakdown" mechanism, solving the integral equation shown in Fig. 3

Fig. 3

approximately but nonperturbatively. There is a problem on the role of γ_5 and conformal invariance which exist in the formal theory but not in the solution. This means the vacuum state is labeled by a phase and a scale, and there is a continuous infinity of possible vacuum states. One may ask two questions here:

a) Is this equivalent to ordinary electrodynamics, where there is presumably only one vacuum state? The answer given is that the different vacua lie in different Hilbert spaces, i.e.,

$$\langle 0' | \psi_1 \ldots \psi_n \ldots A_m | 0 \rangle = 0 \qquad \text{if} \qquad |0'\rangle \neq |0\rangle.$$

At this point there is no *obvious* contradiction with ordinary electrodynamics.

b) Are there zero mass 0^- Goldstone bosons [2] in the theory? The answer here is that the formal arguments involve operator products which are too singular. An example has been given by Johnson. However, to be completely convinced, it would be good to make explicitly the calculation of $e^- - e^+$ scattering in the $^1S_0$ state, as well as some matrix elements of the axial vector current, to see whether they agree with conventional electrodynamics.

3. The most difficult problem concerns vacuum polarization. Here two visions of how this might be made finite were discussed. The first is based on the assertion that if D is replaced by the bare D_0 in calculating vacuum polarization (valid if Z_3 finite), the asymptotic structure of $(q^2 D)^{-1}$ is given to any perturbation order by

$$(q^2 D)^{-1} = 1 + f(\alpha_0) \log \frac{\Lambda^2}{q^2} \qquad q^2 \to \infty$$

Thus if $f(\alpha_0) = 0$, we find a convergent theory, but only for a discrete value of α_0 (or discrete set of values). This is closely connected to the Gell-Mann and Low [3] "renormalization group" argument, which comes essentially to the same conclusion.

Johnson's second vision is based upon a hypothesis of a certain non-perturbative behavior of that part of the vertex Γ which enters into the vacuum polarization. Roughly, the divergence in Fig. 4

Fig. 4

is to be cancelled in the first approximation by diagrams such as Fig. 5

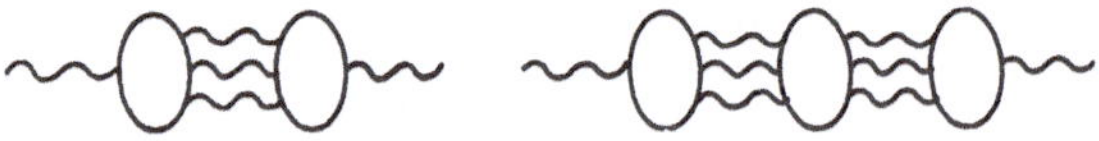

Fig. 5

along with higher iterations. Here the structure of the integral equation for Γ appears to allow the cancellation for any α_0. The calculations implementing this idea are not yet completed. There is an unsolved gauge problem associated with the closed loops. We shall mention this again later.

III. Källén's Approach

The third point of view is, in a word, that the unrenormalized theory is inconsistent, and that $\pi(-q^2) \nrightarrow 0$ as $q^2 \to \infty$. The evidence is divided into two parts:

1. The first point is based on the famous 13-year-old argument that at least one of the renormalization constants is infinite. The idea is based on a proof by contradiction. One assumes that all is well and then on this basis tries to bound $\pi(-q^2)$ from below as $q^2 \to \infty$:

$$\pi(-q^2) = \frac{V}{3\,q^2} \sum_{q_n = q} |\langle 0|j_\mu|n\rangle|^2 \geqslant \frac{V}{3\,q^2} \sum |\langle 0|j_\mu|e^+ e^-\rangle|^2$$

The heart of the argument is that, as $q^2 \to \infty$

$$\langle 0|j_\mu|e^+ e^-\rangle \to \frac{Z_2}{Z_3} \bar{v} \gamma_\mu u = \frac{N^2}{1-L} \bar{v} \gamma_\mu u$$

provided all the Z's are finite. If the above result is true, it follows that the dispersion integral for Z_3 diverges and we get the contradiction. Three objections to this argument were discussed at this meeting:

a) There is a question of rigor, such as interchange of orders of limits and integrations, etc. All agree that the level of rigor is not that of, say, axiomatic field theory, a comment applicable, to be sure, to the Johnson and pragmatic points of view as well. The history and comments on this question are in KÄLLÉN's lectures. There is a modernized version of the argument in preparation which may improve this question of rigor a little. The new version uses the integral representation for the 3-point function derived by KÄLLÉN and TOLL [4].

b) The next objection is that the result is gauge-dependent because Z_2 is gauge-dependent. The answer here is that the calculation is done in a specific gauge, i.e., Gupta-Bleuler gauge, within corrections of order α_0. This raises the question, however, of whether the KÄLLÉN argumentation can be used to refute the Johnson hypothesis of decent behavior at infinite energies (in a different gauge). This question might be answered by repeating the KÄLLÉN argument in an arbitrary covariant gauge. (Johnson indicated that a canonical theory, with indefinite metric, exists for these covariant gauges.) However, this has not been done.

c) Because of the infrared problem, $\langle 0|j_\mu|e^+ e^-\rangle \to 0$ as the photon "mass" $\to 0$. When the infinite number of compensating soft photons are added back in, one gets a factor

$$\sum_{k_{\alpha\,\mathrm{soft}}} |\langle 0|j_\mu|e^+ e^- k_\alpha\rangle|^2 \sim (\mathrm{Born})\, e^{-\frac{\alpha}{\pi}\log\frac{q^2}{m^2}\log\frac{q^2}{(\Delta k)^2}}$$

where Δk is the maximum photon energy included. This expression approaches zero rapidly as $q^2 \to \infty$. However, hard photons must also be included in the sum, so that $(\Delta k)_{\mathrm{effective}} \sim q$. So the result depends sensitively on the hard photons: is $\Delta k_{\mathrm{eff}} > q$ or $< q$? Experience has shown that the soft photon exponential is in most cases almost one when the hard photons are included, a possible exception being the correction [5] to pair annihilation, where $(\Delta k)_{\mathrm{eff}} > q$ and $\sigma_{\mathrm{rad}} > \sigma_0$. So "experience" may perhaps support the KÄLLÉN conclusion, but this is certainly far from a proof.

2. The second part of the KÄLLÉN argument is more heuristic, and is based on the estimate

$$\sum_{q_n = q} |\langle 0|j_\mu{}^{\mathrm{Born}}|n\,\gamma\rangle|^2 \gtrsim \frac{\alpha^n (\log q^2)^{2n}}{n!}$$

If this is a valid estimate and if $\langle 0|j_\mu|n\,\gamma\rangle \to \langle 0|j_\mu{}^{\mathrm{Born}}|n\,\gamma\rangle$ (or some multiple thereof) as $q^2 \to \infty$, then π blows up faster than a power and we obtain the behavior shown in Fig. 1. According to KÄLLÉN, the order of magnitude of q for which these corrections could become important might be [6] as low as $\sim 137\, m_e$ (within factors of 2π, etc.).

If $\pi(-q^2)$ blows up in such a violent way, then the observable part of the vacuum polarization is also no longer computable from the theory. Here the KÄLLÉN point of view is that the theory must at small distances be modified in a profound way, and the present theory bears a relation to the modified theory something like classical to quantum physics.

Although this point of view is in sharp contrast with that put forward by Johnson, in particular the smooth behavior claimed for

$$[q^2 D]^{-1} \sim 1 + f(\alpha_0) \log q^2$$

for large q^2, there is no strict disagreement. Johnson would claim the diagrams (after summing over permutations of photon lines) of Fig. 6

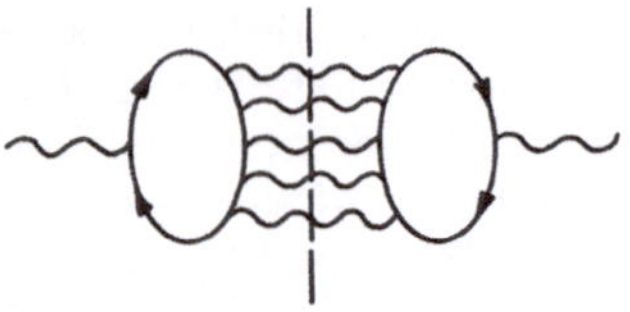

Fig. 6

behave at high energy as $\log q^2$, while Källén argues the absorptive part associated with the photon intermediate state shown by the dotted line behaves as a large power of $\log q^2$. It might be of use to clarify this point further; this should be possible by detailed calculations.

IV. Other Topics

As inevitable as death and taxes, the problem of gauge invariant definitions of the current was discussed at length by various speakers. Now popular is

$$j_\mu = \frac{1}{2} \lim_{\varepsilon \to 0} \left[\bar{\psi}\left(x + \frac{\varepsilon}{2}\right), \gamma_\mu \psi\left(x - \frac{\varepsilon}{2}\right)\right] \exp i e \int_{-\varepsilon/2}^{+\varepsilon/2} d\xi_\mu A^\mu(x + \xi)$$

This seems to ensure the vanishing of the photon mass, but Johnson reported trouble in his nonperturbative approach. This is associated with handling the 6th-order graph of Fig. 5 in an asymmetric way. (In the Dyson equation for the vacuum polarization, there appears a complete Γ at one of the external photon vertices and bare γ at the other.) So perhaps the last word on this subject has not yet been spoken.

In this connection the lectures of Lévy are somewhat relevant. Lévy described a nonlocal version of electrodynamics which allows an arbitrary photon mass, while being formally gauge invariant. This formalism uses gauge invariant fields smeared over timelike paths, along with a Källén-Yang-Feldman integral equation formalism for the dynamics. It is finite to all orders, but there are problems: one is associated with proving the S matrix unitary. Another is associated with getting rid of the dependence of observables on the timelike path; these must be averaged. Furthermore, this averaging must not destroy unitarity, assuming of course that it is there at all. This formalism may be useful as a breakdown-model of electrodynamics, with a breakdown of either covariant or non-covariant nature.

We also heard quantum electrodynamics discussed from the S matrix point of view by Barut, with emphasis on the kinematical and group-theoretical structure. I find it difficult to understand if this S-matrix theory, an open theory (if a theory at all) in which all the basic equations have not even been written down, is supposed to be a replacement for what we have — a theory indeed "analytic" and unitary order by order in α. It is true that for many electrodynamic problems dispersion-relation techniques are powerful and preferable as practical computational tools, but as basic theory the present status of the S-matrix approach leaves much to be desired. At energies $\gtrsim 300$ BeV it may, however, be almost all that is left of conventional electrodynamics.

From Zumino we heard discussed the possibility of generalizations of the electrodynamic formalism to other gauge fields notably of the Yang-Mills type, of possible use in the strong interactions. Here some of the computational techniques developed by Johnson might conceivably find application, since his approximations have the virtue of maintaining analyticity, 2-particle unitarity, and preservation of conservation laws (Ward identities). However, as we heard, there is a long way to go. First, the canonical formalism is more difficult, and secondly there is the problem of the mass of the vector (gauge?) bosons observed in nature, of some difficulty to obtain from the formalism. There are interesting possibilities here of the type discussed by SCHWINGER and seen explicitly in 2-dimensional electrodynamics. Third, there is nonrenormalizability. And, as always, there is the problem, once having this pretty gauge symmetry, of breaking it in accordance with observations.

So, in conclusion, there is a great deal to do, and we may hope to see in the future some of these ideas approach more closely the domain of experimentally accessible phenomena.

References

1. For example, Proceedings 1958, 1959, 1962 Conferences on High Energy Physics; 1961 Aix-en-Provence Conference.
2. GOLDSTONE, SALAM, and WEINBERG, Phys. Rev. **127**, 965 (1962).
3. GELL-MANN and LOW, Phys. Rev. **95**, 300 (1954).
4. KÄLLÉN and WIGHTMAN, Kgl. Danske Vid. Selsk. Mat-Phys. Skr. **1**, 6 (1958). KÄLLÉN and TOLL,
5. ANDREASSI, BUDINI, and FURLAN, Phys. Rev. Letters **8**, 184 (1962).
6. This is based on the idea that α in the above estimate should be replaced by something $\backsim \alpha\,[D(q^2)]/[D(0)] > \alpha$. It is discussed in unpublished CERN lectures (1957—58); CERN preprints 57—43 and Ex. 1476 (in German).